Human Body Dynamics

Springer Science+Business Media, LLC

Aydın Tözeren

Human Body Dynamics

Classical Mechanics and Human Movement

With 177 Illustrations

Springer

Aydın Tözeren
Department of Biomedical Engineering
The Catholic University of America
Washington, DC 20064
USA
tozeren@cua.edu

Illustrations by Dr. Rukmini Rao Mirotznik. Cover photo © copyright Laurie Rubin/The Image Bank.

Library of Congress Cataloging-in-Publication Data
Tözeren, Aydın.
Human body dynamics : classical mechanics and human movement / Aydın Tözeren.
p. cm.
Includes bibliographical references and index.
ISBN 978-1-4757-7390-3 ISBN 978-0-387-21691-1 (eBook)
DOI 10.1007/978-0-387-21691-1

1. Human mechanics. I. Title.
QP303.T69 1999
612.7'6—dc21 99-15365

Printed on acid-free paper.

Originally published by Springer-Verlag New York, Inc. in 2000
Softcover reprint of the hardcover 1st edition 2000

Production coordinated by Chernow Editorial Services, Inc., and managed by Francine McNeill; manufacturing supervised by Erica Bresler.
Typeset by Matrix Publishing Services, Inc., York, PA.

9 8 7 6 5 4 3 2 1

ISBN 978-1-4757-7390-3 SPIN 10715990

To the Memory of My Dad

Preface

"The human body is a machine whose movements are directed by the soul," wrote René Descartes in the early seventeenth century. The intrinsic mechanisms of this machine gradually became clear through the hard work of Renaissance scientists. Leonardo da Vinci is one such scientist from this period of enlightenment. In pursuit of knowledge, Leonardo dissected the bodies of more than 30 men and women. He sawed the bones lengthwise, to see their internal structure; he sawed the skull, cut through the vertebrae, and showed the spinal cord. In the process, he took extensive notes and made carefully detailed sketches. His drawings differentiated muscles that run across several joints from those muscles that act on a single joint. "Nature has made all the muscles appertaining to the motion of the toes attached to the bone of the leg and not to that of the thigh," wrote Leonardo in 1504 next to one of his sketches of the lower extremity, "because when the knee joint is flexed, if attached to the bone of the thigh, these muscles would be bound under the knee joint and would not be able to serve the toes. The same occurs in the hand owing to the flexion of the elbow."

Another Renaissance scholar who made fundamental contributions to the physiology of movement is Giovanni Alfonso Borelli. Born in 1604 in Naples, Borelli was a well-respected mathematician. While teaching at the University of Pisa, he collaborated with the faculty of theoretical medicine in the study of movement. Borelli showed that muscles and bones formed a system of levers. He showed that during some physical activity the hip and the knee transmit forces that are several times greater than the body weight. He spent many years trying to secure funding for the publication of his masterpiece *On the Movement of Animals.* Borelli died in 1679, a few weeks after Queen Catherine of Sweden agreed to pay for the publication costs of the book. The first volume of *On the Movement of Animals* was published the following year.

The advances in the understanding of human body structure and its relation to movement were soon followed by the formulation of nature's laws of motion. In his groundbreaking book *Philosophie Naturalis Principia*

Mathematica, published in 1687, Sir Isaac Newton presented these laws in mathematical language. The laws of motion can be summarized as follows: A body in our universe is subjected to a multitude of forces exerted by other bodies. The forces exchanged between any two bodies are equal in magnitude but opposite in direction. When the forces acting on a body balance each other, the body either remains at rest or, if it were in motion, moves with constant velocity. Otherwise, the body accelerates in the direction of the net unbalanced force.

Newton's contributions to mechanics were built on the wealth of knowledge accumulated by others. In this regard, perhaps the most critical advances were made by Galileo Galilei. Born in Italy on February 15, 1564, Galileo became fascinated with mathematics while studying medicine at the University of Pisa. At the university, he was perceived as an arrogant young man. He made many enemies with his defiant attitude toward the Aristotelian dogma and had to leave the university for financial reasons without receiving a degree. Galileo recognized early on the importance of experiments for advancing science. He observed that, for small oscillations of a pendulum, the period of oscillation was independent of the amplitude of oscillation. This discovery paved the way for making mechanical clocks. One of his stellar contributions to mechanics is the law of free fall. Published first in his 1638 book *Discorsi,* the law states that in a free fall distances from rest are proportional to the square of elapsed times from rest. Although Galileo found recognition and respect in his lifetime, he was nonetheless sentenced to prison at the age of 70 by the Catholic Church for having held and taught the Copernican doctrine that the Earth revolves around the Sun. He died while under house arrest.

Newton's laws were written for so-called particles, however large they may be. A particle is an idealized body for which the velocity is uniform within the body. In the eighteenth century, Leonhard Euler, Joseph-Louis Lagrange, and others generalized these laws to the study of solid bodies and systems of particles. Euler was the first to assign the same gravitational force to a body whether at rest or in motion. In 1760, his work *Thoria Motus Corporum Solidurum seu Rigidorum* described a solid object's resistance to changes in the rate of rotation. A few years later, in 1781, Charles-Augustin de Coulomb formulated the law of friction between two bodies: "In order to draw a weight along a horizontal plane it is necessary to deploy a force proportional to the weight . . . ." Coulomb went on to discover one of the most important formulas in physics, that the force between two electrical charges is inversely proportional to the square of the distance between them. Analytical developments on solid mechanics continued with the publication in 1788 of Lagrange's elegant work *Mechanic Analytique.*

The foundation of classical mechanics set the stage for further studies of human and animal motion. "It seems that, as far as its physique is concerned, an animal may be considered as an assembly of particles separated by more or less compressed springs," wrote Lazare Carnot in 1803. In the 1880s, Eadweard Muybridge in America and Ettiene-Jules Marey in France established the foundation of motion analysis. They took sequential photographs of athletes and horses during physical activity to gain insights into movement mechanics. Today, motion analysis finds particular use in physical education, professional sports, and medical diagnostics. Recent research suggests that the video recording of crawling infants may be used to diagnose autism at an early stage.

The sequential photography allows for the evaluation of velocities and accelerations of body segments. The analysis of forces involved in movement is much more challenging, however, because of the difficult mathematics of classical mechanics. To illustrate the point, scientists were intrigued in the nineteenth century about the righting movements of a freely falling cat. How does a falling cat turn over and fall on its feet? M. Marey and M. Guyou addressed the issue in separate papers published in Paris in 1894. About 40 years later, in 1935, G.G.J. Rademaker and J.W.G. Ter Braak came up with a mathematical model that captured the full turnover of the cat during a fall. The model was refined in 1969 by T.R. Kane and M.P. Schmer so that as observed in the motion of the falling cat the predicted backward bending would be much smaller than forward bending. The mechanism presented by Kane and Schmer is simple; it consists of two identical axisymmetric bodies that are linked together at one end. These bodies can bend relative to each other but do not twist. Space scientists found the model useful in teaching astronauts how to move with catlike ease in low gravity.

Although the mechanical model of a falling cat is simple conceptually, its mathematical formulation and subsequent solution are quite challenging. Since the development of the falling cat model, computational advances have made it easier to solve the differential equations of classical mechanics. Currently, there are a number of powerful software packages for solving multibody problems. Video recording is used to quantify complex modes of movement. Present technology also allows for the measurement of contact forces and the evaluation of the degree of activation of muscle groups associated with motion. Nowadays, the data obtained on biomechanics of movement can be overwhelming. A valid interpretation of the data requires an in-depth understanding of the laws of motion and the complex interplay between mechanics and human body structure. The main goal of this book is to present the principles of classical mechanics using case studies involving human movement. Unlike nonliving objects, humans and animals have the capacity to initiate movement and to modify motion through changes of shape. This capability makes the mechanics of human and animal movement all the more exciting.

I believe that *Human Body Dynamics* will stimulate the interests of engineering students in biomechanics. Quantitative studies of human movement bring to light the healthcare-related issues facing classical mechanics in the twenty-first century. There are already a number of outstanding statics and dynamics books written for engineering students. In recent years, with each revision, these books have incorporated more examples, more problems, and more colored photographs and figures, a few of which touch on the mechanics of human movement. Nevertheless, the focus of these books remains almost exclusively on the mechanics of man-made structures. It is my hope that *Human Body Dynamics* exposes the reader not only to the principles of classical mechanics but also to the fascinating interplay between mechanics and human body structure.

The book assumes a background in calculus and physics. Vector algebra and vector differentiation are introduced in the text and are used to describe the motion of objects. Advanced topics such as three-dimensional motion mechanics are treated in some depth. Whenever possible, the analysis is presented graphically using schematic diagrams and software-created sequences of human movement in an athletic event or a dance performance. Each chapter contains illustrative examples and problem sets. I have spent long days in the library reading scientific journals on biomechanics, sports biomechanics, orthopaedics, and physical therapy so that I could conceive realistic examples for this book. The references included provide a list of sources that I used in the preparation of the text. The book contains mechanical analysis of dancing steps in classical ballet, jumping, running, kicking, throwing, weight lifting, pole vaulting, and three-dimensional diving. Also included are examples on crash mechanics, orthopaedic techniques, limb-lengthening, and overuse injuries associated with running.

Although the emphasis is on rigid body mechanics and human motion, the book delves into other fundamental topics of mechanics such as deformability, internal stresses, and constitutive equations. If *Human Body Dynamics* is used as a textbook for a graduate-level course, I would recommend that student projects on sports biomechanics and orthopaedic engineering become an integral part of the course. The references cited at the end of the text provide a useful guide to the wealth of information on the biomechanics of movement.

Human Body Dynamics should be of great interest to orthopaedic surgeons, physical therapists, and professionals and graduate students in sports medicine, movement science, and athletics. They will find in this book concise definitions of terms such as *linear momentum* and *angular velocity* and their use in the study of human movement.

I wish to acknowledge my gratitude to all authors on whose work I have drawn. My colleagues and students at The Catholic University of America helped me refine my teaching skills in biomechanics. Professor Van Mow provided me with generous resources during my sabbatical at Columbia University where I prepared most of the text. I am deeply indebted to Professor H. Bülent Atabek of The Catholic University of America for his careful reading of the manuscript. Professor Atabek corrected countless equations and figures and provided valuable input to the contents of the manuscript. My teachers, Professors Maciej P. Bieniek and Frank L. DiMaggio of Columbia University, also spent considerable time reviewing the manuscript. I am very grateful to them for their corrections and constructive suggestions. Dr. Rukmini Rao Mirotznik enriched the text with her beautiful sketches and sublime figures. Barbara A. Chernow and her associates contributed to the book with careful editing and outstanding production. Finally, my thanks goes to Dr. Robin Smith and his associates at Springer-Verlag for bringing this book to life.

Washington, D.C. AYDIN TÖZEREN

Contents

Preface vii
Nomenclature xvii

Chapter 1 Human Body Structure
Muscles, Tendons, Ligaments, and Bones 1

1.1 Introduction 1
1.2 Notation for Human Movement 3
1.3 Skeletal Tree 6
1.4 Bone, Cartilage, and Ligaments 10
1.5 Joints of the Human Body 14
1.6 Physical Properties of Skeletal Muscle 17
1.7 Muscle Groups and Movement 21
1.8 Summary 27
1.9 Problems 27

Chapter 2 Laws of Motion
Snowflakes, Airborne Balls, Pendulums 30

2.1 Laws of Motion: A Historical Perspective 30
2.2 Addition and Subtraction of Vectors 33
2.3 Time Derivatives of Vectors 39
2.4 Position, Velocity, and Acceleration 40
2.5 Newton's Laws of Motion and Their Applications 43
2.6 Summary 52
2.7 Problems 53

Chapter 3 Particles in Motion

Method of Lumped Masses and Jumping, Sit-Ups, Push-Ups 56

3.1 Introduction 56
3.2 Conservation of Linear Momentum 57
3.3 Center of Mass and Its Motion 58
3.4 Multiplication of Vectors 64
3.5 Moment of a Force 67
3.6 Moment of Momentum About a Stationary Point 70
3.7 Moment of Momentum About the Center of Mass 77
3.8 Summary 78
3.9 Problems 79

Chapter 4 Bodies in Planar Motion

Jumping, Diving, Push-Ups, Back Curls 84

4.1 Introduction 84
4.2 Planar Motion of a Slender Rod 85
4.3 Angular Velocity 88
4.4 Angular Acceleration 94
4.5 Angular Momentum 97
4.6 Conservation of Angular Momentum 100
4.7 Applications to Human Body Dynamics 103
4.8 Instantaneous Center of Rotation 109
4.9 Summary 111
4.10 Problems 112

Chapter 5 Statics

Tug-of-War, Weight Lifting, Trusses, Cables, Beams 117

5.1 Introduction 117
5.2 Equations of Static Equilibrium 117
5.3 Contact Forces in Static Equilibrium 121
5.4 Structural Stability and Redundance 127
5.5 Structures and Internal Forces 135
5.6 Distributed Forces 144
5.7 Summary 146
5.8 Problems 146

Chapter 6 Internal Forces and the Human Body

Complexity of the Musculoskeletal System 150

6.1 Introduction 150
6.2 Muscle Force in Motion 152
6.3 Examples from Weight Lifting 157
6.4 Moment Arm and Joint Angle 161
6.5 Multiple Muscle Involvement in Flexion of the Elbow 164
6.6 Biarticular Muscles 165
6.7 Physical Stress 169
6.8 Musculoskeletal Tissues 172
6.9 Limb-Lengthening 178
6.10 Summary 182
6.11 Problems 183

Chapter 7 Impulse and Momentum

Impulsive Forces and Crash Mechanics 194

7.1 Introduction 194
7.2 Principle of Impulse and Momentum 194
7.3 Angular Impulse and Angular Momentum 200
7.4 Elasticity of Collision: Coefficient of Restitution 207
7.5 Initial Motion 211
7.6 Summary 213
7.7 Problems 214

Chapter 8 Energy Transfers

In Pole Vaulting, Running, and Abdominal Workout 220

8.1 Introduction 220
8.2 Kinetic Energy 221
8.3 Work 225
8.4 Potential Energy 227
8.5 Conservation of Mechanical Energy 230
8.6 Multibody Systems 232
8.7 Applications to Human Body Dynamics 235
8.8 Summary 246
8.9 Problems 247

Chapter 9 Three-Dimensional Motion
Somersaults, Throwing, and Hitting Motions 256

9.1 Introduction 256
9.2 Time Derivatives of Vectors 257
9.3 Angular Velocity and Angular Acceleration 258
9.4 Conservation of Angular Momentum 264
9.5 Dancing Holding on to a Pole 271
9.6 Rolling of an Abdominal Wheel on a Horizontal Plane 275
9.7 Biomechanics of Twisting Somersaults 280
9.8 Throwing and Hitting Motions 283
9.9 Summary 287
9.10 Problems 289

Appendix 1 Units and Conversion Factors 297

Appendix 2 Geometric Properties of the Human Body 299

Selected References 304

Index 311

Nomenclature

${}^{R}\mathbf{a}^{P}$: Acceleration of point P in reference frame R (m/s^2)
$\alpha^P = {}^{E}\mathbf{a}^{P}$: Acceleration of point P in the reference frame E, which is fixed on earth
$\mathbf{a}^c$: Acceleration of the center of mass of a body in the inertial reference frame E
α: Angular acceleration of body B in reference frame E (1/s)
B: Represents a body with volume V and mass m
$\mathbf{b}_1$, $\mathbf{b}_2$, $\mathbf{b}_2$: Orthogonal unit vectors associated with body B
C: Center of mass
$d\mathbf{a}/dt$: Time derivative of $\mathbf{a}$
$d^2\mathbf{a}/dt^2$: Second time derivative of $\mathbf{a}$
E: Reference frame fixed on earth
E: Young's modulus for elastic materials (N/m^2)
ϵ: Strain, ratio of change in length to stress-free length of a line element
$\mathbf{e}_1$, $\mathbf{e}_2$, $\mathbf{e}_2$: Orthogonal unit vectors defining the reference frame E
$\mathbf{F}$: Force (N)
$\mathbf{f}_{ij}$: Force exerted by mass element j on the mass element i within a body B (system of particles)
$\mathbf{g}$: Gravitational acceleration (m/s^2)
$\mathbf{H}^c$: Moment of momentum of a body (system of particles) about a point C
$\mathbf{H}^0$: Moment of momentum of a body (system of particles) about a point O (kg-m^2/s)
I^c_{ij}: ijth component of mass moment of inertia about the center of mass (kg-m^2)
I^o_{ij}: ijth component of mass moment of inertia about point O
J_x: Axial moment of inertia (m^4)
k: Spring constant (N/m)
k: Radius of gyration (m)
Λ: Angular impulse (N-m-s)
$\mathbf{L}$: Linear momentum of a particle, body, or system of particles (kg-m/s)
$\mathbf{M}^o$: Moment of a force about point O (N-m)

m: Mass of a particle or a body (kg)
μ: Coefficient of friction
P: Mechanical power (rate of work done by a system of forces) (N-m/s)
$\mathbf{r}^{P/O}$: Position vector connecting point O to point P (m)
ρ: Position vector connecting the center of mass of a body to a point of the body
σ: Stress, force intensity, force per unit area (N/m^2)
t: Time
T: Kinetic energy (N-m)
T: Tension in a cable, tendon, or ligament (N)
V: Potential energy (N-m)
V: Rate of shortening, ratio of rate of change of length to the length of a muscle fiber (1/s)
${}^R\mathbf{v}^P$: Velocity of point P in reference frame R (m/s)
$\mathbf{v}^P = {}^E\mathbf{v}^P$: Velocity of point P in the reference frame E, which is fixed on earth
W: Work done on a system by a force (N-m)
w: Load per unit area (length) that is acting on a structure (N/m^2 or N/m)
${}^R\omega^B$: Angular velocity of rigid body B in reference frame R (1/s)
ω: Angular velocity of body B in reference frame E
ζ: Impulse (N-s)

Notes: The terms in parentheses present the units of each variable. The abbreviations kg, m, N, and s stand, respectively, for kilogram, meter, Newton, and second. In general, a left superscript refers to a reference frame under consideration. For simplicity, we omit the superscript when the reference frame is one that is fixed on Earth. A right superscript may indicate a point or a body. Frequently, we omit this superscript when the text clearly indicates which point or body is being referred to. The subscripts typically indicate a component along a certain coordinate axis.

1

Human Body Structure: Muscles, Tendons, Ligaments, and Bones

1.1 Introduction

Humans possess a unique physical structure that enables them to stand up against the pull of gravity. Humans and animals utilize contact forces to create movement and motion. The biggest part of the human body is the trunk; comprising on the average 43% of total body weight. Head and neck account for 7% and upper limbs 13% of the human body by weight. The thighs, lower legs, and feet constitute the remaining 37% of the total body weight. The frame of the human body is a tree of bones that are linked together by ligaments in joints called articulations. There are 206 bones in the human body. Bone is a facilitator of movement and protects the soft tissues of the body.

Unlike the frames of human-made structures such as that of skyscrapers or bridges, the skeleton would collapse under the action of gravity if it were not pulled on by skeletal muscles. Approximately 700 muscles pull on various parts of the skeleton. These muscles are connected to the bones through cable-like structures called tendons or to other muscles by flat connective tissue sheets called aponeuroses. About 40% of the body weight is composed of muscles.

Skeletal muscles act on bones using them as levers to lift weights or produce motion. A lever is a rigid structure that rotates around a fixed point called the fulcrum. In the body each long bone is a lever and an associated joint is a fulcrum. The levers can alter the direction of an applied force, the strength of a force, and the speed of movement produced by a force. The principle of the lever was presented by Archimedes in the third century B.C. Moreover, the practical use of levers is illustrated in the sculptures of Assyria and Egypt, two millennia before the times of Archimedes. The types of levers observed in the human body are sketched in Fig. 1.1. Neck muscles acting on the skull, controlling flexion/extension movements, constitute a first-class lever (Fig. 1.1a). When the fulcrum lies between the applied force and the resistance, as in the case of a seesaw, the

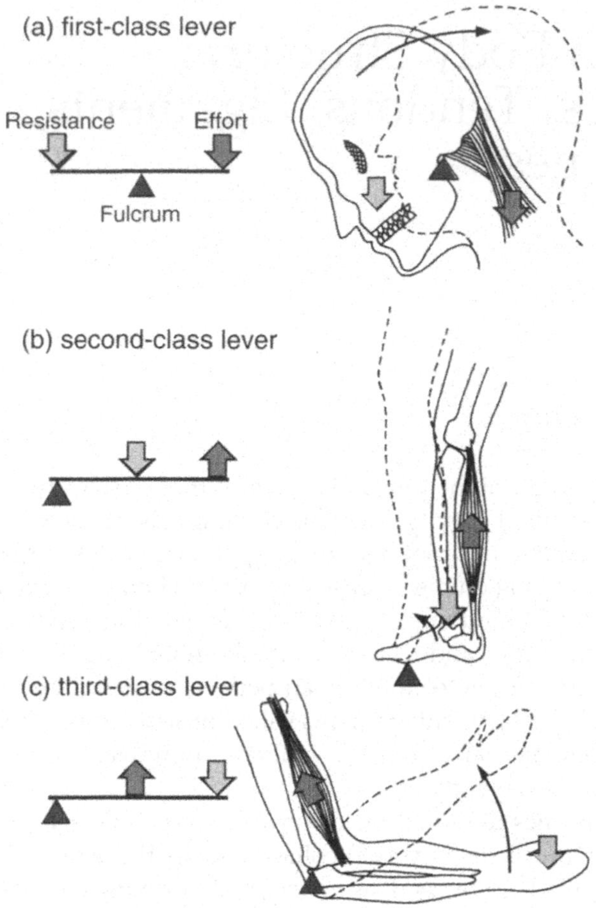

FIGURE 1.1a–c. Three different types of lever systems found in the human body. Bones serve as levers and joints as fulcrums. The resistance to rotation of a bone around a joint comes from two sources: the weight of the part of the body and an external weight to be lifted. The dark *arrows* in the diagram indicate the direction of the muscle pull exerted on the lever. The neck muscles pull on the skull in a first-class lever arrangement **(a)**. The action of the calf muscle acting on the ankle is part of a second-class lever system **(b)**. The action of biceps on the forearm constitutes a third-class lever system **(c)**.

lever is called a first-class lever. The first-class lever alters the direction and the speed of movement and changes the amount of force transmitted to the resistance. In the case shown in the figure, the fulcrum is the joint connecting the atlas, the first vertebra, to the skull. The resultant weight of the head and neck muscle controlling flexion/extension act at opposite sides of the fulcrum. When the muscle pulls down, the head rises up.

Calf muscles that connect the femur of the thigh to the calcaneus bone of the ankle constitute a second-class lever (Fig. 1.1b). A second-class lever magnifies force at the expense of speed and distance. In the case shown in the figure, the fulcrum is at the line of joints between the phalanges and the metatarsals of the feet. The weight of the foot acts as the resistance. In this arrangement, the calf muscle can lift a weight much larger than the tensile force it creates, but in doing so, it has to move a longer distance than the weight it lifts.

An example of a third-class lever in the human body is shown in Fig. 1.1c. In the case of the biceps muscle of the arm shown in the figure, the load is located at the hand and the fulcrum at the elbow. When the biceps contract, they pull the lower arm closer to the upper arm. In this lever system the speed and the distance traveled are increased at the expense of force. Often the various large muscles of the human body produce forces that are multiple times the total body weight. Biceps create movement by their ability to shorten as they continue to sustain tension. Skeletal muscles contract in response to stimulation from the central nervous system and are capable of generating tension within a few microseconds after activation. A skeletal muscle might be able to shorten as much as 30% during contraction.

Among the typical structures built by humans, human body structure most resembles the tensegrity toys in which one forms a three-dimensional body by connecting compression-resistant bars (bones) to tension-resistant cables. However, tensegrity models cannot duplicate the contractility of muscle fibers and therefore cannot generate movement. There is yet another unique feature of the structures of the living, and that is the capacity for self-repair, growth, and remodeling. Almost all structural elements of the human body, the ligaments, tendons, muscles, and bones, remodel in response to applied forces: they possess what are called intrinsic mechanisms of self-repair.

1.2 Notation for Human Movement

Spatial positions of various parts of the human body can be described referring to a Cartesian coordinate system that originates at the center of gravity of the human body in the standing configuration (Fig. 1.2). The directions of the coordinate axis indicate the three primary planes of a standing person. The transverse plane is made up of the x_1 and x_3 axes. It passes through the hip bone and lies at a right angle to the long axis of the body, dividing it into superior and inferior sections. Any imaginary sectioning of the human body that is parallel to the (x_1, x_3) plane is called a transverse section or cross section.

The frontal plane is the plane that passes through the x_1 and x_2 axes of the coordinate system (see Fig. 1.2). It is also called the coronal plane. The

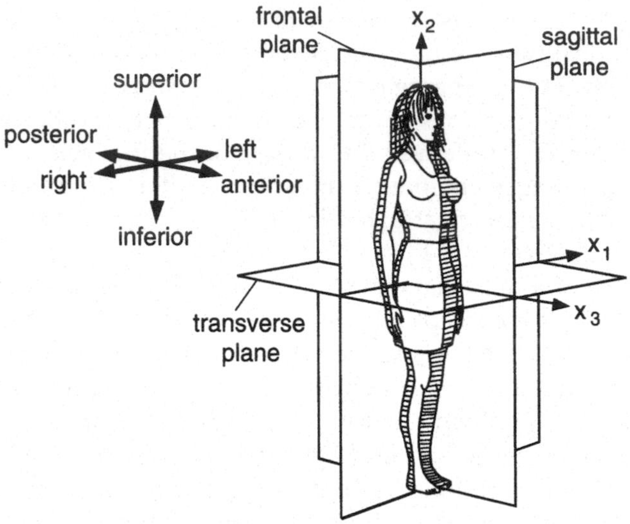

FIGURE 1.2. The three primary planes of a standing person. The sagittal plane is the only plane of symmetry. This plane divides the body into left- and right-hand sides. The frontal plane separates the body into anterior and posterior portions. The transverse (horizontal) plane divides the body into two parts: superior and inferior.

frontal plane divides the body into anterior and posterior sections. The sagittal plane is the plane made by the x_2 and x_3 axes. The sagittal plane divides the body into left and right sections. It is the only plane of symmetry in the human body.

Anatomists have also introduced standard terminology to classify movement configurations of the various parts of the human body (Fig. 1.3). Most movement modes require rotation of a body part around an axis that passes through the center of a joint, and such movements are called angular movements. The common angular movements of this type are *flexion, extension, adduction,* and *abduction.*

Flexion and *extension* are movements that occur parallel to the sagittal plane. Flexion is rotational motion that brings two adjoining long bones closer to each other, such as occurs in the flexion of the leg or the forearm. Extension denotes rotation in the opposite direction of flexion; for example, bending the head toward the chest is flexion and so is the motion of bending down to touch the foot. In that case, the spine is said to be flexed. Extension reverses these movements. Flexion at the shoulder and the hip is defined as the movement of the limbs forward whereas extension means movement of the arms or legs backward. Flexion of the wrist moves the palm forward, and extension moves it back. If the movement of extension continues past the anatomical position, it is called hyperextension.

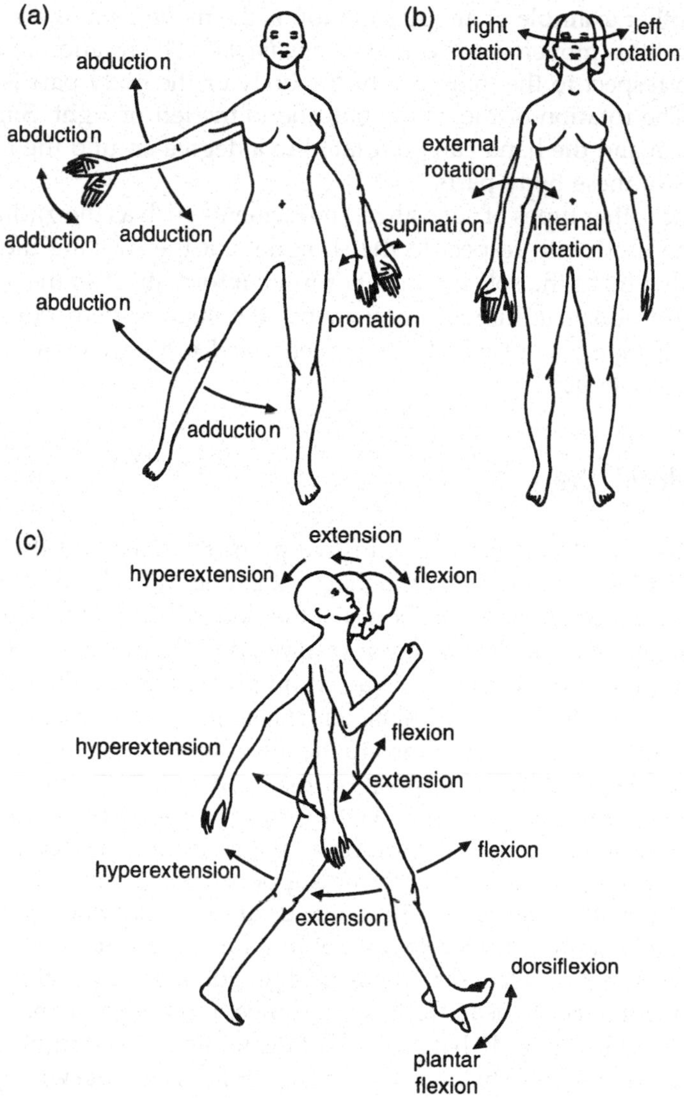

FIGURE 1.3a–c. Anatomical notations used in describing the movements of various body parts: abduction and adduction **(a)**, rotation **(b)**, and flexion and extension **(c)**.

Abduction and *adduction* are the movements of the limbs in the frontal plane. Abduction is movement away from the longitudinal axis of the body whereas adduction is moving the limb back. Swinging the arm to the side is an example of abduction. During a pull-up exercise, an athlete pulls the arm toward the trunk of the body, and this movement constitutes adduction. Spreading the toes and fingers apart abducts them. The act of bringing them together constitutes adduction.

Yet another example of angular motion is the movement of the arm in a loop, and this movement is called *circumduction.* The rotation of a body part with respect to the long axis of the body or the body part is called *rotation.* The rotation of the head could be to the left or right. Similarly, the forearm and the hand can be rotated to a degree around the longitudinal axis of these body parts.

There are other types of specialized movements such as the gliding motion of the head with respect to the shoulders or the twisting motion of the foot that turns the sole inward. For more information on the anatomical classification of human movement, the reader may consult an anatomy book, some of which have been listed in the references at the end of this volume.

1.3 Skeletal Tree

The human skeleton is divided into two parts: the axial and the appendicular (Fig 1.4). The *axial skeleton* shapes the longitudinal axis of the human body. It is composed of 22 bones of the skull, 7 bones associated with the skull, 26 bones of the vertebral column, and 24 ribs and 1 sternum comprising the thoracic cage. It is acted on by approximately 420 different skeletal muscles. The axial skeleton transmits the weight of the head and the trunk and the upper limbs to the lower limbs at the hip joint. The muscles of the axial skeleton position the head and the spinal column, and move the rib cage so as to make breathing possible. They are also responsible for the minute and complex movements of facial features.

The vertebral column begins at the support of the skull with a vertebra called the atlas and ends with an insert into the hip bone (Fig. 1.5a). The average length of the vertebral column among adults is 71 cm. The vertebral column protects the spinal cord. In addition, it provides a firm support for the trunk, head, and upper limbs. From a mechanical viewpoint, it is a flexible rod charged with maintaining the upright position of the body (Fig. 1.5b). The vertebral column fulfills this role with the help of a large number of ligaments and muscles attached to it.

A typical vertebra is made of the vertebral body (found anteriorly) and the vertebral arch (positioned posteriorly). The vertebral body is in the form of a flat cylinder. It is the weight-bearing part of the vertebra. Between the vertebral bodies are 23 intervertebral disks that are made of relatively deformable fibrous cartilage. These disks make up approximately one-quarter of the total length of the vertebral column. They allow motion between the vertebrae. The shock absorbance characteristics of the vertebral disks are essential for physical activity. The compressive force acting on the spine of a weight lifter or a male figure skater during landing of triple jumps peak at many times the body weight. Without shock absorbants, the spine would suffer irreparable damage.

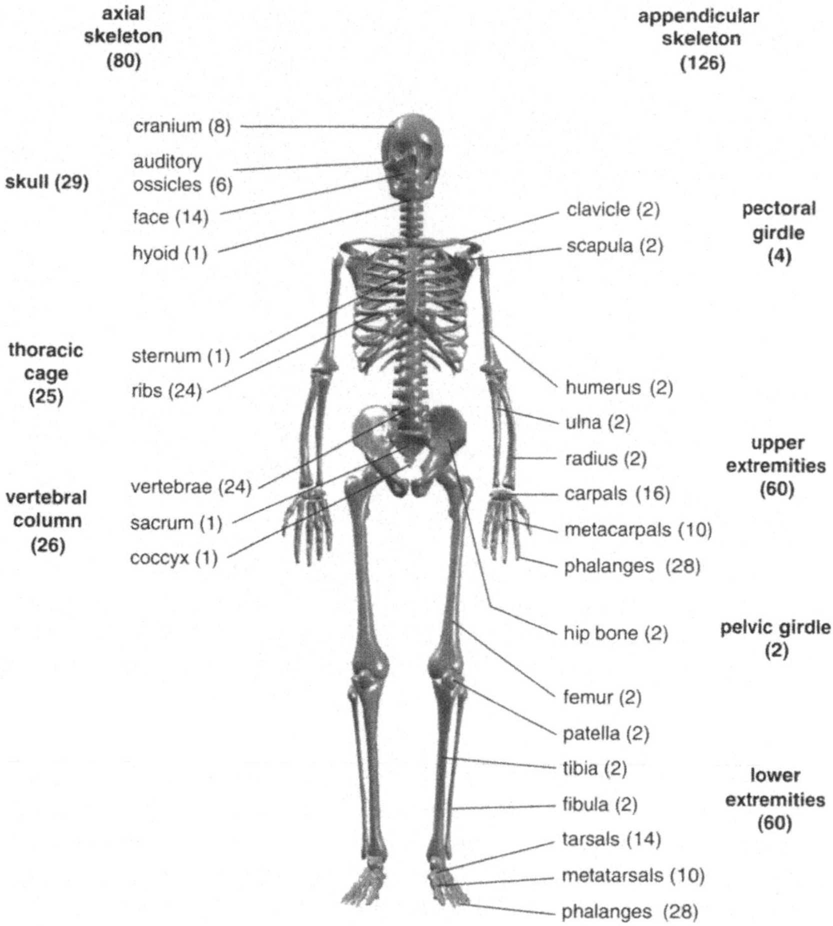

FIGURE 1.4. Frontal view of the human skeleton. The skeleton is composed of 206 bones. It is divided into two parts: the axial skeleton and appendicular skeleton. The numbers in *parentheses* indicate the number of bones of a certain type (or in a certain subgroup). The names of the major bones of the skeleton are identified in the figure.

The vertebral disks are also instrumental in determining the curvature of the spinal column. Most of the body weight lies in front of the vertebral column during standing, walking, and running. Individual disks are not of uniform thickness, but are slightly wedged. The curvatures in the cervical (neck) and lumbar (pelvic) regions are primarily caused by the greater anterior thickness of the disks in that region. The reverse S shape of the vertebral column in the standing position brings the weight in line with the body axis.

The bodies of the vertebrae are held together by longitudinal ligaments that extend the entire length of the vertebral column. There are also a

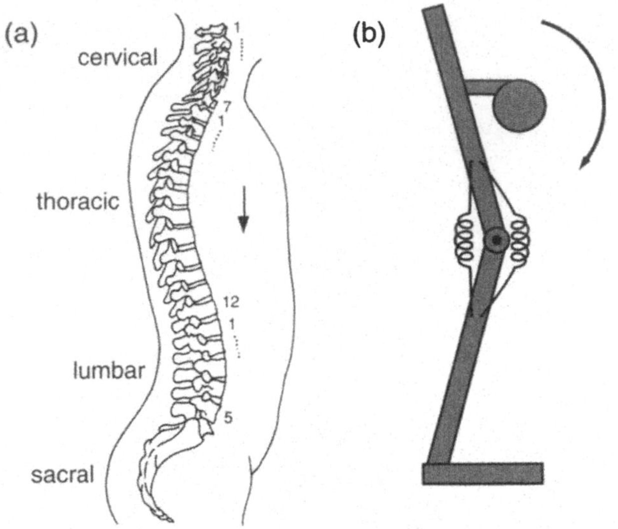

FIGURE 1.5a,b. Side view of the spinal column **(a)**. The spinal column is like a string of beads of irregular shape. It would collapse under its own weight in the absence of the large number of ligaments and muscles that are attached to it. Most of the body weight lies anterior to the spinal column, and to balance it, ligaments and the erector spinae muscles pull the spine to its curved shape **(b)**.

multitude of ligaments that connect arches of the adjacent vertebrae. The supraspinous ligament runs posteriorly along the axis of the vertebral column and plays an important role in restoring the upper body from flexion to a extension. Contractile muscles that are attached to the vertebral column provide mobility as well as stability.

The thoracic cage is directly connected to the vertebral column. The ribs arise on or between thoracic vertebrae and are connected to the sternum by cartilaginous extensions. There are 12 pairs of ribs in the thoracic cage. The joints of the axial skeleton are heavily reinforced by an array of ligaments, and as a result they permit only limited movement.

The *appendicular skeleton* consists of the bones of the upper and lower limbs and the supporting elements (girdles) that connect them to the trunk (see Fig. 1.4). Each arm articulates with the trunk at the shoulder (the pectoral girdle), and the lower extremities are attached to the trunk at the pelvic girdle. There are 126 bones in the appendicular skeleton, and approximately 300 muscles act on them to cause movement or to sustain a certain pause.

The upper limbs are connected to the trunk at the shoulder (pectoral) girdle. The shoulder girdle consists of the S-shaped clavicle (collarbone) and a broad, flat scapula (the shoulder blade). The clavicle joins at one end to the sternum and at the other end meets the scapulae. The only direct connection between the shoulder girdle and the axial skeleton is the joint between the clavicle and sternum. Skeletal muscles support and position the scapula, which has no direct bony or ligamentous connections to the rib cage. Once the shoulder joint is in position, muscles that originate on the pectoral girdle help to move the upper extremity.

The bone of the upper arm, the humerus, articulates with the scapula on the proximal end. At its distal end, it articulates with the bones of the forearm, the radius and ulna. These are parallel bones that support the forearm. Their distal ends form joints with the bones of the wrist. The radius and the ulna are connected through their entire length by a flexible interosseus membrane.

The wrist is composed of eight carpal bones that are arranged in two rows, proximal and distal carpals. In the hand, five metacarpals articulate with the distal carpals of the wrist and support the palm. Distally, the metacarpals articulate with the finger bones or phalanges. There are 14 phalanges bones in each hand.

The pelvic girdle attaches the lower limbs to the axial skeleton. The pelvis is a composite structure that is composed of the hip bone (coxae) of the appendicular skeleton and the sacrum and coccyx, the last two elements of the vertebral column. An extensive network of fibers connect the elements of the pelvis, increasing the stability of this structure under various types of loading conditions. Because the bones of the pelvic girdle bear the weight of the human body, they are more massive than those of the pectoral girdle. Similarly, the bones of the thigh and the lower leg are more massive than those of the arm and the forearm.

The long bone of the thigh, the femur, is the longest and heaviest bone in the body. More than 7% of all stress fractures in the human occur in the femur. The head of the femur joins the pelvis and the other end articulates with the tibia of the leg at the knee joint. The other bone of the lower leg, the fibula, is slender in comparison with the tibia. The fibrous membrane between these two bones stabilizes their position and provides additional surface area for muscle attachment. The fibula is excluded from the knee joint and generally does not transfer weight to the ankle and the foot. However, it is an important site for muscle attachment. In addition, the distal tip of the fibula extends laterally to the ankle joint, providing lateral stability to the ankle. About half of all stress fractures in the human occur in the tibia. These fractures are usually the result of repetitive, cyclic loading of the bone such as occurs during running, ballet, and jumping sports. As we shall see later in the text, high-impact activities drastically increase the loads carried by the bones of the lower leg. The reaction forces at the feet may be 5 to 10 times higher than the body weight during sprinting or jumping. Usually the strong muscles and mobile joints act as shock absorbers, damping the intensity of the peak load transmitted to the bone. Muscle fatigue, and stiff or immobile joints have been implicated in increased load on bone.

The patella (kneecap) is a large sesamoid bone that forms within the tendon of the quadriceps femoris, a group of muscles that extend the leg.

The kneecap prevents the knee from extensive damage caused by an impact force. It also increases the lever arm of the quadriceps muscle group, making the muscle more efficient in extending the knee.

The ankle (also called the tarsus) consists of seven tarsals. The bones of the foot include the five long bones that form the sole of the foot. Tarsals bear 25% of the stress fractures in the human. The phalanges (the toes) have the same anatomical organization as fingers. Together, they contain 14 phalanges in each foot.

1.4 Bone, Cartilage, and Ligaments

Bones are the parts of the human body that are most resistant to deformation. Unless they are broken or fractured, bones do not undergo significant shape changes during short periods. As such, they can be considered as rigid bodies in the analysis of movement and motion. In a rigid body neither the distance between any two points nor the angle between any three points changes during motion.

The bone matrix is composed of collagen fibers and inorganic calcium salts decorating these fibers (Fig. 1.6a). Collagen is the most abundant structural protein in the body. Collagen fibers bend easily when compressed but resist stretching; they have enormous tensile strength. The salts are primarily calcium phosphate and, in lesser amounts, calcium carbonate. The salt crystals can withstand large compressive forces but they are brittle and inflexible. However, when deposited on flexible collagen fibers, the resultant composite behaves differently. The bone composite possesses the best structural features of the collagen and the salt: it can withstand large compressive forces and has considerable strength against tension and torsion.

Bone is not a homogeneous material; that is, its physical properties vary with location. In a long bone, compact bone tissue (relatively dense and solid) forms the walls of the cylindrical shaft (Fig. 1.6b). The compact bone constitutes the surface layer of other bones. An internal layer of spongy bone, an open network of struts and plates, surrounds the marrow cavity. Spongy bone is also present at the expanded areas (heads) of long bones. Both compact and cancellous (spongy) bone have the same matrix composition but they differ in weight density and three-dimensional microstructure. In general, spongy bone is found where bones are not heavily stressed or where stresses arrive in many different directions. On the other hand, compact bone is thickest where the bone is stressed extensively in a certain direction.

Using shape as a criteria, bones of the human body have been classified into six categories. *Long bones* are found in the upper arm and forearm, thigh and lower leg, palms, soles, fingers, and toes. They play a crucial role in movement, functioning as lever systems. *Short bones* such as

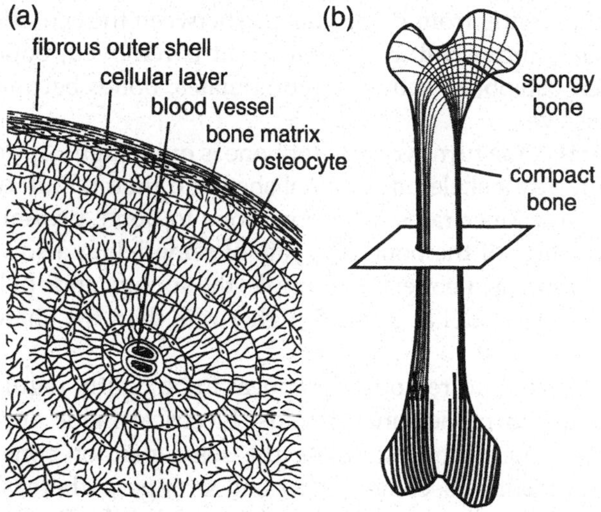

FIGURE 1.6a,b. The microstructure of a thin section of a long bone **(a)**. The thin, branching lines in the figure represent the collagen fibers decorated with calcium salts. Roughly one-third of the matrix of bone consists of collagen fibers. The balance is primarily a mixture of calcium salts. The bone cells called osteocytes are usually organized in groups around a central space that contains blood vessels. The lamellar organization in a long bone **(b)** shows that the walls of the shaft of the femur are of compact bone whereas the heads are composed of spongy bone.

those found in wrists and ankles are boxlike in appearance. *Flat bones* form the roof of the skull, sternum, the ribs, and the scapula. They protect the underlying soft tissues from the forces of impact. They also offer an extensive surface area for the attachment of skeletal muscles. *Irregular bones* such as the vertebrae of the spinal column have complex shapes with short, flat, and irregular surfaces. *Sutural bones* are small, flat, and oddly shaped bones of the skull in the suture line. Finally, *sesamoid bones* such as the patellae are usually small, round, and flat. They develop inside tendons.

Bone is a living tissue. Cells constitute approximately 2% of the mass of a typical bone. Among the bone cells, osteoblasts excrete collagen and control the deposition of inorganic material on them. They are responsible for the production of new bone. Osteoclasts, on the other hand, secrete acids that dissolve the bony matrix and release the stored minerals of calcium and phosphate. During this activity, osteoclasts are tightly sealed to the bone surface. They dissolve bone mineral by active secretion of hydrogen ions. Bone degradation products are then transported within vesicles across the cell and emptied out to the extracellular space. This process, called *resorption,* is fundamental to the regulation of calcium and phosphate concentration in body fluids. In the human body, regardless of age, osteoblasts are adding to the bone matrix at the same time osteoclasts are removing from it. The balance between the activities of these two cell types is important: if too much salt is removed, bones become weaker. When osteoblast activity predominates, bones become stronger and more massive.

On the average, the turnover rate for bone is quite high. Approximately one-fifth of the adult skeleton is demolished and then rebuilt or replaced in a year. The turnover rates vary from bone to bone, possibly depending on the function of the bone. The rate of remodeling also varies with the spatial location on a bone. For example, the spongy ends of long bones of human limbs remodel at a much higher rate than the shaft of a long bone.

The bone growth and remodeling appear to be tightly regulated in the human body by hormones and steroids. Electrical fields are known to stimulate bone repair and stimulate the self-repair of bone fractures. Heavily stressed bones become thicker and stronger, whereas bones not subjected to ordinary stresses become thin and brittle. Regular exercise serves as a stimulus that maintains normal bone structure.

Growth plates are the sites of bone growth during childhood and early adulthood. They are positioned at the spongy ends of the long bones. Osteoblasts proliferate on the surface of the growth plate and make new bone. The long bones of the average infant lengthen by 50% during the first year after birth. The bone growth rate drops to about 7% per year by age 3. The bone growth stops around 30 years of age, and between 35 and 40 the osteoblast activity begins to decline gradually while osteoclast activity continues at previous levels. Nevertheless, among all the mature tissues and organs of adult body, only one has the ability to remake itself and that is bone. When broken, bone reconstructs itself by triggering biological processes reminiscent of those that occur in the embryo. The repair begins when a class of stem cells travel to the damaged site and undertake specific tasks such as producing a calcified scaffolding around the break. Thus, a break or a fracture uncovers the remaking characteristics of bone tissue in adulthood. As discussed in Chapter 6, surgeons have utilized this capacity to lengthen limbs in people with limb abnormalities.

Cartilage is a gelatinous matrix that covers bone surfaces at a large number of articulations. It is glassy smooth, glistening, and bluish-white in appearance. It is found in the connections between the ribs and the sternum, and on the surface of articulating bones of the shoulder and hip joints, elbow, knee, and the wrist. Cartilage pads are positioned between spinal vertebrae. One important function of cartilage is to absorb compressive shocks and thereby prevent bone damage. Cartilage drastically reduces friction between opposing bony surfaces and enables rotation of one surface over the other. The only cell type found within the cartilage matrix are chondrocytes. These cells live in small pockets known as lacunae, and all nutrient and waste product exchange occur by diffusion through the matrix. Cartilage is avascular because chondrocytes produce a chemical that discourages the formation of blood vessels. The outer layer of cartilage is composed of a dense irregular connective tissue providing mechanical support and protection. Cartilage does not grow in adults, and in fact decreases in thickness with aging. Unlike other components of the skeletal system, cartilage has a poor self-repair mechanism; most cartilages cannot repair themselves after a severe injury. This is one reason why so many middle-aged runners have "bad knees."

Ligaments connect one bone to another (Fig. 1.7). These are cable-like structures consisting primarily of collagen fibers. Another fibrous protein found in ligaments is elastin. While collagen acts to oppose tensile forces, elastin acts to increase flexibility. A ligament is slightly more compliant

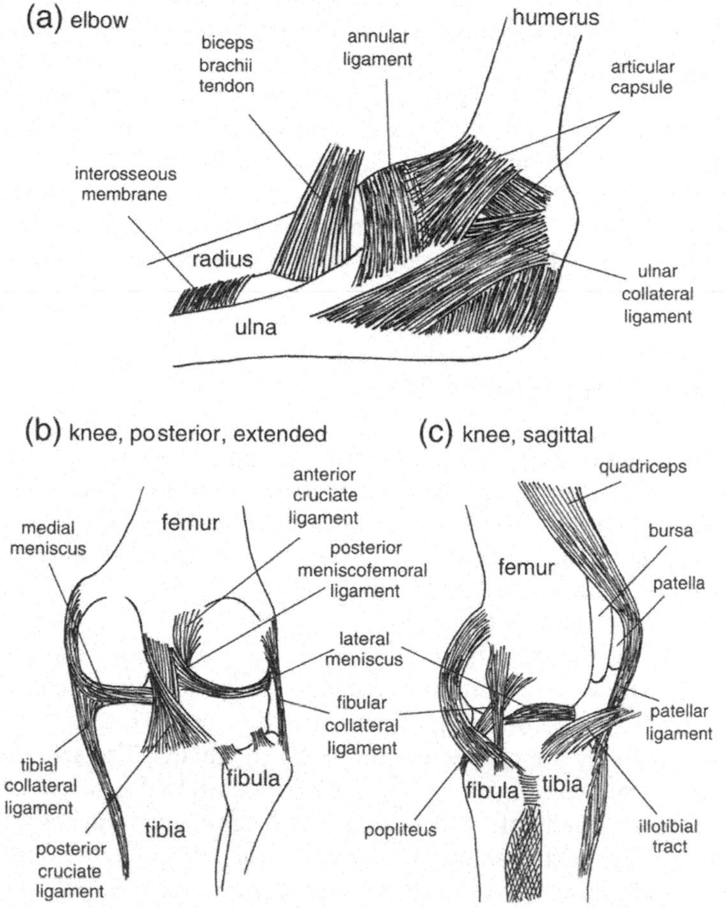

FIGURE 1.7a–c. The ligaments of the elbow **(a)** and the knee **(b,c)**. A large number of ligaments are necessary to keep multiple bone segments in place.

than a tendon but is stiffer than a muscle. Ligaments support the joints by holding the ends of bones together. Ligaments also support body organs such as the liver and hold the teeth in the jawbone.

Fibroblasts are the most abundant cells found in ligaments. These cells manufacture and secrete protein subunits to form collagen-rich extracellular fibers. Fibroblasts also secrete hyaluronic acid, a substance that gives tissue matrix its syrupy consistence. Also present are a number of immune system cells and stem cells that respond to local injury by dividing to produce additional cells for self-repair.

The joints of the upper and lower limbs contain an abundance of ligaments positioned in various directions. Seven major ligaments stabilize the knee joint (Fig. 1.7b,c). Tearing of one or two of these ligaments result in increasing mobility and instability of the knee. A news article in the November 15, 1998, issue of the *New York Times* illustrates this point. The article tells the story of Jason Sehorn, an emerging star of the Giants football team who had landed awkwardly on one knee during an exhibition game with the Jets on August 20. His lower leg lay flat on the ground sideways and his thigh was perpendicular to it. Here is how Sehorn describes the diagnosis: "The doctor came over and grabbed my knee and twisted it in one way. He looked me right in the eye and said, 'They took your a.c.l. (anterior cruciate ligament).' Then he turned it the other way. He said, 'They got your m.c.l. (medial collateral ligament)'." Ligaments are crucial for the strength of the body structure and the control of movement.

1.5 Joints of the Human Body

Human joints can be classified into three groups based on the range of motion permitted at the joint. An immovable joint is called *synarthrosis* in anatomy. These are the joints found between the bones of the skull and between teeth and the surrounding bone of the jaw. In the skull, the edges of the bones are interlocked and bound together by dense connective tissue. These joints are called sutures.

The second group of joints, such as the distal articulation between tibia and fibula, allow for slight movements. Such a joint is called *amphiarthrosis*. The bones forming these joints do not have to be touching each other but they are connected tightly by ligaments. The articulations between adjacent vertebrae form this type of a joint. In this case the bones are separated by pads of fibrocartilage. The slight movements allowed between adjacent vertebrae permit the vertebral column to bend forward and backward and to the sides as well as rotate to some extent about its longitudinal axis.

The joints that allow considerable motion of the articulating bones are called freely moving joints (*diarthrosis* or synovial joints). These joints are typically found at the end of the long bones, such as those of the leg and arm. Under normal conditions, the bony surfaces do not contact one another because articulating surfaces are covered by cartilage and because there is a layer of fluid (synovial fluid) between the opposing surfaces. The matrix of the cartilage contains water and is squeezed out in compressive loading, creating a lubrication layer on the surface of the interface. This thin layer of fluid reduces the frictional forces and help distribute the compressive stress more uniformly along the surfaces of the articulating bones. The hip, knee, and ankle joints are all examples of synovial joints.

A joint is called a monoaxial joint when rotation is allowed only on one axis. An example of a monoaxial joint is the one that attaches the two vertebrae which are most proximal to the skull. This joint allows rotation of the head to the left or to the right. Because it acts like a pivot, it is called a *pivot joint* (Fig. 1.8a). Another example of pivot joint is the articulation between the forearm bones, the radius and ulna, at the elbow. These bones have the capacity to rotate relative to each other along the long axis of the forearm.

The elbow and the knee are called *hinge joints* because they permit flexion and extension in the sagittal plane (Fig. 1.8b). In terms of their physical function, these joints are much like door hinges. The elbow and the knee joints are monoaxial joints. As shown in Fig. 1.7, in the knee joint the rounded surface of the distal end of the femur and the flatter surface of the tibia do not fit together. Collateral ligaments on either side hold the bones together while allowing the knee to bend. But because these ligaments run along the axis of the leg, they cannot prevent small movements of one bone on the other. Angled ligaments found within the capsule of the knee joint, the anterior cruciate ligament and posterior cruciate ligament, also connect the femur to the tibia. These ligaments are crucial for the stability of the knee. Also contributing to stability are the two crescent-shaped wedges of fibrous cartilage (the menisci) that lie in the gap between the articulating surfaces of the femur and the tibia. The menisci are not freestanding but are held together by ligaments. They help distribute the contact force between the femur and tibia over the surface of articulation.

The ankle joint permits sole elevation and sole depression. The articulation between the lower end of the tibia and the talus of the foot is responsible for bending the ankle toe up and toe down. The ankle allows rotations in other directions too, thanks to the movement of the small bones of the ankle relative to each other. The complexity of the articulating surfaces and the abundance of ligaments of this joint can be a challenge even to the leading experts of the biology of movement. "I do not fully understand the complicated array of ligaments that hold the ankle bones together (and I do not think any one else does, either) so I will not try to explain them," writes R. McNeill Alexander in his illuminating book, *The Human Machine*.

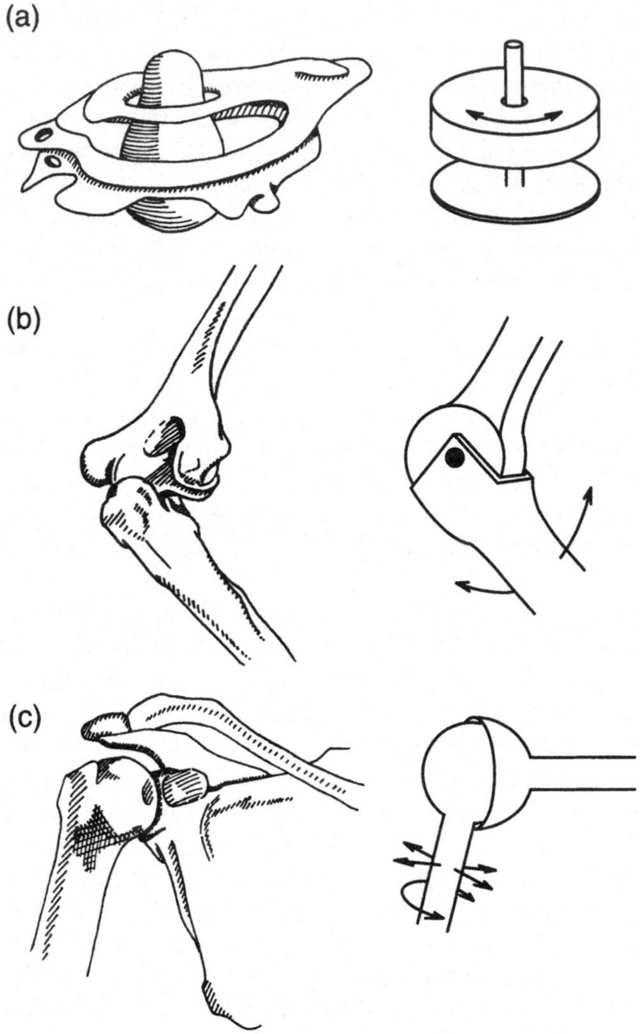

FIGURE 1.8a–c. Engineering representations of the synovial joints of the human body. The figure shows the pivot joint between axis and atlas, the two vertebrae closest to the head **(a)**, the elbow **(b)**, and the shoulder joint **(c)**. The pivot joint between atlas and axis allows the head to rotate to the left and to the right. A hinge joint allows relative rotation in the plane of two articulating long bones. The knee and elbow are examples of hinge joints. A ball-and-socket joint allows rotation in three directions. This type of joint has the most degrees of freedom in movement. Examples are the shoulder joint and the pelvic girdle.

Ball-and-socket joints are articulations in which the round head of one bone rests within a cup-shaped depression in another (Fig. 1.8c). These joints are called multiaxial because they permit rotation on more than one plane or axis. Examples of these joints include the *shoulder joint* and the *hip joint.* The shoulder joint is formed by the head of the humerus and the small, shallow pear-shaped cavity of the scapula. This joint allows the greatest range of motion of any joint in the body, mainly because the cavity of the scapula is shallow in depth and also the articular capsule enclosing it is remarkably loose. Perhaps because of the greater degrees of freedom, the shoulder joint is also the most frequently dislocated joint.

1.6 Physical Properties of Skeletal Muscle

Muscles are composed of bundles of long and thin cells that are called muscle fibers. Bundles of skeletal muscle fibers are encased by a dense fibrous connective tissue layer called the epimysium. Bundles are separated from each other by connective tissue fibers of the perimysium, and within each bundle the muscle fibers are surrounded by a delicate network of reticular fibers called the endomysium. Scattered satellite cells lie between the endomysium and the muscle fibers. These cells function in the repair of the damaged muscle tissue. The connective tissue fibers of the endomysium and perimysium are interwoven. These fibers converge at each end of the muscle to form tendons. Tendons are a bit stiffer than ligaments. They are woven into the fibrous outer layer of bone. This meshwork provides an extremely strong bond. As a result, any contraction of the muscle exerts a pull on the bone to which it is attached. Tendons not only transmit the muscle force to the ends of bones but they also have a stabilizing influence on articulations. In the case of the shoulder joint, the tendons that cross the shoulder joint reinforce the joint more effectively than the ligament reinforcements. The tendon of the long head of the biceps secures the head of the humerus tightly against the glenoid cavity of the scapula.

Approximately 70% of all tendon injuries are sports related. If the elongation of tendon is less than 4% in an activity, the tendon will return to its original length when unloaded. Above this strain level, however, cross-links between the fibers of a tendon may begin to fail and some fibrils may rupture. Higher loads such as occur in an accidental fall result in the macroscopic rupture of the tendon. Examples of tendon disorders include achilles tendinitis, caused by running, patellar tendinitis, the result of running and jumping, and carpal tunnel syndrome of the tendons of the wrist and the fingers and tennis elbow, from the overuse of wrist extensor muscles. In most cases, resting the tendon may alleviate the problem. The treatment of tendon injuries may prove difficult because of the need to preserve the balance between resting the injured tendon and preventing atrophy of the surrounding muscles and joints.

Muscle fibers measure as much as 30 cm in length. In the contracted state, a muscle fiber can produce a tensile force of 50 N/cm^2 of cross-sectional area. The force of contraction depends strongly on the temperature.

For a given stimulus, the peak force that can be developed increases with increasing temperature. One purpose of the warm-up period preceding athletic activity is to increase the temperature of the muscles so they produce greater forces during the subsequent athletic activity. Another purpose of warm-up is to stretch the muscle, as the peak tensile force a muscle fiber can generate also depends on the length of the fiber at the initiation of the contraction. A fiber that contracts from a prestretched condition develops forces greater than that of a fiber contracting from resting length.

A muscle fiber can produce tensile force while shortening. However, the faster a fiber shortens, the less force it can exert. Experiments with single muscle fibers indicate that there is a maximum speed of shortening above which a muscle cannot shorten fast enough to keep up with the apparatus driving the shortening. The single muscle fiber experiments also show that when a muscle is forcibly stretched it can exert very large forces, up to a limit; excessive forces acting on the muscle can cause it to tear.

Muscle cells convert chemical energy found in fatty acids and blood sugar glucose into movement and heat. Muscle fibers contain thousands of smaller strands called myofibrils. The smallest contractile unit of a myofibril is called a sarcomere. As shown in Fig. 1.9, a sarcomere is composed of thin and thick filaments; the actin-rich filament is called the thin filament and the myosin-rich filament is called the thick filament. Actin and myosin are protein molecules that are associated with motility in living systems. Relative translation between thick and thin filaments is responsible for much of the change in length of a muscle during contraction (Fig. 1.9). According to the sliding filament theory of muscle contraction, myosin heads on the thick filaments (crossbridges) interact with actin-binding sites on the thin filaments. The crossbridges are presumed to generate force only when they are attached to actin.

Skeletal muscle must be stimulated by the central nervous system before it contracts. Messages to activate a muscle travel from the brain to the nervous system and to individual muscle fibers. These signals are carried by nerve cells called motor neurons. The amount of tension produced by a skeletal muscle depends on both the frequency of stimulation and the number of motor units involved in the activation. In the muscles of the eye, a motor neuron might control only two or three fibers because precise control is extremely important. On the other hand, in leg muscles more than 2000 fibers are controlled by a single neuron. All the fibers in a motor control unit contract at the same time. Even when a muscle appears to be at rest, some motor units in the muscle may be active. The contraction of the activated muscle fibers does not produce enough pull to cause movement but they do tense the muscle. The resting tension in a skeletal muscle is called muscle tone. Resting muscle tone stabilizes the positions of bones and joints and maintains body position.

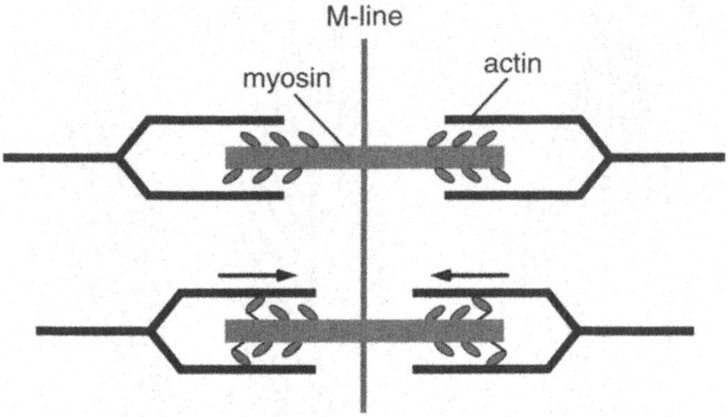

FIGURE 1.9. A simplified schematic diagram of thick and thin filaments during two stages of contraction. According to the sliding filament model of muscle contraction, muscle force is generated by the interaction of myosin heads on the thick filament with the actin sites on the thin filament. This interaction becomes biochemically favorable immediately after the stimulation of the muscle by the central nervous system.

The shape of a muscle provides clues to its function. In a *parallel* muscle, muscle fibers are parallel to the long axis of the muscle (Fig. 1.10a). Most of the muscles in the human body are parallel muscles. While some parallel muscles (abdominal muscles) form flat bands, others are spindle shaped with cordlike tendons at both ends. Such a parallel muscle has a belly. When it contracts it gets shorter and the belly increases in diameter to keep the muscle volume constant. When muscle fibers are parallel to the long axis of the muscle, all the fibers contract the same amount.

In a *pennate muscle*, one or more tendons run through the body of the muscle, with fibers attached to them at an oblique angle (Fig. 1.10b). Pennate muscles do not move their tendons as far as parallel muscles do because the fibers pull on the tendon at an angle less than 90°. On the other hand, pennate muscle contains more muscle fibers than a parallel muscle of the same size. Depending on the pennate angle, a pennate muscle has the potential of generating larger levels of tension than a parallel muscle of the same size. If all the muscle cells in a pennate muscle are found on the same side of the tendon, the muscle is called unipennate. If the tendon branches within the pennate muscle, then the muscle is said to be multipennate.

In a *convergent* muscle, muscle fibers are based over a broad area, but all the fibers come together at the common insertion site (Fig. 1.10c). In this muscle the direction of the pull can be changed by activating one group of muscle cells at any one time. When all the cells in this muscle group are activated at once, they generate less force than a parallel muscle of the same size. This is because muscle fibers on opposite sides of the

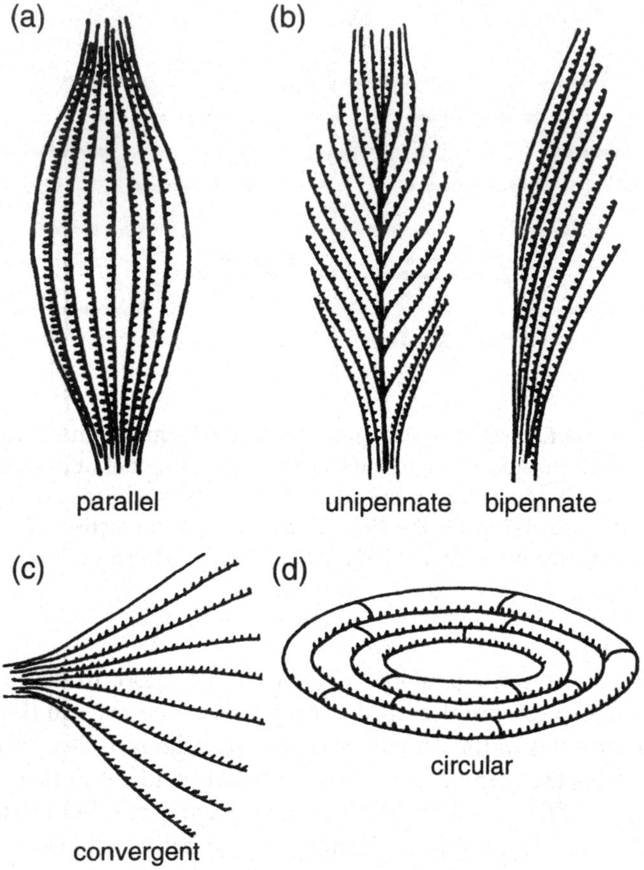

FIGURE 1.10a–d. Various types of muscle organization in the human body. The muscle fibers are aligned in parallel to the axis of the parallel muscle **(a)**. In pennate muscles, the tendon and the muscle fibers are oriented at an oblique angle **(b)**. In convergent muscle, muscle fiber direction varies within the muscle but all the fibers converge at a point **(c)**. Circular muscles contract to control the size of an orifice of the human body **(d)**.

tendon pull in different directions, and thus the resultant force is smaller than the absolute sum of its parts. In a *circular* muscle (Fig. 1.10d), the fibers are concentrically arranged around an opening such as the mouth. When the muscle contracts, the diameter of the opening decreases.

According to their primary functions, muscles are grouped into three categories. A *prime mover (agonist)* is a muscle whose contraction is chiefly responsible for producing a particular movement. For example, the *biceps* is made up of two muscles on the front part of the upper arm that are responsible for flexing the forearm upward toward the shoulder. A *synergist* muscle contracts to help the prime mover in performing the movement of a bone. Synergist muscles may assist the prime mover at the initiation of motion or help stabilize the point of origin. *Antagonists* are muscles whose action opposes that of the agonist; that is, if the agonist produces flexion, the antagonist will produce extension. When an agonist contracts to produce a particular movement, the corresponding antagonist will be stretched. The tension in the antagonist is adjusted by the nervous system to control the speed of the movement and ensure the smoothness of the motion. The triceps of the upper arm act as antagonist to the biceps. In this capacity, they play a role in stabilizing the flexion movement.

1.7 Muscle Groups and Movement

There are layers of muscles in the muscular system. Muscles visible at the body surface are often called *externus* and *superficialis*, and they typically serve important functions to stabilize a joint or cause movement. With the naked eye it is often possible to identify the muscle group responsible for a certain action.

Major muscle groups of the body are shown in Fig. 1.11. The *axial musculature* begins and ends on the axial skeleton. Belonging to the group of axial musculature are the muscles of the head and neck that move the face, tongue, and larynx. The muscles of the spine include flexor and extensor muscles of the head, neck, and spinal column. The oblique and rectus muscles form the muscular walls of the trunk. In the chest area these muscles are partitioned by the ribs, but over the abdominal surface, they form broad muscular sheets. Trunk muscles keep the internal organs of the body intact, and in that function, they are similar to the corset that nineteenth-century women were obliged to wear in the Western world.

The muscles that stabilize the shoulder, hip, and the limbs are called the *appendicular musculature*. These muscles account for approximately 40% of the human musculature. The appendicular musculature is divided into two groups: (1) the muscles of the shoulders and upper extremities (arm, forearm, hand) and (2) muscles of the pelvic girdle (hip joint) and lower extremities (thigh, leg, foot).

Some of the muscles of the appendicular musculature act on a single joint. These are called monoarticular muscles. Gluteus maximus, the major muscle group of the buttocks, is a monoarticular muscle; it only acts on the hip joint. Other muscles may act at two or more joints. For example, the hamstring muscle, the semitendinosus and biceps femoris, traverses two joints and acts both on the hip and the knee. These muscles have the capacity to extend at the hip and flex at the knee. The quad muscle, rectus femoris, and the calf muscle, gastrocnemius, also act on two joints and as such are called biarticular muscles. What is the advantage of having polyarticular muscles in the human body? A plausible answer to this question may be that biarticular muscle, by affecting two joints at

(a)

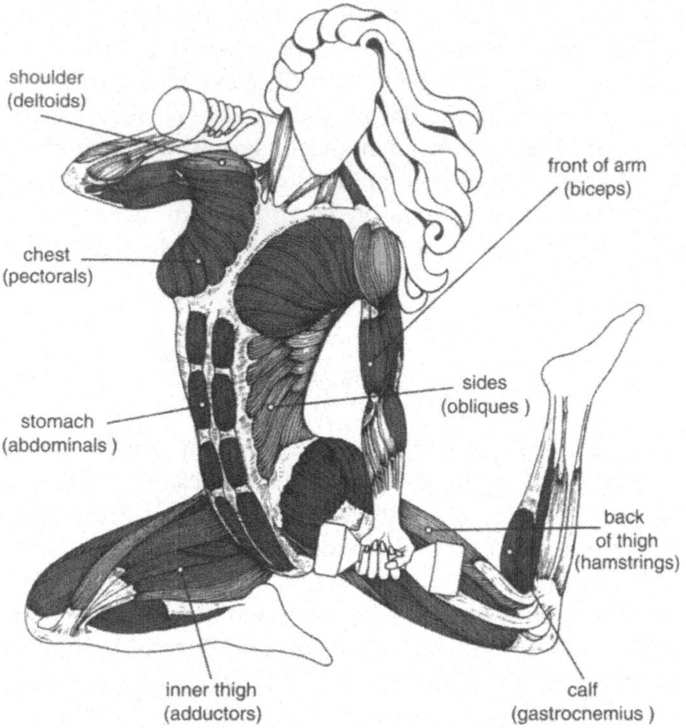

FIGURE 1.11a,b. Major muscle groups of the human body: front view **(a)** and back view **(b)**. (Copied with permission from Bruce Algra Fitness Chart series).

the same time, helps prevent ligament injury in sudden acceleration and deceleration. The rate of rotation of a joint has to be zero at the instant the joint is fully extended. Otherwise, structures traversing the joint run the risk of damage. It has been suggested that biarticular muscles prevent the rotational energy of segments from reaching levels that could lead to injury to the ligaments and tendons.

Of the nine muscles that cross the shoulder joint to insert on the humerus, only the superficial pectoralis major, latissimus dorsi, and deltoid muscles are prime movers of the arms. The remaining six are synergists and fixators.

The *pectoralis major* muscle group of the chest originates in the cartilages of ribs two through six in the front, on the body of the sternum, and the medial portion of the clavicle, and inserts on the humerus (Fig.1.12). This fanlike muscle is of the *convergent* type and belongs to the appendicular musculature. It is responsible for pulling the upper arm across the body. It is a major climbing muscle in the sense that if the arms are fixed above the head, the massive power of the muscle can be used to pull the trunk upward. Because it is a triangular muscle, the line of action of the

(b)

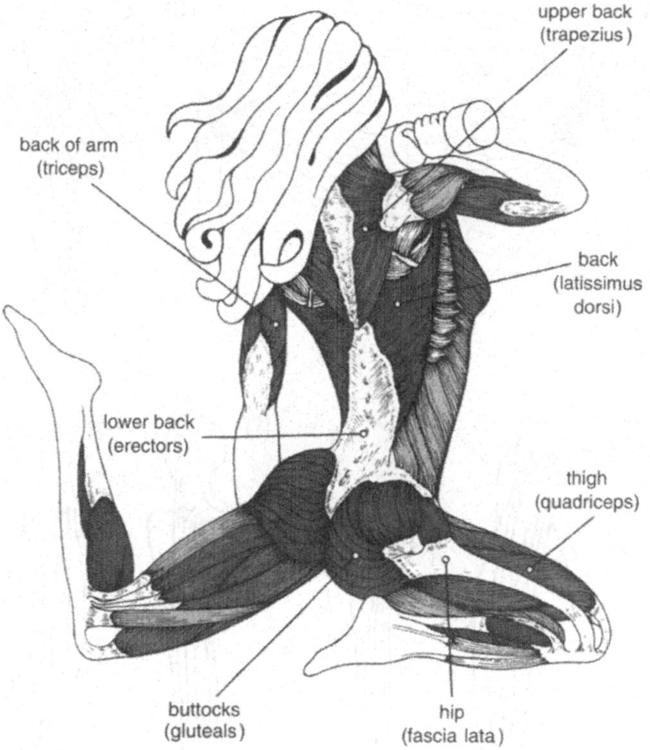

FIGURE 1.11a,b. *Continued*

force it generates may vary depending on which fibers are activated. As such the pectoralis flexes, adducts, and medially rotates the humerus. The muscle group under pectoralis major is the *pectoralis minor*. Its function is to move the shoulder girdle. It originates in ribs one through three and inserts at the scapula. *Pectoralis minor* depresses and protracts the shoulder, rotates the scapula, and elevates ribs if the scapula is made stationary with the use of other muscles that move the shoulder girdle.

Trapezoid muscles are located on the upper region of the back just below the neck. They originate along the middle of the neck and back and insert upon the clavicle and the scapula. The left and the right trapezoid muscles form a broad diamond. They are responsible for elevating the shoulders (shrugging) and for extending the head backward. Similar to the pectoralis major, this muscle group is innervated by more than one nerve, and specific regions can be made to contract independently. As a result, both the direction and the magnitude of the trapezoid muscle force vary greatly.

Latissimus dorsi is the largest muscle group of the upper body (Fig. 1.12). It is located on the back side of the body below the shoulder blades,

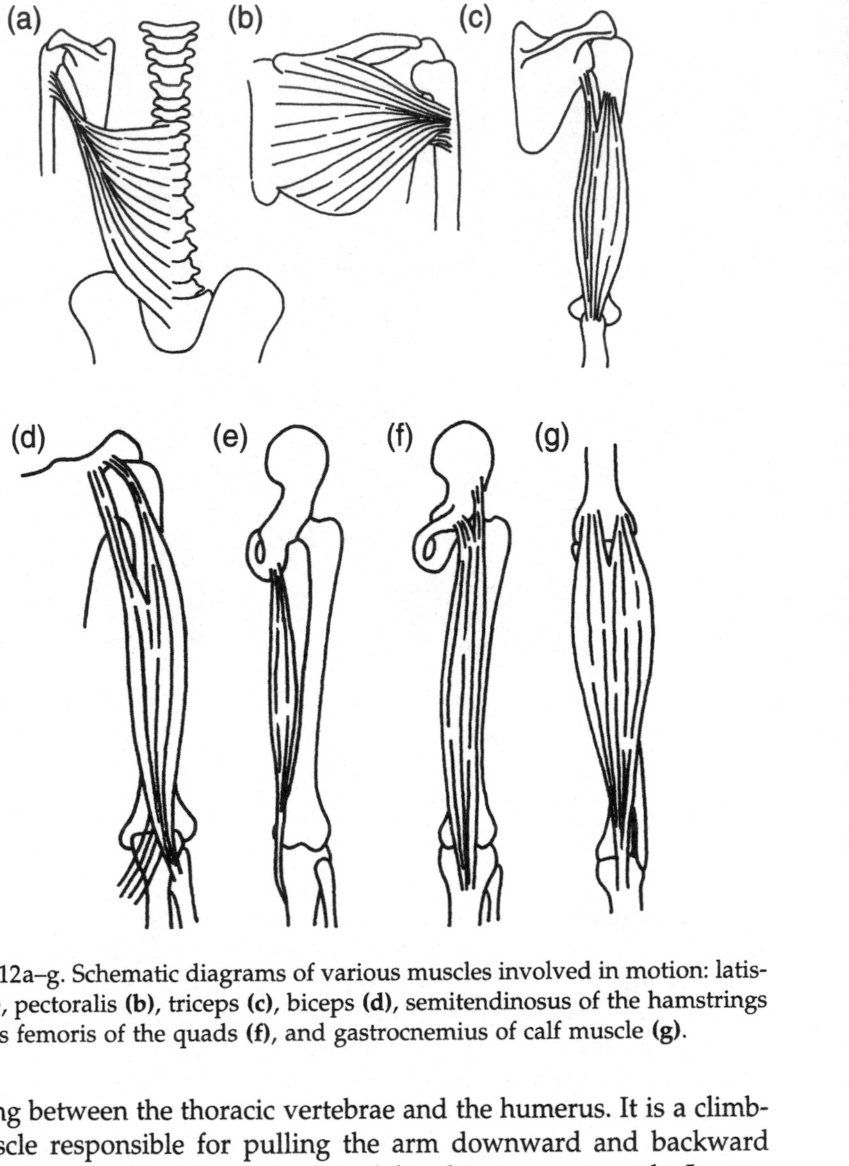

FIGURE 1.12a–g. Schematic diagrams of various muscles involved in motion: latissimus **(a)**, pectoralis **(b)**, triceps **(c)**, biceps **(d)**, semitendinosus of the hamstrings **(e)**, rectus femoris of the quads **(f)**, and gastrocnemius of calf muscle **(g)**.

stretching between the thoracic vertebrae and the humerus. It is a climbing muscle responsible for pulling the arm downward and backward against resistance. This is a very powerful and important muscle. In people using crutches, the latissimus dorsi pulls the trunk forward relative to the arms. Because this muscle attaches to the pelvis, in patients with paralysis of the lower half of the body it can be used to produce movement of the pelvis and the trunk. Patients wearing calipers and using crutches can produce a modified gait by fixing the arms and hitching the hips by the alternate contraction of right and left latissimus dorsi.

Deltoids of the shoulder, a multipennate muscle group, consists of 11 muscles located on the upper side of the arms. Deltoids originate on both the clavicle and scapula and end on the humerus. This muscle group is responsible for raising the upper arm forward, lifting it sideways away from the body, and for rotating the arms front and back.

Rectus abdominis is an axial muscle group that is arranged in parallel between the chest and the pelvis. It originates at the hip bone and inserts at ribs five through seven and at the lower tip of the sternum. It is responsible for spinal forward flexion and is used to contract the upper body toward the lower body. *Obliques,* located on both sides of the abdomen, are also part of the axial muscle group. They originate at the vertebrae and insert at the rib cage. They are responsible for moving the upper body from side to side.

Erector spinae is the main muscle group in the lower back. The erector spinae muscles include superficial and deep layers that align approximately along the long axis of the body. When contracting together, they extend the spinal column. When only the muscles on one side contract, the spine is bent laterally. Overall this muscle is responsible for extending or straightening the upper body from a bent-over position.

Biceps of the upper arms originates on the scapula and ends on the radius of the forearm. It is a parallel muscle (Fig. 1.12). Its major responsibilities are to supinate the forearm (as in inserting a corkscrew) and to flex the forearm upward toward the shoulder (as in pulling out the cork). Because the biceps is biarticular, it assists pectoralis major in the flexion of the upper arm. However, in the midst of other strong muscle groups that move the shoulder, its effect on the shoulder is of secondary importance. The *brachialis* and *brachioradialis* act as synergizers of biceps in flexing the lower arm.

Triceps, a group of three muscles, is located on the back of the upper arm (Fig. 1.12). The long head of triceps brachii originates on the scapula and the lateral and the medial heads on the humerus. All three insert on the ulna. Triceps function as agonists to biceps; they are responsible for straightening the arm at the elbow. The triceps extend the forearm during push-ups, and in various forms of pushing and punching. They are essential to execute a karate chop. A person in a wheelchair uses triceps to push the wheel around and propel the chair forward. In extending the forearm, triceps are aided by synergizers such as *anconeus.* In addition to these major muscle groups that move the shoulder, upper arm, and forearm, there are a large number of muscles that are specialized to flex and extend the wrist and to move the hand and the fingers. In fact, the arm contains 72 muscles, most of which are found in the lower arm. Their functions are to move the hand and the wrist. Small muscles in the hand control the finger and thumb movements.

Gluteus maximus, gluteus medius, and *gluteus minimus* are the main muscle groups that make up the buttocks. They originate at different locations on the hip bone and insert on the femur. Gluteus maximus extends the thigh in such activities as stepping up onto a stool, climbing stairs, and running. With the hamstrings it raises the trunk from a flexed position. Gluteus medius and gluteus minimus abduct and medially rotate the thigh and support the pelvis in walking and running. *Adductors* are five muscles that are located on the inside area of the upper leg; their function is to bring the legs together. They are monoarticular muscles.

Quadriceps (the vastus lateralis, the vastus intermedius, the vastus medialis, and the rectus femoris) are located at the front of the leg above the knee (see Fig. 1.12). They are responsible for straightening the lower leg at the knee joint. We use quads in stair climbing, squats, and walking and running. In the standing position this muscle group performs very little action because the knees are in the locked position. That is why a person collapses when their knees are knocked from behind.

Hamstrings are the primary muscles located at the back of the thighs (Fig. 1.12). They are the semimembranosus, the semitendinosus, and the biceps femoris, which is of the bipennate type. These are responsible for contracting and extending the lower leg and for raising the heel toward the buttocks. Another important function is raising the trunk from a flexed position. This action requires a great deal of power because the weight of the trunk acting on the other side of the hip joint produces a large moment and these muscles work with a very short lever arm in the hip joint. This difficult task explains why hamstrings typically are quite bulky.

Gastrocnemius, the muscles of the calves, are located on the back side of the lower leg below the knee (Fig. 1.12). They are responsible for raising up onto the toes. These muscles are essential for such activities as running, walking, and jumping.

According to the rate at which a muscle generates force or shortens following activation, it falls into one of these three categories: fast, slow, and intermediate. Fast fibers are large in diameter. The fast muscles have the capacity to contract within 0.01 seconds (s) or less following stimulation. However, because readily available chemical energy within a muscle cell can be rapidly exhausted in sustaining contraction, fast fibers, fatigue easily. Slow fibers are about half the diameter of fast fibers, and they contract three times slower than a fast muscle. Intermediate fibers contract at intermediate rates. The percentage of fast versus slow fibers in each muscle is genetically determined, and there are significant individual differences. Marathon runners with a high proportion of slow muscle fibers in their leg muscles outperform those with faster muscle fibers.

The proportion of intermediate fibers within a muscle tissue changes with physical conditioning. Repeated and exhaustive contraction of a skeletal muscle leads to the hypertrophy or enlargement of the muscle. Following repeated stimulations that cause near maximum tension, the cross-sectional area of the fiber as well as the number of contractile fibrils increase. If used repeatedly for endurance events, fast fibers begin to function as intermediate fibers. On the other hand, when a skeletal muscle is not stimulated by a motor neuron on a regular basis, it loses muscle tone and mass. Muscle atrophy is initially reversible, but dying muscle fibers are not replaced. When a muscle tears or strained, it was believed until recently that the damaged muscle had to depend on local cells to repair injury. However, recent research indicates that bone marrow contains cells that repair damaged muscles. This finding raises the exciting possibility that it may one day be possible to replenish degenerating muscles with fresh cells from the patient's bone marrow.

1.8 Summary

The structural frame of the human body is a tree of bones that are linked together by ligaments in joints called articulations. There are 206 bones in the human body. Bone is a facilitator of movement and protector of the soft tissues of the body. Approximately 700 muscles pull on various parts of the skeleton, using the bones as levers to preserve a certain posture or to produce movement. These muscles are connected to the bones through cable-like structures called tendons or to other muscles by flat connective tissue sheets called aponeuroses. About 40% of the body weight is composed of muscles. Often the various large muscles of the human body produce forces that are multiples of the total body weight. Skeletal muscles contract in response to stimulation from the central nervous system and are capable of generating tension within a few microseconds after activation. A skeletal muscle might be able to shorten as much as 30% during contraction. The condensed outline of human anatomy presented in this chapter may be sufficient to follow the literature on human body mechanics. Nevertheless, the reader is recommended to consult an anatomy book for details and further insights. The references listed at the end of the book contain a number of books on human anatomy and movement science.

1.9 Problems

Problem 1.1. Consult an anatomy book to identify the primary functions of muscles and other anatomical structures of the shoulder, arm, forearm, thigh, and leg.

Problem 1.2. Build a model of a spinal column using a nylon string, beads, rubber bands, and corks. How does the bead shape affect the curvature of the model spine? Is it possible to construct a column using beads of spherical shape? Note that the rubber bands, if they were to simulate muscles, would have to be prestretched before being hooked onto the beads.

Problem 1.3. The primary component of the shoulder girdle, the scapula, is not connected to any bone in the human body but it is positioned by a number of muscles acting on it. Take a piece of cardboard or styrofoam and position it in space by using hooks and strings. How many different strings do you need to fix the position of an object in space? Could you fix the object in space by attaching all the strings to the same point? Note that the position of a point in space can be defined by specifying its three coordinates. The spatial configuration of a solid body can be specified in terms of three space coordinates of any one particle of the body and the three angles that determine the orientation of the body in space. Thus, a solid body that is free to move in space is said to have six degrees of freedom.

Problem 1.4. Consult an anatomy book to reason why the upper arm and thigh contain a single bone each (humerus and femur) whereas the forearm and the leg have two bones: radius and ulna of the forearm, and tibia and fibula of the leg, respectively.

Problem 1.5. Consult an anatomy book to explain the functions of various neck muscles.

Problem 1.6. Think of analogies between human body structures and suspension bridge cables, beams, columns, domes, arches, styrofoam packing material, heel cups prescribed by orthopaedists, tent sheets, and corsets.

Problem 1.7. Suppose a parallel muscle of the arm such as the biceps weighs 4 kg and the total length of the muscle–tendon complex is 32 cm. Explain how the muscle force and the extent of shortening depend on the length of tendon. Note that the longer is the tendon, the shorter is the muscle and the bulkier it is at the midsection. Assume that the maximum tension exerted by the biceps is proportional to the maximum diameter of this muscle in the resting state.

Problem 1.8. Search the anatomy literature for a discussion of the types of synovial joints. Provide definitions for plane (gliding), condyloid, and saddle joints, and cite examples.

Problem 1.9. Summarize the mechanism of the sliding filament theory of muscle contraction. What are the basic data supporting this theory? Why haven't researchers been able to obtain photomicrographs of crossbridges pulling on actin filaments at sequential instances of time? Summarize Hill's force–velocity relation of skeletal muscle. Why is this relation called phenomenological?

Problem 1.10. Why is the sole of the foot covered by a thick fatty connective tissue (pad)? Which bones are subjected to the greatest impact force during running?

Problem 1.11. Identify biarticular muscles of the leg. Speculate what roles they would play during vertical jumping.

Problem 1.12. Which muscle group of the upper arm do we use to get out of a armchair?

Problem 1.13. When a muscle shortens while producing movement, it is said to contract concentrically. If the muscle extends while contracting, it is contracting eccentrically. Determine which phase of the push-ups (upward phase or downward phase) the triceps contract concentrically.

Problem 1.14. Which biarticular muscle group of the thigh controls the lowering forward of the trunk?

Problem 1.15. A tendon can withstand tensile forces of 10,000 N/cm^2 of cross-sectional area without tearing. The tendon is thicker than it needs to be to transmit muscle force to the bone. Why do tendons grow so thick? In his book entitled *The Human Machine*, R. McNeill Alexander attributes this to the elastic properties of the tendon and uses the analogy of bicycle brakes. If the cables that work the bicycle brakes were made of rubber rather than steel wire, one would have to move the brake levers a long way before stopping, resulting in longer braking times. Discuss this topic within the context of the tendons of the hand muscles and playing piano.

2

Laws of Motion: Snowflakes, Airborne Balls, Pendulums

2.1 Laws of Motion: A Historical Perspective

All living as well as nonliving objects that are large enough to be visible through a light microscope obey the laws of motion first formulated in mathematical terms by Sir Isaac Newton in 1687 in his book *Philosophica Naturalis Principia Mathematica.*

The first of the three laws Newton formulated is about the resting state: an object (body) remains at rest unless it is compelled to move by a force exerted on it. The first law stems from Galileo's assertion that "a moving body thrown on a horizontal plane, without any obstacle, will remain in uniform motion indefinitely if the plane extends to infinity." The first law is easy to grasp when we consider the behavior of an uncooperative, sulking dog. No amount of enticement, throwing a frisbee or offering a bone cookie, will make the dog move an inch; she pretends to be dead. An option is to pull on her leash and drag her. To do that, one has to overcome the frictional force exerted on the dog by the ground. Once the dog realizes she will not be able to resist the force of the leash as a paralyzed object, she might well reposition herself on the floor and try to balance the force of the leash with the contact forces acting on her paws.

As this example shows, a number of forces may act on an object that is at rest. According to the first law, a building is at rest because its weight is balanced vertically by the upward force exerted on it by the ground. A building will not move unless it is acted upon by wind or earthquake forces that cannot be balanced (Fig. 2.1a). A ballerina can stand still on the tiptoes of one foot by balancing the weight of her body with the contact force exerted by the ground. She is at rest, unless of course she loses her balance and the contact force exerted by the ground is no longer equal and opposite to the force of gravity (Fig. 2.1b). In the diagrams of Fig. 2.1 we have identified all the forces that act on a given body. Such diagrams are called free-body diagrams.

Newton's second law is the fundamental law of motion. According to the second law, an object will accelerate in the direction of the unbalanced

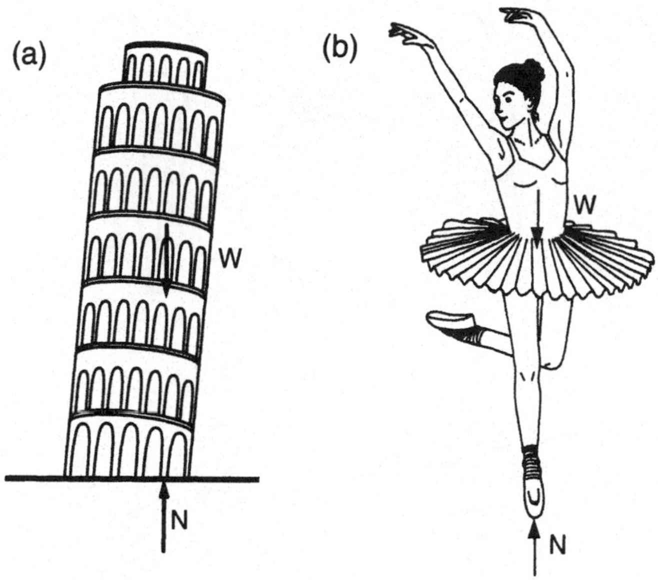

FIGURE 2.1a,b. Objects in static equilibrium: the Pisa tower **(a)**, and a ballerina **(b)** holding a delicate balance on the toes of her one foot. The *arrows* in the figure indicate the forces acting on each object. The symbol **W** usually denotes the weight of a body and **N** is the ground force exerted on the body.

force. The magnitude of acceleration will be equal to the magnitude of the resultant (unbalanced) force divided by the mass of the object. If an object is at rest or moving with constant velocity, the resultant force acting on the object must be equal to zero.

The path toward the precise formulation of the second law was torturous. Could bodies exert force on each other without establishing contact? Could the gravitational force acting on an object depend on the motion of the object? How would one measure gravitational force acting on a body in motion? These are some of the pertinent questions physicists had to consider before formulating the laws of motion. There was a lot of ambiguity about the concept of force. In his famous book *Principia* (1687), Newton wrote that he considered forces mathematically and not physically. But later on, he was compelled to admit their physical reality, for no true physics could be constructed without them.

Newton's third law states that for every action, there is an equal and opposite reaction. That is, a force on object 1 caused by interactions with object 2 is equal and opposite to the force on object 2 caused by the interactions with object 1. The third law may appear counterintuitive at first glance. It is difficult to imagine, when a boxer hits a slender person half his weight, that he is automatically hit back with the same intensity of force (Fig. 2.2a). When a car hits a tree, the tree hits the car back with the

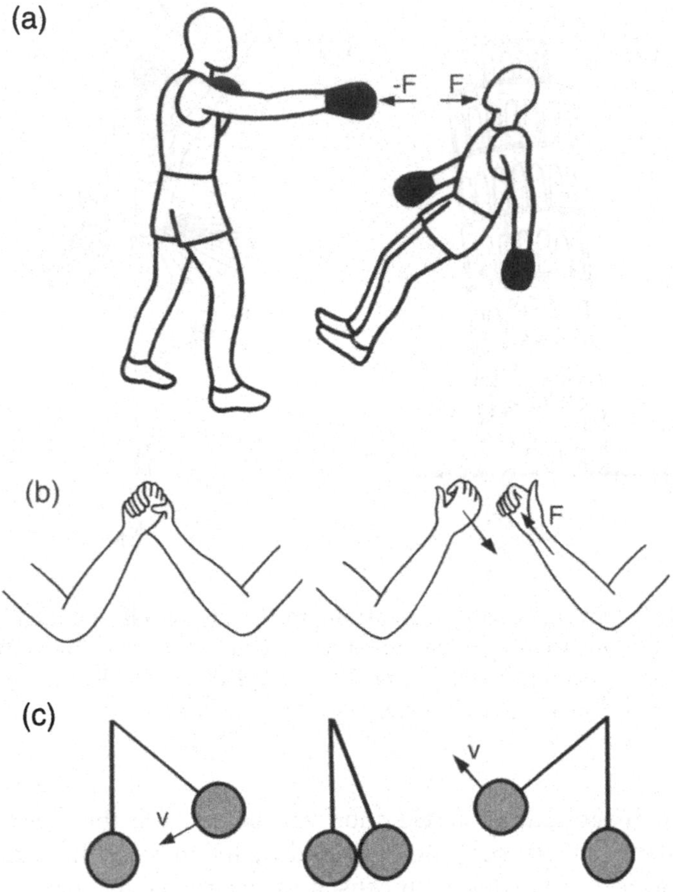

FIGURE 2.2a–c. Newton's third law states that the force of reaction is equal in magnitude and opposite in direction to the force of action. A boxer who hits an ordinary man is hit back with a force of the same intensity but opposite in direction, regardless of the size and the strength of the man **(a)**. Two people who are arm wrestling exert on one another forces of equal magnitude but opposite direction **(b)**. When two pendulums collide, they exert on each other forces of equal magnitude **(c)**.

same intensity. When a person beats another in arm wrestling, the force he exerts on the opposing party has the same level of intensity as the force the losing party exerts on him (Fig. 2.2b). It is just that the winning party is able to continue to contract his biceps muscles while the biceps of the opposing person is yielding to the external load. The third law may be counterintuitive also because we rarely observe equality in nature—things are either bigger or smaller, heavier or lighter, and so on. Newton arrived at this counterintuitive law by considering the data on the impact of two pendulums (Fig. 2.2c). The motion of the bobs after collision could only be explained by the validity of the third law, that action is equal to reaction. Hence, the discovery that in our universe equality exists, at least within the realm of contact forces.

Magnificent structures built by men thousands of years ago suggest that ancient civilizations were at least intuitively aware of many of the subtle features of the laws of motion. However, it took many millennia for a human to state these laws in an explicit and concise manner. Greek philosophers, among them Aristotle, had attempted to formulate the physical laws of motion but they all failed. They had held to the belief that the fundamental principles of nature could be deduced only by rational thinking and not by experimentation. As a result, they did not realize that quantities such as force, velocity, and acceleration have both magnitude and direction and as such they differ fundamentally from scalar entities such as mass and temperature.

The importance of empirical observation in the discovery of physical laws was appreciated much later during the Renaissance Period by Galileo and others. It was Galileo in the seventeenth century who first formulated the parallelogram law for combining vectors such as forces acting on a particle. It was Kepler who first observed a clear illustration of Newton's third law while investigating the motion of stars. Kepler concluded that gravitational force between two stars was proportional to the mass of each star and inversely proportional to the square of the distance between them. The gravitational force had to be aligned on the straight line connecting the centers of the two stars. And, regardless of their mass, any two stars exerted on each other the same amount of force, with direction reversed. Development of calculus and vector differentiation in the seventeenth century led finally to the emergence of the branch of science that we call today classical mechanics.

2.2 Addition and Subtraction of Vectors

Mechanics of human movement can best be explained with the use of vector notation and calculus. We present next a brief review of vector mathematics. A vector is a quantity that has both a magnitude and direction. Perhaps the vector most commonly known is body weight, which acts always in the direction pointing to the center of earth. Graphically, a vector is shown as a directed segment of a straight line. The length of the segment represents the magnitude of the vector. The direction of the line segment is the direction of the vector. The sense of direction is from the tail end of the segment to the end capped by an arrow. In this text we use **boldface** type when referring to a vector. As illustrated in Fig. 2.3, vectors are added by bringing them together arrow to tail and then by connecting the free tail with that of the free arrow. This is called the parallelogram law. The parallelogram law can be used to evaluate the re-

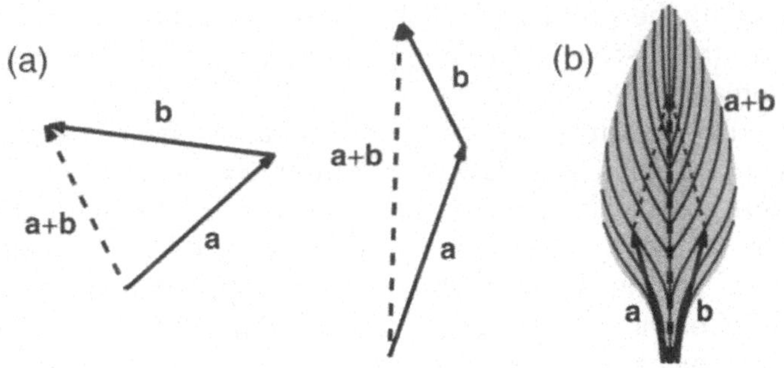

FIGURE 2.3a,b. The parallelogram law of addition of two vectors. Two vectors are added by bringing them together arrow to tail and then by connecting the free tail with that of the free arrow **(a)**. The application of this so-called parallelogram law to the calf muscle is shown in **(b)**.

sultant force acting on a joint or a limb. For example, as shown in Fig. 2.3b, the resultant force exerted by the calf muscle on the ankle is the sum of the muscle force produced by the two fleshy bellies of this muscle. This resultant acts as the propelling force during jumping and sprinting.

Both the commutative law and the associative law hold for vector addition:

$$\mathbf{a} + \mathbf{b} = \mathbf{b} + \mathbf{a} \tag{2.1a}$$

$$\mathbf{a} + (\mathbf{b} + \mathbf{c}) = (\mathbf{a} + \mathbf{b}) + \mathbf{c} \tag{2.1b}$$

Vector subtraction can be thought of as a special case of addition:

$$\mathbf{a} - \mathbf{b} = \mathbf{a} + (-\mathbf{b}) \tag{2.1c}$$

in which $-\mathbf{b}$ is equal to vector $\mathbf{b}$ in magnitude but has the opposite sense of direction. Thus vector subtraction has all the properties associated with vector addition.

Vectors can also be added algebraically. For this purpose, one draws three mutually perpendicular straight lines protruding from a point O in space. This is called a Cartesian reference frame. The point O is termed the origin of the coordinate system and the three straight lines stemming from it are the coordinate axes. Let E (x_1, x_2, x_3) be a Cartesian coordinate system and let $\mathbf{e}_1$, $\mathbf{e}_2$, and $\mathbf{e}_3$ denote unit vectors along x_1, x_2, and x_3 directions as shown in Fig. 2.4. A unit vector is defined as a vector with a unit magnitude. Any vector $\mathbf{b}$ can be expressed in this coordinate system as the sum of three components:

$$\mathbf{b} = b_1\,\mathbf{e}_1 + b_2\,\mathbf{e}_2 + b_3\,\mathbf{e}_3 \tag{2.2a}$$

in which the symbols b_1, b_2, and b_3 represent the projections of vector $\mathbf{b}$ on x_1, x_2, and x_3 axes. These projections are visualized in Fig. 2.4. To determine projection b_1, one draws perpendicular lines from each end of the

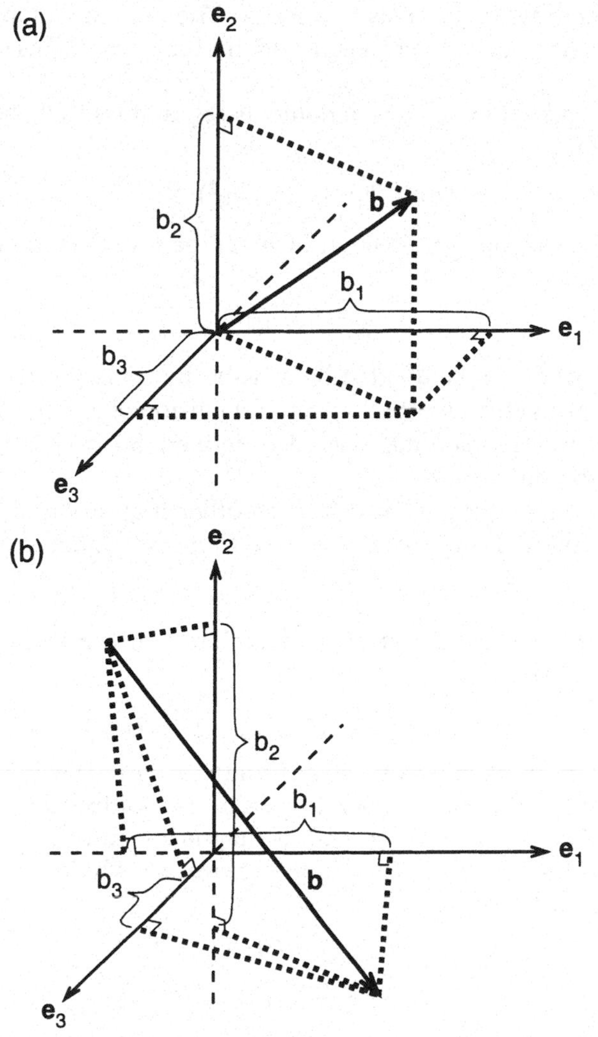

FIGURE 2.4a,b. Representation of a vector in a Cartesian reference frame E. The three mutually perpendicular lines protruding from point O compose the reference frame E. The symbols $\mathbf{e}_1$, $\mathbf{e}_2$, and $\mathbf{e}_3$ denote unit vectors along the coordinate axes. The symbols b_1, b_2, and b_3 denote, respectively, the projections of **b** onto the directions specified by the unit vectors $\mathbf{e}_1$, $\mathbf{e}_2$, and $\mathbf{e}_3$. Shown in **(a)** is a vector originating from the origin of the reference frame. Fig. 2.6b illustrates a vector whose origin does not coincide with the origin of the reference frame. The parameter b_i is negative if the projection is in the opposite direction of the unit vector $\mathbf{e}_i$.

vector **b** to the x_1 axis. The line segment that remains between the two intersections represents b_1. Projections b_2 and b_3 are determined similarly. A projection can be positive or negative. A projection is positive if it points along one of the unit vectors $\mathbf{e}_1$, $\mathbf{e}_2$, and $\mathbf{e}_3$; otherwise, it is negative. Note that in most undergraduate texts on dynamics, the vectors $\mathbf{e}_1$, $\mathbf{e}_2$, and $\mathbf{e}_3$

are represented by the symbols **i**, **j**, and **k**. The notation adopted in this text is especially useful when considering multibody systems such as the human body.

The magnitude of vector **b** is denoted as $\|\mathbf{b}\|$, and by Pythagoras' theorem it is equal to

$$\|\mathbf{b}\| = (b_1{}^2 + b_2{}^2 + b_3{}^2)^{1/2} \tag{2.2b}$$

The unit vector along the direction of **b** can be found by dividing **b** by its magnitude

$$\mathbf{e}^b = \mathbf{b}/\|\mathbf{b}\| \tag{2.2c}$$

Note that a vector can be divided by a scalar by dividing its projections along the coordinate axes with that scalar. Division of a vector by another vector is not an operation that has been defined. So, a vector cannot be divided by another vector.

Two vectors **a** and **b** are equal to each other if and only if their three components along the axes of a reference frame are equal to each other:

$$\mathbf{a} = \mathbf{b} \text{ if and only if } a_1 = b_1;\ a_2 = b_2;\ a_3 = b_3 \tag{2.3}$$

In a Cartesian coordinate system, two vectors **a** and **b** are added algebraically as follows:

$$\mathbf{a} + \mathbf{b} = (a_1 + b_1)\,\mathbf{e}_1 + (a_2 + b_2)\,\mathbf{e}_2 + (a_3 + b_3)\,\mathbf{e}_3 \tag{2.4}$$

A vector is multiplied by a positive number (a scalar) when its magnitude is multiplied by that number. If a vector is multiplied by a negative number, its direction is reversed and its magnitude multiplied by the absolute value of the number. The following equations hold for multiplication of vectors by a scalar:

$$s\,(t\mathbf{a}) = (st)\,\mathbf{a} \tag{2.5a}$$

$$(s + t)\,\mathbf{a} = s\mathbf{a} + t\mathbf{a} \tag{2.5b}$$

$$s\,(\mathbf{a} + \mathbf{b}) = s\mathbf{a} + s\mathbf{b} \tag{2.5c}$$

Example 2.1. Elbow Force During Baseball Pitching. Baseball players, especially pitchers, are prone to overuse injuries associated with throwing. These injuries result from accumulated microtrauma developed during repetitive use. To test the hypothesis that the microtrauma in the throwing arm of a pitcher is caused by the large forces and torques exerted at the shoulder and elbow joint during pitching, a large number of studies have investigated the mechanics of pitching using high-speed motion analysis. The forces applied by the ligaments and tendons on the elbow joint during baseball pitching were measured in the medial (M), anterior (A), and compression (C) directions (Fig. 2.5). The magnitudes of these forces were found to be

$$F^{\mathrm{M}} = 428\text{ N},\ F^{\mathrm{A}} = 101\text{ N},\ F^{\mathrm{C}} = 253\text{ N}.$$

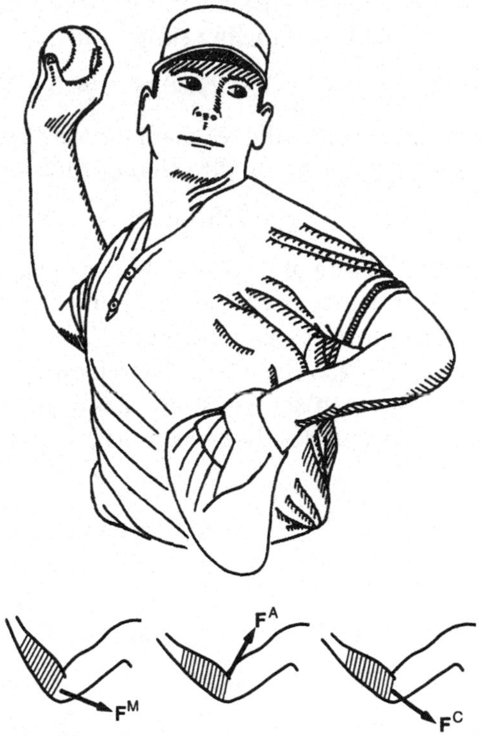

FIGURE 2.5. Forces exerted at the elbow by the rest of the body in the medial (M), anterior (A), and compressive (C) directions during pitching. These forces have been estimated from the three-dimensional videoanalysis of a pitching event using the method of inverse dynamics.

where N denotes the force unit Newton. A force that accelerates 1 kg at a rate of 1 m/s^2 is 1 N.

The unit vectors in the medial, anterior, and compression directions were expressed in unit vectors fixed on earth ($\mathbf{e}_1$, $\mathbf{e}_2$, $\mathbf{e}_3$,):

$$\mathbf{e}^M = 0.79\ \mathbf{e}_1 + 0.17\ \mathbf{e}_2 + 0.59\ \mathbf{e}_3$$

$$\mathbf{e}^A = 0.21\ \mathbf{e}_1 - 0.98\ \mathbf{e}_2$$

$$\mathbf{e}^C = -0.58\ \mathbf{e}_1 - 0.12\ \mathbf{e}_2 - 0.81\ \mathbf{e}_3$$

Using these data, compute the resultant force acting on the elbow.

Solution: The resultant force acting on the elbow is equal to the sum of the three forces acting on the elbow:

$$\mathbf{F} = (F^M)\ \mathbf{e}^M + (F^A)\ \mathbf{e}^A + (F^C)\ \mathbf{e}^C$$

in which $\mathbf{F}$ stands for the resultant force.

Substituting the expressions given for the magnitude of the forces and their directions into this expression, we obtain:

$$\mathbf{F} = 428\ (0.79\ \mathbf{e}_1 + 0.17\ \mathbf{e}_2 + 0.59\ \mathbf{e}_3) + 101\ (0.21\ \mathbf{e}_1 - 0.98\ \mathbf{e}_2) + 253\ (-0.58\ \mathbf{e}_1 - 0.12\ \mathbf{e}_2 - 0.81\ \mathbf{e}_3)$$

Summing the coefficients in front of the unit vectors, this expression can be put into the following simpler form:

$$\mathbf{F} = 212.8\ \mathbf{e}_1 - 56.6\ \mathbf{e}_2 + 47.6\ \mathbf{e}_3\ \text{(N)}$$

We can express the resultant force as the product of its magnitude times the unit vector in the direction of the resultant force:

$$\mathbf{F} = 225.3\ (0.94\ \mathbf{e}_1 - 0.25\ \mathbf{e}_2 + 0.21\ \mathbf{e}_3)\ \text{N}$$

The magnitude of the resultant force was determined by using Eqn. 2.2b. The unit vector along the direction of force was obtained by dividing the resultant force by its magnitude.

Example 2.2. Resultant Force Exerted by the Pectoralis on the Upper Arm. The man shown in Fig. 2.6 is performing lateral flies to work his pectoralis muscles. The pectoralis is a triangular muscle of the upper chest;

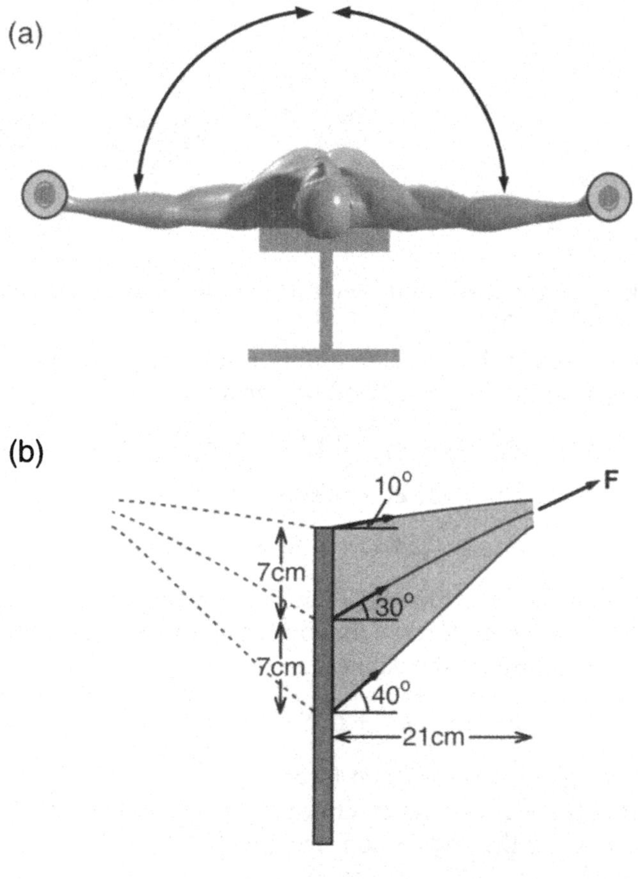

FIGURE 2.6a,b. A man exercising his pectoralis muscle group **(a)**. The schematic diagram of the pectoralis as composed of three distinct sets of muscle fibers **(b)**.

assume that it can be represented as composed of three sets of muscle fiber groups connecting the sternum to the humerus. Determine the resultant force exerted by the pectoralis on the humerus. At the position shown in the figure, the muscle acts in a plane, and the magnitude of the force produced by each set of fibers is 75 N.

Solution: The resultant pectoralis force **F** is the sum of the forces produced by the three sets of fibers

$$\mathbf{F} = 75\text{ N }(\cos 40° + \cos 30° + \cos 10°)\ \mathbf{e}_1 + 75\text{ N }(\sin 40° + \sin 30° + \sin 10°)\ \mathbf{e}_2$$

$$\|\mathbf{F}\| = [3 + 2(\cos 10° + \cos 20° + \cos 30°)]^{1/2}\ 75\text{ N}$$

In deriving the equation for the magnitude of the pectoralis force, we used the trigonometric relation

$$\cos (a - b) = \cos (a) \cos (b) + \sin (a) \sin (b)$$

In this example, we have assumed that all fibers of the pectoralis were activated by the central nervous system. The pectoralis is capable of exerting forces on the humerus in wide-ranging directions. This force is accomplished by varying the spatial activation pattern of the muscle itself.

2.3 Time Derivatives of Vectors

Newton's second law relates the resultant force acting on a particle to the acceleration of the particle. Acceleration is the time derivative of velocity. How do we take the time derivative of a vector?

The time derivative of a vector **b** in the reference frame E is defined as

$$d\mathbf{b}/dt = (db_1/dt)\ \mathbf{e}_1 + (db_2/dt)\ \mathbf{e}_2 + (db_3/dt)\ \mathbf{e}_3 \qquad (2.6)$$

in which *db/dt* denotes the time derivative of **b**, and ($\mathbf{e}_1$, $\mathbf{e}_2$, $\mathbf{e}_3$) are the three orthogonal unit vectors associated with reference frame E. Parameters b_1, b_2, and b_3 are the projections of the vector **b** on the unit vectors of the reference frame E.

In this definition, the time derivative of a vector depends on the coordinate system in which it is taken. To illustrate this point, let us visualize a youngster drawing with red ink a line of 5 cm on his abdomen along the axis of his body. Let us denote by **b** the vector joining the ends of this red line. Suppose that the youth proceeds to perform somersaults. The time derivative of **b** will be equal to zero with respect to a coordinate system embedded on the trunk of the youth. In that coordinate system, **b** is not a function of time. On the other hand, when expressed in terms of unit vectors fixed on earth, **b** is a function of time and, therefore, its time derivative is not equal to zero. The term acceleration in Newton's second law is the acceleration with respect to a Cartesian coordinate system fixed on earth.

2.4 Position, Velocity, and Acceleration

Position of a particle P moving in space is identified by a vector connecting the particle P with the origin O of reference frame E. This vector is called the position vector, and it is denoted by the symbol **r**. Because the position of the terminal point of **r** depends on time, **r** is a vector function of time t: $\mathbf{r} = \mathbf{r}(t)$. The vector **r** can be written as

$$\mathbf{r} = (x_1\, \mathbf{e}_1 + x_2\, \mathbf{e}_2 + x_3\, \mathbf{e}_3) \tag{2.7a}$$

in which x_1, x_2, and x_3 are all functions of time t:

$$x_1 = x_1(t);\ x_2 = x_2(t);\ x_3 = x_3(t) \tag{2.7b}$$

The velocity **v** and the acceleration **a** of particle P in the reference frame E are then defined as follows:

$$\mathbf{v} = d\mathbf{r}/dt = (dx_1/dt\ \mathbf{e}_1 + dx_2/dt\ \mathbf{e}_2 + dx_3/dt\ \mathbf{e}_3) \tag{2.8a}$$

$$\mathbf{a} = d\mathbf{v}/dt = (dv_1/dt\ \mathbf{e}_1 + dv_2/dt\ \mathbf{e}_2 + dv_3/dt\ \mathbf{e}_3) \tag{2.8b}$$

These equations indicate that to compute the velocity and acceleration of a point P with respect to a reference frame fixed on earth, we need to express the position vector of P in terms of the unit vectors fixed on earth and then take the time derivatives of vector projections on the coordinate axes. This procedure is illustrated next with an example.

Example 2.3. Particle Path, Velocity, and Acceleration. The position vector connecting a fixed point O in the reference frame E to a moving point P in space is given by the expression:

$$\mathbf{r}^{P/O} = (1.67 + 3\ t^2)\ [\cos\ (2t^2)\ \mathbf{e}_1 + \sin\ (2t^2)\ \mathbf{e}_2]$$

Determine the velocity and acceleration of point P in reference frame E.

Solution: The unit vectors $\mathbf{e}_1$, $\mathbf{e}_2$ are constants in E so their time derivatives will be zero. Using differentiation by parts, we find

$$\begin{aligned}
\mathbf{v} &= d\mathbf{r}/dt \\
&= 6t\ [\cos\ (2t^2)\ \mathbf{e}_1 + \sin\ (2t^2)\ \mathbf{e}_2] + (1.67 + 3\ t^2)\ (4t)\ [-\sin\ (2t^2)\ \mathbf{e}_1 \\
&\quad + \cos\ (2t^2)\ \mathbf{e}_2] \\
&= [6t\ \cos\ (2t^2) - (6.7t + 12t^3)\ \sin\ (2t^2)]\ \mathbf{e}_1 + [6t\ \sin\ (2t^2) \\
&\quad + (6.7t + 12t^3)\ \cos(2t^2)]\ \mathbf{e}_2
\end{aligned}$$

$$\begin{aligned}
\mathbf{a} &= d\mathbf{v}/dt \\
&= [6\cos\ (2t^2) - 24t^2\sin\ (2t^2) - (6.7 + 36t^2)\sin\ (2t^2) \\
&\quad + (26.8t^2 + 48t^4)\ \cos\ (2t^2)]\ \mathbf{e}_1 \\
&\quad + [6\sin\ (2t^2) + 24t^2\cos\ (2t^2) + (6.7 + 36t^2)\ \cos\ (2t^2) \\
&\quad - (26.8t^2 + 48t^4)\ \sin\ (2t^2)]\ \mathbf{e}_2
\end{aligned}$$

Note that in this book we typically refer to velocity (acceleration) with respect to a reference frame fixed on earth as simply velocity (acceleration).

Velocity and Acceleration in Polar Coordinates

During an arm wrestle, the forearm of the man who is at the brink of defeat will begin to draw a circle whose center is his elbow pushed against a table. In this case, in the computation of velocity and acceleration of the forearm, it may be easier to use polar coordinates rather than Cartesian coordinates.

In polar coordinates, we define $\mathbf{e}_r$ to be the unit vector in the direction of the position vector connecting origin O of the coordinate system to a moving point P. Consider, for example, the case of abduction of the arm as shown in Fig. 2.7a. The unit vector along the line of the arm $\mathbf{e}_r$ is given by the equation:

$$\mathbf{e}_r = \cos\theta\, \mathbf{e}_1 + \sin\theta\, \mathbf{e}_2$$

Then the position vector connecting the shoulder to the elbow can be written as

$$\mathbf{r} = L\, \mathbf{e}_r$$

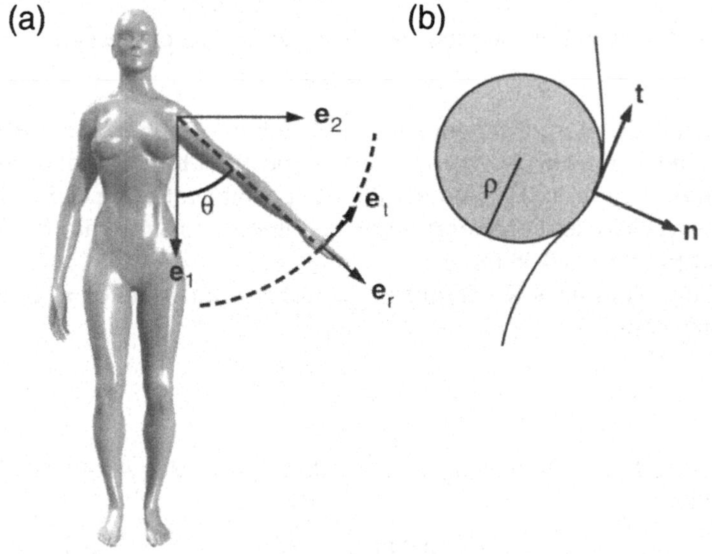

FIGURE 2.7a,b. Polar and path coordinates. The unit vectors associated with polar coordinates are $\mathbf{e}_r$ and $\mathbf{e}_t$. The vector $\mathbf{e}_r$ is in the radial direction pointing outward whereas $\mathbf{e}_t$ is tangent to the circle and points in the direction of increasing θ **(a)**. In the case of path coordinates, the unit vector $\mathbf{n}$ is normal and $\mathbf{t}$ is tangent to the trajectory **(b)**. The symbol ρ denotes the radius of curvature; it is the radius of the largest circle that is tangent to the particle path at the location shown.

in which L denotes the length of the upper arm. Taking the time derivative of the position vector, we determine the velocity of the elbow:

$$\mathbf{v} = L\, d\mathbf{e}_r/dt$$
$$= L\,(d\theta/dt)\,(-\sin\theta\,\mathbf{e}_1 + \cos\theta\,\mathbf{e}_2)$$
$$= L\,(d\theta/dt)\,\mathbf{e}_t$$

in which $\mathbf{e}_t$ is perpendicular to $\mathbf{e}_r$ as shown in Fig. 2.7.

Next, let us determine acceleration by taking the time derivative of velocity $\mathbf{v}$:

$$\mathbf{a} = d\mathbf{v}/dt$$
$$= L\,(d^2\theta/dt^2)\,(-\sin\theta\,\mathbf{e}_1 + \cos\theta\,\mathbf{e}_2) - L\,(d\theta/dt)^2\,(\cos\theta\,\mathbf{e}_1 + \sin\theta\,\mathbf{e}_2)$$
$$= L\,(d^2\theta/dt^2)\,\mathbf{e}_t - L\,(d\theta/dt)^2\,\mathbf{e}_r$$

Thus, when a particles traverses a circular path, its position, velocity, and acceleration can be expressed as follows:

$$\mathbf{r} = L\,\mathbf{e}_r \qquad (2.9a)$$
$$\mathbf{v} = L\,(d\theta/dt)\,\mathbf{e}_t \qquad (2.9b)$$
$$\mathbf{a} = L\,(d^2\theta/dt^2)\,\mathbf{e}_t - L\,(d\theta/dt)^2\,\mathbf{e}_r \qquad (2.9c)$$

Note that the speed of the particle v is given by the expression

$$v = L\,(d\theta/dt) \qquad (2.9d)$$

Example 2.4. Arm Movements in Aerobics. An aerobic instructor abducts her arm from downward vertical position to horizontal position at shoulder length in 0.6 seconds (s), at constant rate (Fig. 2.7a). Determine the velocity and acceleration of her elbow. Assume that the length of her upper arm is 0.38 m.

Solution: Since 0.6 s was required to traverse an angle of $\pi/2$ radians at constant rate

$$d\theta/dt = \pi/1.2$$
$$d^2\theta/dt^2 = 0$$

Thus, using Eqn. 2.9, velocity and acceleration of the elbow can be expressed as

$$\mathbf{v} = 0.99\ (\mathrm{m/s})\ \mathbf{e}_t$$
$$\mathbf{a} = -2.60\ (\mathrm{m/s^2})\ \mathbf{e}_r$$

Velocity and Acceleration in Path Coordinates

Even when a particle draws a planar curved path that is not a circle, it is possible to describe its velocity and acceleration in terms of unit vectors that are tangential and normal to the path. Consider a particle traversing a complex planar curve. Let **t** be the unit vector tangent to the particle path and let **n** be the unit normal vector drawn outward as shown in Fig. 2.7b. The velocity and acceleration of particle P can then be written as

$$\mathbf{v} = v\,\mathbf{t} = (ds/dt)\,\mathbf{t} \tag{2.10a}$$

$$\mathbf{a} = (dv/dt)\,\mathbf{t} - (v^2/\rho)\,\mathbf{n} \tag{2.10b}$$

in which s is the arc length along the particle path, $v = ds/dt$ is the speed of the particle, and ρ is the radius of curvature. It is defined as the radius of the largest circle that has the same tangent with the particle path at point P. In a sense, Eqn. 2.10 can be considered as a generalization of Eqn. 2.9. It is used quite often in the analysis of motion of satellites and stars because of the elliptical nature of their particle path.

2.5 Newton's Laws of Motion and Their Applications

We have explored the physical content of the laws of motion in the introduction of this chapter. The mathematical operations presented in the preceding sections now allow us to write these laws in mathematical language. According to Newton's first law, the resultant force acting on a particle must be equal to zero when the particle is at rest or moving with constant velocity in an inertial reference frame:

$$\Sigma\mathbf{F} = 0 \tag{2.11a}$$

in which $\Sigma\mathbf{F}$ denotes the sum of all forces acting on the particle. Newton's second law relates the resultant force to the acceleration of the particle

$$\Sigma\mathbf{F} = m\,\mathbf{a} \tag{2.11b}$$

in which m is the mass of the particle and **a** is its acceleration with respect to an inertial reference frame.

According to the third law, the force of action is equal in magnitude and opposite in direction to the force of reaction

$$\mathbf{f}_{1-2} = -\mathbf{f}_{2-1} \tag{2.11c}$$

in which $\mathbf{f}_{1-2}$ represents the force exerted on particle 1 by particle 2.

The term particle is used in the description of these mathematical formulations to represent an object, small or large, with the stipulation that the variation of velocity (acceleration) within the object is negligible compared to the mean velocity (acceleration) of the object. In his studies, Newton considered earth and the stars as particles. When considering the trajectory of a football in air, it is appropriate to consider the ball as a particle. However, when studying the spin of the ball as it traverses the air, the size and the shape of the ball must be taken into account.

Example 2.5. Free Fall of an Object: An Experiment by Galileo. The legend is that Galileo determined the law of gravity by conducting an experiment in which iron balls of different sizes were dropped from the top of the Pisa tower and a group of his friends at different floors of the tower measured the time of the free fall at specified heights. Iron balls of varying diameter fall toward earth with constant acceleration (Fig. 2.8). This acceleration is called gravitational acceleration. Its magnitude is usually denoted by g. The force that causes gravitational acceleration is the gravitational force (the weight of the object).

With respect to a Cartesian coordinate system fixed on earth, the position of a free-falling iron ball varies with time in accordance with the following equation:

$$\mathbf{r} = 0\ \mathbf{e}_1 + (h_o - gt^2/2)\ \mathbf{e}_2 + 0\ \mathbf{e}_3 = (h_o - gt^2/2)\ \mathbf{e}_2 \tag{2.12a}$$

$$x_1 = 0,\ x_2 = h_o - gt^2/2,\ x_3 = 0 \tag{2.12b}$$

where x_1, x_2, and x_3 are the projections of $\mathbf{r}$ along the coordinate axes, h_o is the height of the Pisa tower (~80 m), and t denotes time in seconds (Fig. 2.8).

Taking the time derivative of Eqn. 2.12, one can compute the velocity and the acceleration of an iron ball falling under the action of gravity:

$$\mathbf{v} = -gt\ \mathbf{e}_2 \text{ and } \|\mathbf{v}\| = gt \tag{2.13a}$$

$$\mathbf{a} = -g\ \mathbf{e}_2 \text{ and } \|\mathbf{a}\| = g \tag{2.13b}$$

According to Eqn. 2.13, the speed of an iron ball falling in air ($\|\mathbf{v}\|$) increases with time whereas its acceleration remains constant. Furthermore, acceleration is also not dependent on the mass of the ball.

Example 2.6. Terminal Velocity of a Snowflake. When a body moves in a medium such as air or water, the medium exerts a resistance force on

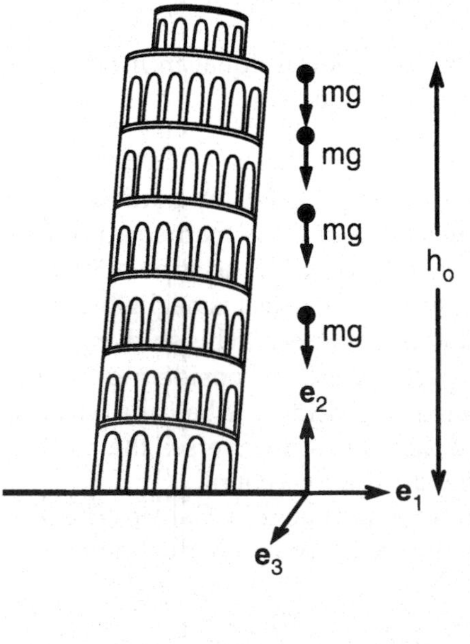

FIGURE 2.8. A steel ball dropped from the Pisa tower. The *black circles* indicate the position of a steel ball at equal time intervals. The only force acting on the ball is the force of gravity.

the body. The resistance force increases with the velocity of the motion relative to the fluid. It is also dependent on the shape of the moving body and on its orientation relative to the direction of motion. Resistance force depends nonlinearly on the velocity and geometry of the body. However, for very small relative velocities (to 1 m/s), resistance force is proportional to the first power of velocity. Consider the case of a snowflake falling under gravity and retarded by an air resistance. Let u be the downward velocity of the snowflake and let $(-k\,u)$ be the frictional resistance per unit mass of the snowflake, k being a constant resistance coefficient. Determine the trajectory of the snowflake. What is its terminal velocity?

Solution: In one-dimensional motion where velocity, force, and acceleration all point in the same direction, it is not necessary to use the vector notation. The equation of motion for the snowflake in the direction of gravity can be written as follows:

$$(du/dt) = g - k\,u \tag{2.14}$$

where (du/dt) is the acceleration of the snowflake, considered positive downward. This differential equation can be integrated analytically with respect to time to yield

$$u = (dx/dt) = g/k - (C/k)\, e^{-k\,t} \tag{2.15a}$$

$$x = (g/t)/k - (1 - e^{-k\,t})\,(C/k^2) \tag{2.15b}$$

where x is the position of the snowflake, measured downward from a reference point in air $(dx/dt = u)$, and C is an arbitrary constant to be determined by the initial conditions at $t = 0$. The exponential term in Eqn. 2.15 rapidly decreases toward zero with increasing time. Thus, the speed u of the falling snowflake becomes (g/k) for sufficiently large times. This constant velocity is termed the terminal velocity of the snowflake. Suppose the terminal velocity of a snowflake is 1.2 m/s; then the resistance coefficient k must be equal to 8.175 s^{-1}. For this k value the term e^{-kt} is practically equal to zero when $t = 1$ s. The velocity of the snowflake approaches its terminal value within a fraction of a second. Experiments indicate that the resistance coefficient k increases with the mass density of air. This finding is consistent with the common observation that the snowflakes fall with smaller velocities in colder climates.

Example 2.7. Motion of Spherical Balls on an Inclined Plane. In a free-fall experiment, a heavy body rapidly gains speed so that it becomes difficult to record the falling distance as a function of time. To bypass this difficulty, Galileo designed an experiment in which the iron balls were set free on top of a smooth plane that made an angle α with the horizontal plane (Fig. 2.9). The iron balls slid down the inclined plane much slower than the velocities observed during a free fall.

Two distinct forces act on an iron ball moving down an inclined plane: the gravitational force $m\mathbf{g}$ pulling the ball downward and the force $\mathbf{N}$ ex-

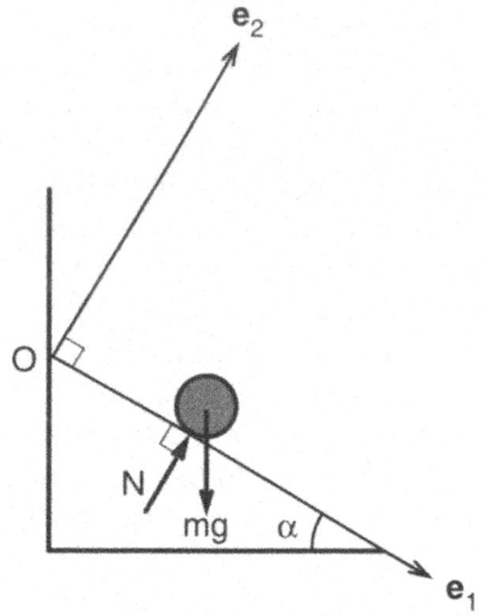

FIGURE 2.9. A steel ball sliding down a smooth (frictionless) plane. The diagram shows the forces acting on the ball as it speeds down the inclined plane.

erted by the smooth plane on the steel ball. This latter force is called the contact force. Its function is to keep the ball from cutting into the inclined platform. Galileo polished the surface of the inclined plane as well as that of the spherical balls to reduce the frictional resistance to motion. Therefore, the contact force **N** acted in the direction normal to the plane of motion as shown in Fig. 2.9. Let us denote the magnitude of this contact force by N. Newton's second law for the iron balls moving down an inclined plane can be written as

$$mg \sin \alpha \, \mathbf{e}_1 - mg \cos \alpha \, \mathbf{e}_2 + N \, \mathbf{e}_2 = m \, a_1 \, \mathbf{e}_1 \tag{2.16a}$$

This vectorial equality is equivalent to the following two scalar equations:

$$a_1 = g \sin \alpha \tag{2.16b}$$

$$N = mg \cos \alpha \tag{2.16c}$$

According to Eqn. 2.16b, acceleration of an iron ball sliding down an inclined plane is sin α times smaller than the acceleration during free fall. For example, if α is chosen as 5°, then $\sin \alpha = 0.087$, and the acceleration of the sliding ball would be equal to 0.85 m/s^2 as opposed to 9.81 m/s^2 for the free fall.

Integrating acceleration a_1 with respect to time, we find the following expressions for velocity (v_1) and distance (s_1) traveled:

$$v_1 = v_{10} + g \, t \sin \alpha \tag{2.16d}$$

$$s_1 = s_{10} + v_{10} \, t + g \, (t^2/2) \sin \alpha \tag{2.16e}$$

in which v_{10} and s_{10} are arbitrary constants to be determined by initial conditions. The velocity of an iron ball is zero at the instant it is released from rest ($t = 0$) and thus $v_{10} = 0$. The parameter s_{10} can be set equal to zero if the distance s is measured from the point where the ball is released from rest. Thus:

$$v_1 = g\, t \sin \alpha \tag{2.16f}$$

$$s_1 = g\, (t^2/2) \sin \alpha \tag{2.16g}$$

Example 2.8. Oscillation of Pendulum. Galileo understood that to discover the laws of motion, one would have to be able to measure time with better accuracy than permitted by an hourglass. He used his pulse to observe that the period of oscillation of a given pendulum was approximately constant. His interest in the properties of pendulums arose while looking at a chandelier that swung from the roof of the baptistry in Pisa. He attached a bob to the end of a string and swung it. He determined that the period of the pendulum, the time it takes to swing from extreme right to extreme left and back, did not change with the weight of the bob. The maximum angle of swing only had a minor effect on the period of the pendulum. On the other hand, the period varied in proportion with the square root of the length of the pendulum. Galileo used pendulums to measure time in his famous experiments on mechanics.

An equation for the period of a pendulum can be obtained by using Newton's equations of motion. Consider a pendulum with length L and mass m (Fig. 2.10). The forces acting on the bob are the weight $m\mathbf{g}$ and the tension $\mathbf{T}$ in the string. Newton's second law in the direction of the string (shown by the unit vector $\mathbf{e}_r$) and in the direction normal to the string (shown by the unit vector $\mathbf{e}_t$) result in the following equations:

$$m\, g \cos \theta - T = -m\, (d\theta/dt)^2\, L \tag{2.17a}$$

$$-m\, g \sin \theta = m\, L\, (d^2\theta/dt^2) \tag{2.17b}$$

where θ is the angle the pendulum makes with the vertical axis and T is the magnitude of the tension (pulling force) applied by the string on the

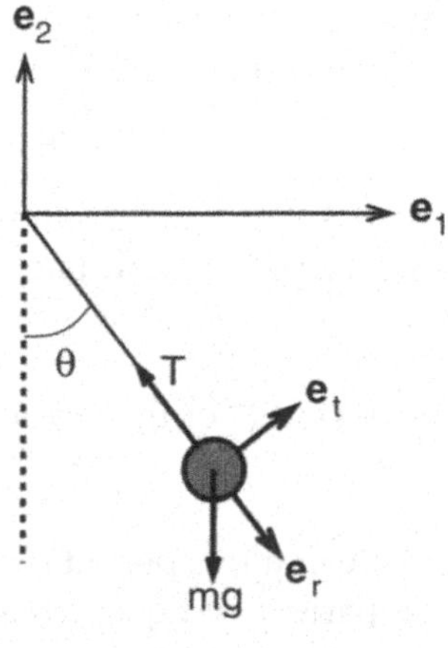

FIGURE 2.10. A pendulum of mass m and length L. The symbol θ denotes the angle the pendulum makes with the horizontal axis; T is the tension applied by the thin rod on the bob.

mass m. The acceleration of the bob was determined by using Eqn. 2.9c, specific to circular paths.

There are two unknowns in Eqn. 2.17a: the tension T and the angle of swing θ. So, we begin with Eqn. 2.17b in which time t is the independent variable and θ is the only dependent variable. This is a second-order ordinary differential equation. For this differential equation to have a unique solution, a set of initial conditions must be satisfied. As initial conditions, we specify that the mass m is held at its maximum elevation ($\theta = \theta^*$) and is let go at $t = 0$ with zero velocity. In mathematical language, the initial conditions are

$$\theta = \theta^*, \text{ and } (d\theta/dt) = 0 \text{ at } t = 0 \tag{2.17c}$$

where θ^* is the maximum angle of swing.

Equation 2.17b can be solved analytically by noting the following equality:

$$(d^2\theta/dt^2) = d(d\theta/dt)/dt = [d(d\theta/dt)/d\theta]\,(d\theta/dt) \tag{2.17d}$$

Substituting this equation into Eqn. 2.17b and integrating with respect to θ one obtains:

$$(d\theta/dt)^2 = (2g/L)\,(\cos\theta^* - \cos\theta) \tag{2.17e}$$

Subsequent integration of Eqn. 2.17e with respect to time involves an elliptic integral. The result can be found in some integral tables. For $\theta^* = \pi/2$, the period of the pendulum (the time it takes for the pendulum to complete a whole swing and reach the same spatial point) can be shown to be equal to

$$t^* = 7.45\,(L/g)^{1/2} \tag{2.18a}$$

For small angles of swing, Eqn. 2.17b governing the swing of the pendulum can be simplified by using the approximation $\sin\theta = \theta$:

$$d^2\theta/dt^2 + (g/L)\,\theta = 0 \tag{2.18b}$$

This is a second-order linear and homogeneous differential equation. Its solution can be written as

$$\theta = A\sin(g/L)^{1/2}t + B\cos(g/L)^{1/2}t \tag{2.18c}$$

where A and B are arbitrary constants to be determined by initial conditions. For $\theta = \theta^*$ and $(d\theta/dt) = 0$ at $t = 0$:

$$\theta = \theta^*\cos(g/L)^{1/2}t \tag{2.18d}$$

The period t^* of this equation can be found by dividing 2π by $(g/L)^{1/2}$:

$$t^* = 2\pi\,(L/g)^{1/2} \tag{2.18e}$$

Note that this period is independent of the amplitude of oscillation (θ^*). Comparison of Eqns. 2.18a and 2.18e shows that when the angle of swing is increased from 0 to $\pi/2$, the period increases by only 18%. That is why, when we observe children swing, that the period of the swing does not seem to change with the maximum angle of swing.

Example 2.9. Trajectory of a Golf Ball. A man hits a golf ball with initial speed V_o and at an angle of α from the horizontal plane as shown in Fig. 2.11. Determine the trajectory of the ball.

Solution: Neglecting air friction, the only force acting on the ball once it is off the ground is the gravitational force $m\mathbf{g}$. With respect to the coordinate system E that is fixed on earth, the equation of motion reduces to

$$-mg\ \mathbf{e2} = m\ (dv_1/dt\ \mathbf{e}_1 + dv_2/dt\ \mathbf{e}_2 + dv_3/dt\ \mathbf{e}_3)$$

This is, in effect, three scalar equations:

$$dv_1/dt = 0 \qquad (2.19a)$$

$$dv_2/dt = -g \qquad (2.19b)$$

$$dv_3/dt = 0 \qquad (2.19c)$$

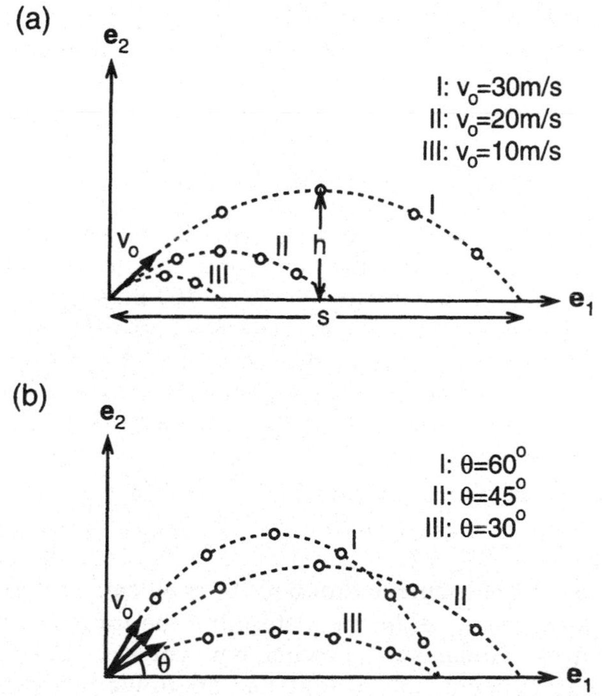

FIGURE 2.11a,b. Parabolic trajectory of a golf ball in air. The *circles* indicate the position of a golf ball at various time intervals. The direction of initial velocity is the same for the trajectories shown in **(a)**. Figure 2.11b illustrates trajectories with the same initial speed.

Integrating these equations with respect to time we find

$$v_1 = v_{10};\ v_2 = -gt + v_{20};\ v_3 = v_{30} \tag{2.20a}$$

$$x_1 = v_{10}\,t + x_{10};\ x_2 = -gt^2/2 + v_{20}\,t + x_{20};\ x_3 = v_{30}\,t + x_{30} \tag{2.20b}$$

The symbols v_{10} and x_{10} refer to components of velocity and position at $t = 0$. To determine the trajectory of the golf ball completely, we need to specify these constants. We assume the ball is at the origin of the reference frame E and its initial velocity is in the (x_1,x_2) plane:

$$x_{10} = x_{20} = x_{30} = 0;\ v_{10} = V_o \cos\alpha,\ v_{20} = V_o \sin\alpha,\ v_{30} = 0 \text{ at } t = 0$$

where V_o is the speed of the ball at $t = 0$ and α is the angle that the initial velocity makes with the $\mathbf{e}_1$ axis. Then, for all latter times, the following equations determine the trajectory traversed by the golf ball:

$$x_1 = V_o\,t \cos\alpha;\ x_2 = V_o t \sin\alpha - gt^2/2;\ x_3 = 0$$

Hence, the golf ball moves with constant speed along the $\mathbf{e}_1$ direction, whereas its velocity decreases at a constant rate in the $\mathbf{e}_2$ direction and the golf ball has no velocity in the $\mathbf{e}_3$ direction. The velocity of the golf ball in the $\mathbf{e}_2$ direction becomes equal to zero when the ball reaches the maximal height h, and that occurs at $t^* = (V_o/g)\sin\alpha$. The maximum height reached by the golf ball (h) and the horizontal distance traveled (s) are given by the following equations:

$$h = (V_o^2/2g)\sin^2\alpha \tag{2.21a}$$

$$s = (2V_o^2/g)\sin\alpha\cos\alpha \tag{2.21b}$$

From Eqn. 2.21, it is clear that the initial velocity of the golf ball determines the height it reaches and the horizontal distance it covers during free fall. For example, if the initial velocity is 100 m/s and $\alpha = 30°$, the ball climbs as high as 12.5 m in 5 s and covers a horizontal distance of 86 m before it touches the ground. The centroid of the ball traverses a parabola, and this trajectory is independent of the size or shape of the particle so long as the air friction is negligible in comparison to the force of gravity.

Example 2.10. Machine Curls and the Biceps Muscle. A woman performs seated machine curls to strengthen her biceps muscles (Fig. 2.12). She uses a weight of 10 kg. Horizontal distance between her elbows (A) and the bottom of the frictionless pulley (B) is 60 cm. Vertical distance between the same points is 20 cm. Determine the force exerted by the holding bar on the woman's hands when her arm is at 45° with the horizontal. At that moment, the weight has an upward acceleration of 3 m/s^2.

Solution: Let us first consider the forces exerted on the 10-kg weight. Tension T in the cable pulls it up whereas the gravity pulls it down, and Newton's second law dictates that

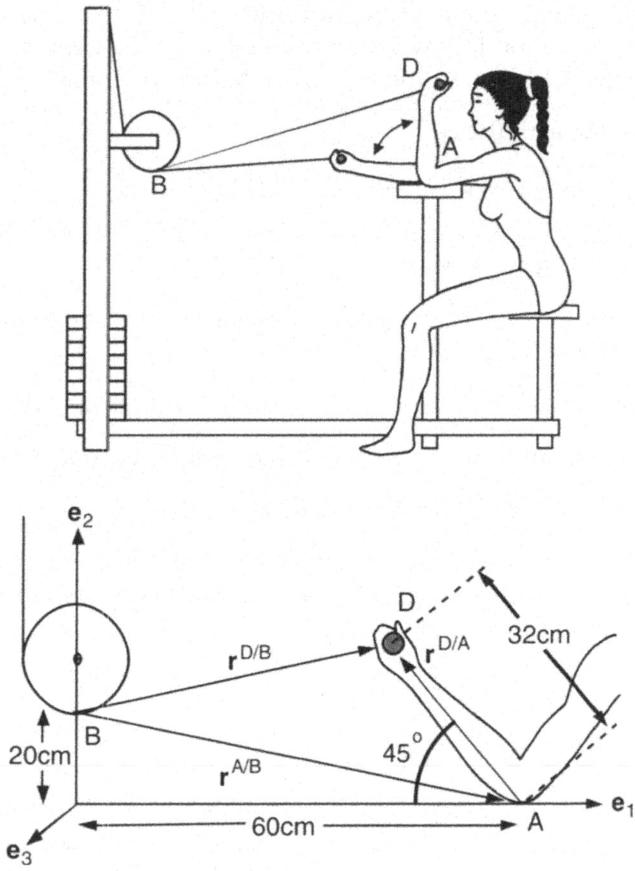

FIGURE 2.12. A woman performs seated machine curls to strengthen her biceps muscles. The points A, B, and D denote, respectively, her elbow, the lowest point of the pulley, and the holding bar. The pulley is frictionless.

$$T - 10 \times 9.81 = 10 \times 3 \Rightarrow T = 128.1 \text{ N}$$

The tension T in the cable is uniform since the pulley is frictionless.

Next, let us consider the forces acting on the holding bar connected to the cable (see point D in Fig. 2.12). If we neglect the weight of this lightly weighted bar, we find that the tension in the cable must be equal in magnitude but opposite in direction to the force exerted by the woman on the bar. Newton's third law then dictates that the tension in the cable is equal to the force exerted by the bar on the hands of the woman holding the bar.

Now that we know the magnitude of the force, we need to determine its direction. To do that we need to compute the orientation of the unit vector in the direction of the cable connecting points B and D. Using the parallelogram law we can show that

$$\mathbf{r}^{D/B} = \mathbf{r}^{A/B} + \mathbf{r}^{D/A}$$

in which $\mathbf{r}^{D/B}$ denotes the position vector from B to D. All other position vectors in the equation follow the same notation. With respect to the coordinate system E shown in the figure, it can be shown that

$$\mathbf{r}^{A/B} = 0.6\,\mathbf{e}_1 - 0.2\,\mathbf{e}_2$$

$$\mathbf{r}^{D/A} = -0.32 \times \cos 45^\circ\,\mathbf{e}_1 + 0.32 \times \sin 45^\circ\,\mathbf{e}_2$$

$$\mathbf{r}^{D/B} = (0.6 - 0.32 \times \cos 45^\circ)\,\mathbf{e}_1 + (-0.2 + 0.32 \times \sin 45^\circ)\,\mathbf{e}_2 = 0.37\,\mathbf{e}_1 + 0.03\,\mathbf{e}_2$$

The unit vector $\mathbf{e}^{D/B}$ along the position vector $\mathbf{r}^{D/B}$ can be found by using Eqns. 2.2b and 2.2c:

$$\mathbf{e}^{D/B} = (0.37\,\mathbf{e}_1 + 0.03\,\mathbf{e}_2)/(0.37^2 + 0.03^2)^{1/2} = 0.99\,\mathbf{e}_1 + 0.08\,\mathbf{e}_2$$

Thus the force exerted by the woman on the holder is equal to

$$\mathbf{F} = T\,\mathbf{e}^{D/B} = 128.1\,(0.99\,\mathbf{e}_1 + 0.08\,\mathbf{e}_2)\ \text{N}$$

Note that this force is approximately horizontal. If the woman were to perform arm curls with free weights, the force exerted on her hands would always be directed vertical downward.

2.6 Summary

Newtons's laws describe the interaction between forces and motion. Newton's first law states that a particle will remain at rest unless it is acted on by an unbalanced (resultant) force. Newton's second law is about how forces acting on a particle affect its motion:

$$\Sigma\mathbf{F} = m\,\mathbf{a}$$

where $\Sigma\mathbf{F}$ denotes the resultant force acting on a particle, m is the mass of the particle, and $\mathbf{a}$ is its acceleration, measured with respect to a coordinate system fixed on earth. According to this law, a particle will accelerate in the direction of the resultant force acting on the particle. The magnitude of the acceleration will be equal to the magnitude of the resultant force divided by the mass of the particle. Newton's first law is merely a subset of the second law. A particle is an object, large or small, in which the variations in velocity on its body is negligibly small at any instant when compared to the mean velocity of the object at the same instant. A particle could be a star, a baseball, or even an ice-skater gliding on ice.

Newton's third law describes how objects, living as well as nonliving, interact with each other:

$$\mathbf{f}_{1-2} = -\mathbf{f}_{2-1}$$

in which $\mathbf{f}_{1-2}$ is the force exerted on particle 1 by particle 2. This law states that reactive force to an action is equal in magnitude to the force of action but is in the reverse direction. Contrary to common belief, when a bull hits a bullfighter and sends him off into the air, the fighter actually hits the bull with the same intensity; it is just that the bull has a much larger mass and therefore can easily withstand the force.

2.7 Problems

Problem 2.1. Consider a cable connecting two points A and B that are fixed in space (Fig. P.2.1). The cable is in tension so that it forms a straight line passing through A and B. Why is the tension in the cable uniform between these points? Take a small segment of the cable and draw all the forces that act on it. Use Newton's second law for this segment of the cable.

Problem 2.2. Why does the tension in an inextensible cable remain uniform when the cable goes over a frictionless pulley?

Problem 2.3. The coordinates of a point with respect to a Cartesian coordinate system E with unit vectors ($\mathbf{e}_1$, $\mathbf{e}_2$, and $\mathbf{e}_3$) can be written as A (x_1, x_2, x_3) where A denotes the point under question. Let A (1, 5, −2) and B (3, −4, 6) be two points in space. Determine the position vector connecting A to B. Also determine the unit vector along the straight line from A to B.

Answer: $\mathbf{r}^{B/A} = 2\,\mathbf{e}_1 - 9\,\mathbf{e}_2 + 8\,\mathbf{e}_3$; $\mathbf{e} = 0.16\,\mathbf{e}_1 - 0.74\,\mathbf{e}_2 + 0.66\,\mathbf{e}_3$.

Problem 2.4. Determine the velocity and acceleration of a small child in a swing of length 3.5 m at a time when the swing is at 30° with the ver-

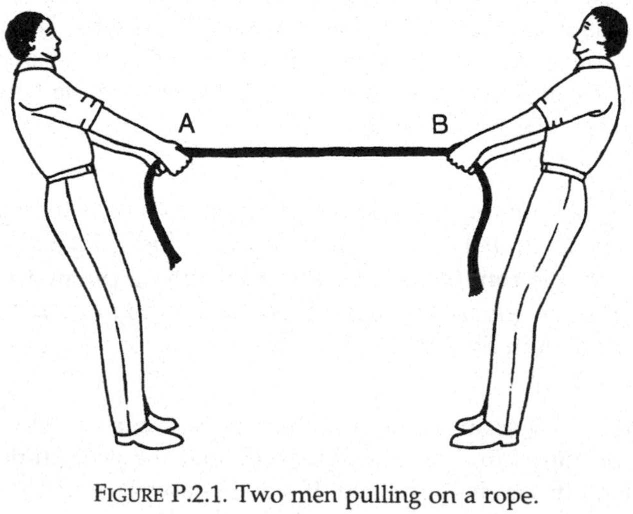

FIGURE P.2.1. Two men pulling on a rope.

tical axis. At that instant, the swing was moving up at a constant rate of $(\pi/8)$ rad/s.
Answer: $\mathbf{v} = 1.2\ \mathbf{e}_1 + 0.7\ \mathbf{e}_2$ (m/s); $\mathbf{a} = 0.27\mathbf{e}_1 + 0.47\mathbf{e}_2$ (m/s^2) where $\mathbf{e}_1$ and $\mathbf{e}_1$ are unit vectors in the plane of motion in the horizontal and vertical directions, respectively.

Problem 2.5. A volleyball, thrown from position $\mathbf{r} = 2\ \mathbf{e}_2$ at time $t = 0$, occupies the position $\mathbf{r} = 6\ \mathbf{e}_1 + 4\ \mathbf{e}_2$ at time $t = 2$ s. Determine the initial velocity of the ball. Determine the highest elevation (y) the ball reaches while airborne.
Answer: $\mathbf{v}_o = 3\ \mathbf{e}_1 + 10.8\ \mathbf{e}_2$ (m/s); $y = 7.9$ m.

Problem 2.6. In baseball, a pitcher standing 60 ft from the home plate needs about 0.4 s to raise his glove in defense of a hard line drive hit at him. In March 1996, a batter hit the pitcher in the face with a ball that traveled back in 0.29 s. Five of the pitcher's teeth were shattered, and he needed 65 stitches to mend his face. Determine the horizontal component of the velocity of the baseball as it hit the pitcher.
Answer: $v_1 = 207$ ft/s.

Problem 2.7. A person drags a sulking dog (who is pretending to be dead) on a polished wood floor along a straight line at a rate of 0.50 m/s. The angle the taut leash makes with the horizontal is 30°. The dog weighs 20 kg and the coefficient of dynamic friction $\mu = 0.2$. Determine the tensile force exerted on the dog by the leash. What force does the dog exert on her 'inconsiderate' owner? Note that the frictional force acting on the dog is equal to the contact force normal to the floor times the coefficient of dynamic friction.
Answer: $T = 40.6$ N.

Problem 2.8. A person drags a sulking dog on a polished wood floor along a straight line at an acceleration of 0.40 m/s^2. The dog pretends to be dead as if she were incapable of moving her limbs. The angle the taut leash makes with the horizontal is 60°. The dog weighs 20 kg and the coefficient of dynamic friction $\mu = 0.2$. Determine the tensile force exerted on the dog by the leash.
Answer: $T = 70.2$ N.

Problem 2.9. A woman doing seated machine curls rests her upper arm on an arm pad, parallel to the floor as shown in Fig. 2.12. She then pulls on the bar that is connected through a cable and a pulley to a weight of 10 kg. The weight moves up with an acceleration of 2 m/s^2. Determine the tension in the cable.
Answer: $T = 118$ N.

Problem 2.10. A woman performs seated machine curls as shown in Fig. 2.12. She performs the exercise slowly so that the acceleration of the weights she lifts can be neglected. If she lifts a weight of 15 kg, deter-

mine the force exerted by the bar on the woman when her forearm makes 30° with the horizontal axis. Hint: first determine the unit vector along line BD.
Answer: $\mathbf{F} = 147\ (0.997\ \mathbf{e}_1 - 0.07\ \mathbf{e}_2)$.

Problem 2.11. Experiments on a wind tunnel showed that the frictional force exerted by air on an object can be estimated by the following formula:

$$F = (1/2)\rho A C_d u^2 \tag{2.22}$$

where ρ is the mass density of air ($\rho = 1.2\ \mathrm{kg/m^3}$), A is the area of projection of the object normal to the direction of motion, C_d is the dimensionless drag coefficient ($C_d = 0.8$), and u is the speed of the moving object. Provide an upper-bound estimate for the maximum resistance force exerted by air on a 5-kg steel ball falling from a distance of 60 m. How big is the air resistance in comparison with the force of gravity? The diameter of the ball is 6 cm. Hint: Determine the velocity of the ball at $y = 0$ m by neglecting the effect of air resistance on the fall, and then use this velocity to compute the resistance force.
Answer: $F = 1.6$ N, $W = 49.1$ N.

Problem 2.12. Determine the arm length and the period of swing of 15 subjects (students in a classroom) and see whether the data are consistent with the predicted dependence of the period of the pendulum on the length of the pendulum (Eqn. 2.18e).

Problem 2.14. Provide a proof for the velocity and acceleration expressions in the path coordinates. The proof can be found in most books on rigid body dynamics.

3

Particles in Motion: Method of Lumped Masses and Jumping, Sit-Ups, Push-Ups

3.1 Introduction

During movement, the velocity in a human body is not uniform. For example, during running, the arms and legs rotate relative to the trunk, and therefore the velocity varies from point to point in each of the limbs. What are nature's laws of motion for objects in which the velocity is not uniform? In this chapter, we address this question. Our starting point is that any object in space can be thought of as a collection of small mass elements. Therefore, any object can be idealized as a system of particles. For each particle in the system, Newton's laws of motion hold. Summing over these equations, we arrive at an important principle in mechanics that is called the conservation of linear momentum. The linear momentum was characterized as "the quantity of motion of a body" in Newton's *Principia*. The linear momentum of a system of particles is the product of the mass of the system and the velocity of its center of mass. The center of mass is defined as the point of support (pivot, fulcrum) at which an object would be in a delicate balance under the action of gravity. In 1589, at the age of 25, Galileo published a treatise on the center of mass of solids. Almost a century later, Borelli devised an experiment to determine the position of center of mass of humans. His book *On the Movement of Animals* (1680) includes a sketch of a man lying on a seesaw in perfect balance. Borelli correctly predicted that the center of mass of the man to be placed right on top of the fulcrum to achieve static equilibrium.

The concept of center of mass (or center of gravity) has inspired the use of similar concepts in fields other than mechanics. For example, the U.S. Census Bureau assigned Butte County, South Dakota, as the geographic center of the United States. The geographic center is defined as the point at which the surface of a geographic entity would balance if the surface were a plane of uniform weight per unit area. According to the Census Bureau, the population center of the United States is Crawford County, Missouri. This is the point at which a flat, weightless map of the United States would balance if weights of equal value were placed on it with each weight representing the location of one person at a specified date. Unlike the geographic center of the United States, the position of the center of mass of a human body is not stationary but varies with body movement. Aware of this fact, dancers learn to manipulate the movement of the body around the center of mass. Once airborne during a jump, dancers reposition their arms and legs to give the impression they can defy gravity and remain suspended in air.

The conservation of linear momentum yields no information about the rate of rotation of a solid body or the average rate of rotation of a system of particles. How do we obtain an equation that relates the rate of rotation to the external forces acting on a body? To obtain such an equation, we multiply the equations of motion for each particle in the body with an appropriate weight and sum over all particles. Similar procedures are employed in statistics to evaluate the extent of scattering of data. The concepts presented in this chapter are those most fundamental to mechanics. We illustrate the use of the conservation principles with examples involving human movement and motion.

3.2 Conservation of Linear Momentum

The human body is a collection of solid objects arranged in treelike configuration. Each object or tree of objects is composed of large number of particles with small mass and small volume (Fig. 3.1). Newton's laws presented in the previous chapter hold for each of these particles. According to the second law:

$$\mathbf{F}^i + \Sigma \mathbf{f}_{ij} = m^i \, \mathbf{a}^i \tag{3.1}$$

where $\mathbf{F}^i$ denotes the net force exerted on particle i by particles external to the system B of particles under consideration. The gravitational force is an example of an external force. It is acted on particle i by earth, which is outside the collection of particles under study. The internal force $\mathbf{f}_{ij}$ is exerted by particle j in the system of particles B on particle i in B. The symbol m^i is the mass of particle i, and $\mathbf{a}^i$ is its acceleration. Note that, in accordance with Newton's third law, the force of reaction is equal to the negative of the force of action; that is:

$$\mathbf{f}_{ij} = -\mathbf{f}_{ji} \tag{3.2}$$

Because a particle cannot exert a force on itself, $\mathbf{f}_{ii}$ is always equal to zero.

The motion of each particle i in B is described by Eqn. 3.1. Let us sum the left- and the right-hand side of Eqn. 3.1 from $i = 1$ to $i = n$ where i takes the integer values and n is the number of particles in B:

$$\Sigma \mathbf{F}^i + \Sigma\Sigma \mathbf{f}_{ij} = \Sigma m^i \, \mathbf{a}^i \tag{3.3}$$

In this equation, $\Sigma \mathbf{F}^i$ denotes the sum of external forces acting on B (external contact forces plus the gravitational body force). The term $\Sigma\Sigma \mathbf{f}_{ij}$ is

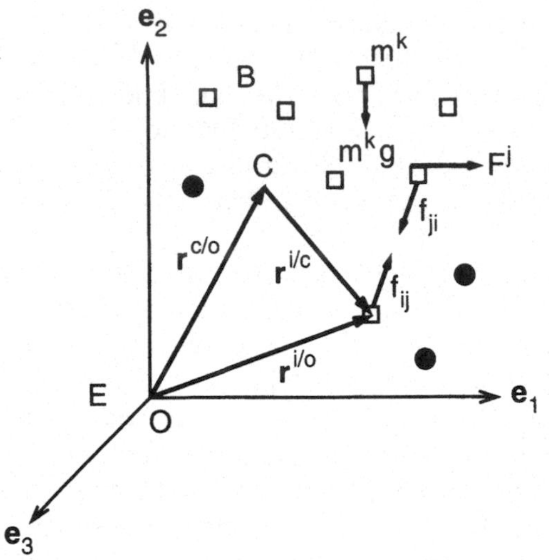

FIGURE 3.1. A system of particles (*B*) moving with respect to reference frame *E*. Particles in B were marked with *rectangles* and particles outside of B with *filled circles*. The position vectors $\mathbf{r}^{i/o}$, $\mathbf{r}^{i/c}$, and $\mathbf{r}^{c/o}$ connect, respectively, the point *O* to particle *i*, the center of mass C to point *i*, and point O to the center of mass C. Also shown in the figure are internal forces $\mathbf{f}_{ij}$ and $\mathbf{f}_{ji}$ and external forces $\mathbf{F}^j$ and m^k $\mathbf{g}$.

the sum of all internal forces acting on particles in B. Newton's third law requires that this latter sum be equal to zero. Thus, internal forces do not contribute to the acceleration of the system at all and hence Eqn. 3.3 assumes the following form:

$$\Sigma \mathbf{F}^i = \Sigma m^i \, \mathbf{a}^i \tag{3.4}$$

Linear momentum **L** of a system of particles B is defined as

$$\mathbf{L} = \Sigma m^i \, \mathbf{v}^i \tag{3.5}$$

where $\mathbf{v}^i$ denotes the velocity of particle *i*. Combining Eqns. 3.4 and 3.5, we obtain

$$\Sigma \mathbf{F}^i = d\mathbf{L}/dt \tag{3.6}$$

This equation is called the conservation of linear momentum. According to this equation, the time rate of change of linear momentum of a system of particles is equal to the resultant external force acting on the system. When a body is at rest or moving with constant speed, its linear momentum remains constant and hence the sum of all forces acting on the body must vanish. The branch of mechanics that considers bodies in equilibrium is called statics. Later in the text, in Chapters 5 and 6, we discuss statics as it relates to human movement.

3.3 Center of Mass and Its Motion

The center of mass of a body B, living or nonliving, is defined by the following equation:

$$(\Sigma m^i) \, \mathbf{r}^c = \Sigma m^i \, \mathbf{r}^i \tag{3.7}$$

Σm^i is the total mass in B, $\mathbf{r}^c$ is the position vector for the center of mass of B, m^i is the mass of the ith element in B, and $\mathbf{r}^i$ is its position vector. These entities are shown in Fig. 3.1. Note that the center of mass is also commonly known as the center of gravity.

Let us now perceive center of mass as if it were a particle in space. In reality, the center of mass may not correspond to any point of the object B. The position of center of mass may be occupied by different particles of B at different times during motion. The time rate of change of position vector $\mathbf{r}^c$ (derivative of $\mathbf{r}^c$ with respect to time t) is equal to the velocity of the center of mass, which we denote by $\mathbf{v}^c$. The acceleration of the center of mass $\mathbf{a}^c$ is the time derivative of $\mathbf{v}^c$:

$$\mathbf{r}^c = [\Sigma(m^i\, \mathbf{r}^i)]/(\Sigma m^i) \tag{3.8a}$$

$$\mathbf{v}^c = [\Sigma(m^i\, \mathbf{v}^i)]/(\Sigma m^i) \tag{3.8b}$$

$$\mathbf{a}^c = [\Sigma(m^i\, \mathbf{a}^i)]/(\Sigma m^i) \tag{3.8c}$$

Using Eqn. 3.8b, the linear momentum $\mathbf{L}$ of a system of particles can be written as

$$\mathbf{L} = \Sigma(m^i\, \mathbf{v}^i) = (\Sigma m^i)\, \mathbf{v}^c$$

Substituting Eqn. 3.8c into Eqn. 3.6, we obtain an equation governing the motion of center of mass of an object:

$$\Sigma \mathbf{F}^i = (\Sigma m^i)\, \mathbf{a}^c \tag{3.9}$$

According to this equation, the net force acting on a system of particles is equal to the mass of the system times the acceleration of the center of mass. For a sphere of uniform mass density, the center of mass is positioned at the center of the sphere. As in the case of an L-shaped body, the center of mass may lie outside the body. The tables at the end of the text provide information about the position of the center of mass for solid bodies of a variety of geometric shapes as well as human body configurations (see Appendix 2).

Example 3.1. The Center of Mass of a Human Body as Represented by Two Rods. Consider a body consisting of two slender rods ab and bc of length L and mass m that are connected with a pin at b. The bar ab is tilted 45° from the horizontal axis. Determine the center of mass of B.

Solution: We draw a reference frame whose origin O is at the pin b connecting the two rods (Fig. 3.2). In a uniform slender rod, the center of mass occupies the midpoint along the axis of the rod. Thus, the centers of mass of the two rods ab and bc are located at the following positions:

$x_1 = -(L/2) \cos 45°$; $x_2 = (L/2) \sin 45°$; $x_3 = 0$ for the rod ab

$x_1 = (L/2)$, $x_2 = 0$, $x_3 = 0$ for the rod bc

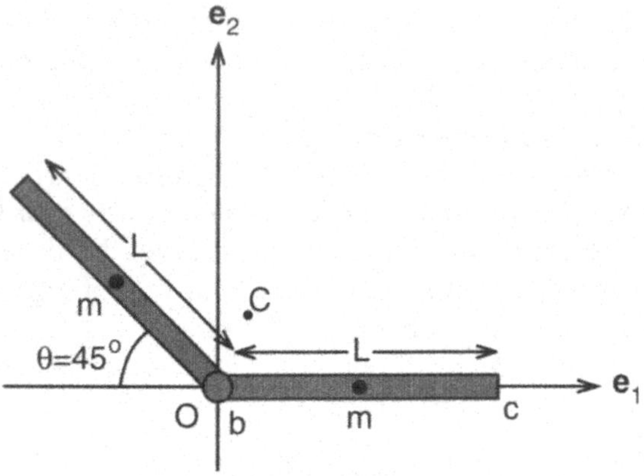

FIGURE 3.2. The location of the center of mass of a system composed of two slender rods that are linked together at point b.

Next, we insert these values into Eqn. 3.8a and determine the center of mass of the serially linked rods:

$$(2m)\, x_1{}^c = -m\,(L/2)\cos 45° + m\,(L/2) \Rightarrow x_1{}^c = 0.07\,L$$

$$(2m)\, x_2{}^c = m\,(L/2)\sin 45° m\,(0) \Rightarrow x_2{}^c = 0.18\,L$$

$$(2m)\, x_3{}^c = -m\,(0) + m\,(0) \Rightarrow x_3{}^c = 0$$

Example 3.2. The Raising of Arms Changes the Position of the Center of Mass. An athlete weighs 163 lb and has a height of 69 in. His center of mass lies 38 in. above the ground during the standard standing position shown in Fig. 3.3. Determine the change in the elevation of his center of mass when he raises his arms from the sides of his body 180° over his head. His arms weigh 7 lb each and the center mass of his arms moves upward 25 in. when he raises them over his head.

Solution: We will consider the athlete to be composed of three parts: two arms each having a mass m and the rest of his body (M). The location of the center of mass in the vertical x_2 direction before and after raising the arms is given by the following equations:

$$(M + 2m)\,(38) = Md + 2m\,h \tag{3.10a}$$

$$(M + 2m)\,(38 + y) = Md + 2m\,(h + 25) \tag{3.10b}$$

where d denotes the distance between the origin of the coordinate system and the center of mass of the body without the arms, and h is the distance between the origin and the center of mass of the arms when they lie on the sides. The symbol y denotes the unknown additional elevation of the center of mass. When Eqn. 3.10a is subtracted from Eqn. 3.10b, one finds that $y = 2.15$ in.

Note that one can easily estimate the mass of a body part by immersing it in a container full of water and measuring the amount of water displaced. The mass of the body part, say a foot or an arm, is equal to the volume of water displaced times the average specific density of human body, which is about 0.9 g/cm^3.

Example 3.3. Chin-ups. A 75-kg man is doing chin-ups to work out his latissimus muscle (Fig. 3.4). If during the rising phase his acceleration reaches a peak of 4 m/s^2, determine the contact force exerted by the horizontal bar on the athlete at that instant.

Solution: The forces acting on the athlete are the gravitational force ($-mg\mathbf{e}_2$) pointing downward and the contact forces **F** exerted by the bar on the athlete, pointing upward. The resultant of these forces must be equal to the mass times the acceleration of center of mass. This can be written in the vector notation in the following form:

$$-75 \times 9.81\ \mathbf{e}_2 + 2\mathbf{F} = 75 \times 4\ \mathbf{e}_2 \Rightarrow \mathbf{F} = 518\ \text{N}\ \mathbf{e}_2$$

Note that the contact force exerted by the bar is equivalent to a weight of $(2 \times 518/9.81) \cong 105.6$ kg. This example shows that when the man accelerates upward, the force the bar exerts on the man is an upward force that is greater than the man's body weight. Because action equals to reaction, this is also the force the athlete exerts on the bar as he pulls on the bar.

Example 3.4. Vertical Jumping. Animals and humans jump to reach higher places, to leap over obstacles, or for competition. Volleyball and basketball players must react instantaneously to the ball and jump within

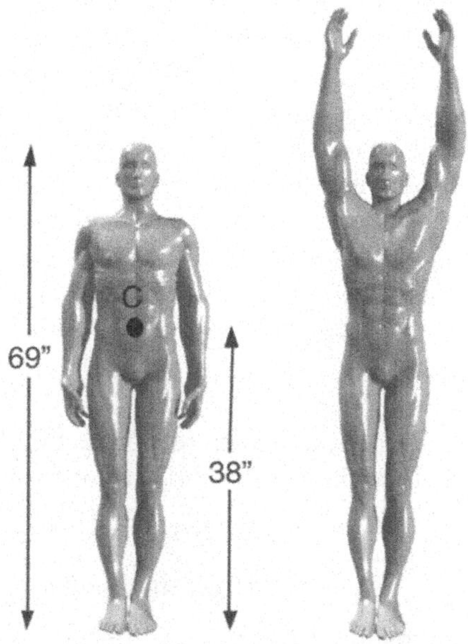

FIGURE 3.3. A man raises his arms from the side of his body to over his head. The center of mass (C) of the man also moves up, albeit a much smaller amount.

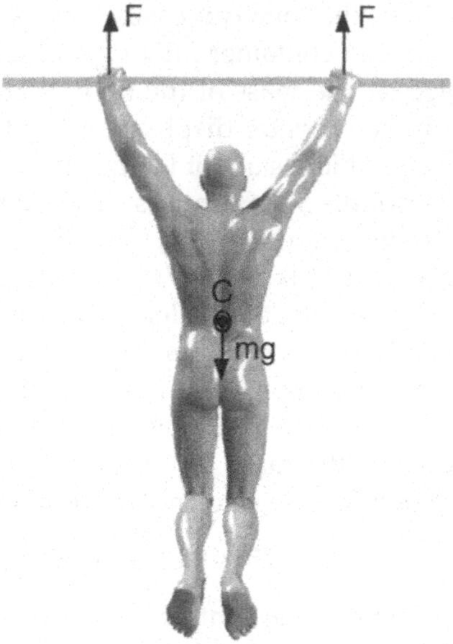

FIGURE 3.4. A man doing chin-ups to exercise his latissimus muscle. The diagram shows the external forces acting on the man.

a fraction of a second. Videotapes of jumping events show that the entire duration of the propulsive stage of the human vertical jump, from backward rotation of the trunk to toe-off, lasts about 0.3 s. In this brief period of time, the angles between various segments of the lower body (feet, shank, thigh, and hips) change to lift the upper body (70% of the body weight) vertically. The mechanics of vertical jumping can be captured with reasonable accuracy by using a four-segment model of a human body composed of foot, shank, thigh, and the upper body (Fig. 3.5a).

Here we consider a much simpler model in which a mass M (representing the upper body) is attached to two slender bars of length L (weightless legs) as shown in Fig. 3.5b. The bars ab and bc must be connected by a muscle–tendon system that enables the bars to change the angle between them. If there was no such mechanism, the two bars would collapse onto the floor under the weight of mass M. Our aim is to determine the conditions of jumping for this two-segment model.

Solution: The geometry of the assumed structure dictates that the spatial position of the point mass M at any time t is given by the equation:

$$\mathbf{r} = 2L \sin\theta \, \mathbf{e}_2 \tag{3.11a}$$

Differentiating $\mathbf{r}$ with respect to time twice, we determine the velocity and acceleration of mass M as a function of time:

$$\mathbf{v} = 2L \cos\theta \, (d\theta/dt) \, \mathbf{e}_2 \tag{3.11b}$$

$$\mathbf{a} = [-2L \sin\theta \, (d\theta/dt)^2 + 2L \cos\theta \, (d^2\theta/dt^2)] \, \mathbf{e}_2 \tag{3.11c}$$

Substituting Eqn. 3.11c into the law of motion for the center of mass

(Eqn. 3.9), we arrive at the following vectorial equation for the force **P** exerted by the ground on the two-segment jumper:

$$\mathbf{P} - Mg\,\mathbf{e}_2 = M\,[-2L \sin\theta\,(d\theta/dt)^2 + 2L \cos\theta\,(d^2\theta/dt^2)]\,\mathbf{e}_2 \quad (3.12a)$$

$$\mathbf{P} = M\,[g - 2L \sin\theta\,(d\theta/dt)^2 + 2L \cos\theta\,(d^2\theta/dt^2)]\,\mathbf{e}_2 \quad (3.12b)$$

The ground can only exert an upward force on the jumper; it has no capability to pull the jumper toward earth. At the instant the jumper leaves the ground, the ground reaction force **P** must be equal to zero. This gives us

$$g = 2L \sin\theta\,(d\theta/dt)^2 - 2L \cos\theta\,(d^2\theta/dt^2) \quad (3.13)$$

Let us explore the physical implications of this equation. Let the lower limbs of the jumper be represented by two linked rods of length $L = 0.5$ m. Furthermore, let us assume that at the time of the takeoff, $\theta = 60°$ and $(d^2\theta/dt^2) = 0$. The rate of rotation $(d\theta/dt)$ can then be found by using Eqn. 3.13 to be equal to 3.25 rad/s. This means that, for the ground force to diminish to zero, the angle between the shank and thigh must increase at a rate greater than 180°/s. If the legs straighten at lesser speed, jumping cannot occur because the ground will continue to exert some finite level of upward force.

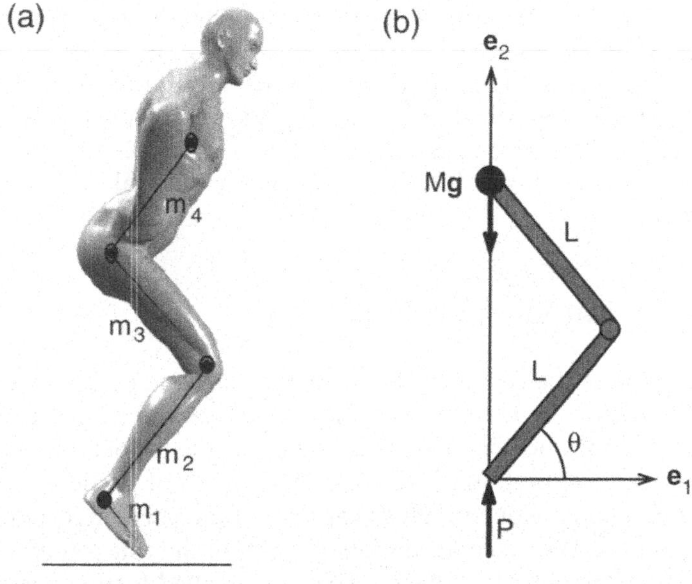

FIGURE 3.5a,b. Schematic representation of a four-segment model of an athlete performing a vertical jump **(a)**. The athlete performing the jump is represented by even a simpler model in **(b)**: a mass representing the weight of the upper body is connected to two slender rods. The external forces acting on the athlete are also shown in **(b)**.

This simple analysis implies that action of hip, thigh, and calf muscles on the lower limbs must be fast enough for jumping to occur. Additionally, the analysis presented here shows that the upward velocity increases with increasing length of the slender bars ab and bc. According to the model, athletes with long legs reach higher velocities during vertical jumping than athletes with shorter legs, when both the short and the tall athletes have the same capacity for quick rotation of the thigh over the lower leg.

Once the two-segment model lifts itself off the ground, how far will it travel in air? The only force acting on the airborne jumper is gravity, and therefore its center of mass will follow the equations of free fall presented in the previous chapter. To determine how far up the center of mass will reach, we need to determine its velocity at the time of takeoff. The vertical velocity at the time of the takeoff can be obtained by substituting L = 0.5 m, $\theta = 60°$, and $(d\theta/dt) = 3.25$ rad/s in Eqn. 3.11b. This initial velocity is found to be 1.625 m/s. After the takeoff, the path of the center of mass is determined by the following equations:

$$\mathbf{a} = -g\ \mathbf{e}_2 \tag{3.14a}$$

$$\mathbf{v} = (1.625 - gt)\ \mathbf{e}_2 \tag{3.14b}$$

$$\mathbf{r} = (1.625\ t - 0.5\ gt^2)\ \mathbf{e}_2 \tag{3.14c}$$

The center of mass will reach its highest point when **v** becomes equal to zero. Eqn. 3.14b shows that $t = 1.625/g$ at this point. Substituting this value of t into Eqn. 3.14c, we find that the maximum distance h that the center of mass of the athlete rises during the jump must be equal to $(1.625)^2/2g$ or 13 cm. In general, vertical distance traveled in air during jumping is equal to $h = V_o^2/2g$ where V_o is the vertical velocity at the time of takeoff.

3.4 Multiplication of Vectors

The description of the motion of the center of mass is neccessary but not sufficient to define the motion of a body or a system of particles. Specifically, Eqn. (3.9) provides no information on the rate of rotation of a solid body or the relative motion of particles in a system of particles. Clearly, a mere summation is not enough to capture all the very important parameters of motion. What is nature's law governing the rotation of a body? Intuitively, we know that whenever a force is applied further away from a pivot point, the better is its capacity to induce rotation. A good example of this is the positioning of door knobs. They are located at the free edge of the door rather than being close to the edge hinged to the wall. This way the distance between the axis of rotation and the force one exerts on the door when one wants to swing it open is maximized.

It turns out that a vector equation governing the rate of rotation of a solid body can be obtained by writing Newton's second law for each particle in the body, multiplying it vectorially with the position vector of that particle, and summing over all particles in the system under consideration. To present this procedure in more detail, we introduce a brief synopsis of the vector multiplication.

Vectors may be multiplied with other vectors in two distinct ways. The scalar product (also called the dot product) of **a** and **b** is defined as the projection of one of the vectors onto the other:

$$\mathbf{a} \cdot \mathbf{b} = \|\mathbf{a}\| \, \|\mathbf{b}\| \cos \theta \tag{3.15}$$

where $\mathbf{a} \cdot \mathbf{b}$ is the dot product, and $\|\mathbf{a}\|$ and $\|\mathbf{b}\|$ are the magnitudes of **a** and **b**. The symbol θ represents the angle between **a** and **b**. Note that the dot product can be positive or negative but has no direction, and therefore it is a scalar.

An important example of dot product is the mechanical work done by a force. When a force acting on a particle is multiplied scalarly with the displacement of the particle, the product W is called the work done by that force. When a person resists a large force to remain at rest, the work done by this force is equal to zero because there is no displacement. Resisting a force statically requires caloric expenditure but produces no mechanical work. In a pendulum, the tensile force exerted by the cord on the bob does no work because this force is always perpendicular to the path traversed by the bob. Work done by a force is positive if the projection of the force on the displacement vector is in the same direction as the displacement. When a particle falls toward earth, gravity does positive work on the particle. On the other hand, when an object is raised vertically the work done by gravity on the object is negative.

According to Eqn. 3.15, the dot product of two vectors that are perpendicular to each other is zero. The following results then hold for the unit vectors $\mathbf{e}_1$, $\mathbf{e}_2$, and $\mathbf{e}_3$:

$$\mathbf{e}_1 \cdot \mathbf{e}_2 = \mathbf{e}_1 \cdot \mathbf{e}_3 = \mathbf{e}_2 \cdot \mathbf{e}_3 = \mathbf{e}_2 \cdot \mathbf{e}_1 = \mathbf{e}_3 \cdot \mathbf{e}_1 = \mathbf{e}_3 \cdot \mathbf{e}_2 = 0 \tag{3.16a}$$

$$\mathbf{e}_1 \cdot \mathbf{e}_1 = \mathbf{e}_2 \cdot \mathbf{e}_2 = \mathbf{e}_3 \cdot \mathbf{e}_3 = 1 \tag{3.16b}$$

Scalar product of two vectors can also be written in terms of the projections of vectors on a Cartesian coordinate system. Let **a** and **b** be two vectors whose components with respect to a Cartesian coordinate system E are given by the following equations:

$$\mathbf{a} = a_1\mathbf{e}_1 + a_2\mathbf{e}_2 + a_3\mathbf{e}_3 \tag{3.17a}$$

$$\mathbf{b} = b_1\mathbf{e}_1 + b_2\mathbf{e}_2 + b_3\mathbf{e}_3 \tag{3.17b}$$

Using Eqns. 3.16 and 3.17, the scalar product of **a** and **b** can be shown to obey the following relationship:

$$\mathbf{a} \cdot \mathbf{b} = a_1b_1 + a_2b_2 + a_3b_3 = \mathbf{b} \cdot \mathbf{a} \tag{3.17c}$$

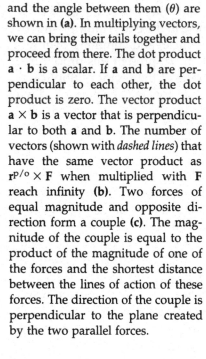

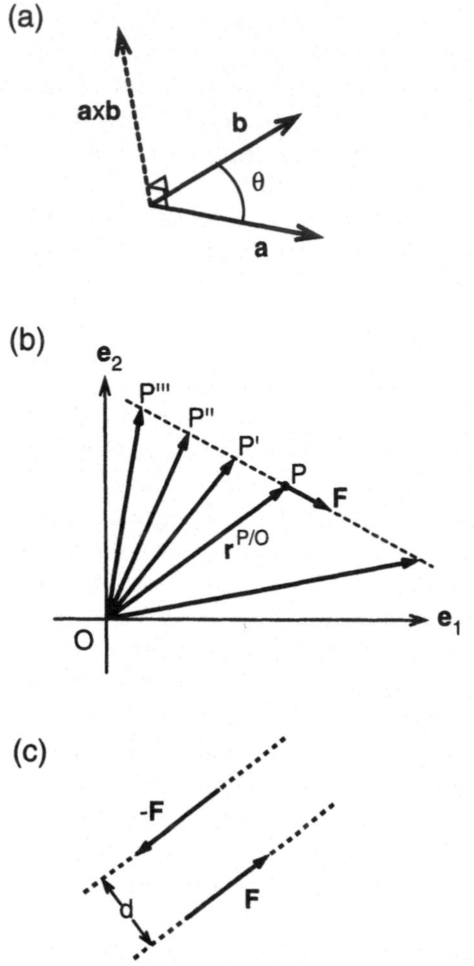

FIGURE 3.6a–c. Two vectors **a** and **b** and the angle between them (θ) are shown in **(a)**. In multiplying vectors, we can bring their tails together and proceed from there. The dot product $\mathbf{a} \cdot \mathbf{b}$ is a scalar. If **a** and **b** are perpendicular to each other, the dot product is zero. The vector product $\mathbf{a} \times \mathbf{b}$ is a vector that is perpendicular to both **a** and **b**. The number of vectors (shown with *dashed lines*) that have the same vector product as $\mathbf{r}^{P/o} \times \mathbf{F}$ when multiplied with **F** reach infinity **(b)**. Two forces of equal magnitude and opposite direction form a couple **(c)**. The magnitude of the couple is equal to the product of the magnitude of one of the forces and the shortest distance between the lines of action of these forces. The direction of the couple is perpendicular to the plane created by the two parallel forces.

Two vectors **a** and **b** can be multiplied vectorially to give a third. The vector product of **a** and **b** is called the cross product and is shown as $\mathbf{a} \times \mathbf{b}$. The resulting product is a vector that is perpendicular to both **a** and **b**:

$$\mathbf{a} \times \mathbf{b} = \|\mathbf{a}\| \, \|\mathbf{b}\| \sin \theta \, \mathbf{e} \tag{3.18a}$$

where θ is the angle between **a** and **b**, and **e** is a unit vector that is normal to both **a** and **b** (Fig. 3.6a). The magnitude of $\mathbf{a} \times \mathbf{b}$ is the area of the parallelogram traversed by **a** and **b**. The sense of direction of **e** is determined by the right-hand rule: point the fingers of the righthand in the direction of **a**, then turn the fingers toward **b**, and the thumb will point to the right sense of direction of **e**. The vector product plays a most important role in mechanics. The lever systems of the human body discussed in Chapter 1 are direct consequences of the vector product of force and the lever arm.

According to Eqn. 3.18a, if two vectors are perpendicular to each other, the magnitude of the cross product is equal to the product of the magnitudes of the vectors. Also, if two vectors are parallel to each other, their cross product is equal to zero. This leads to the following vector products between the unit vectors $\mathbf{e}_1$, $\mathbf{e}_2$, and $\mathbf{e}_3$:

$$\mathbf{e}_1 \times \mathbf{e}_2 = \mathbf{e}_3 = -\mathbf{e}_2 \times \mathbf{e}_1 \tag{3.18b}$$

$$\mathbf{e}_2 \times \mathbf{e}_3 = \mathbf{e}_1 = -\mathbf{e}_3 \times \mathbf{e}_2 \tag{3.18c}$$

$$\mathbf{e}_3 \times \mathbf{e}_1 = \mathbf{e}_2 = -\mathbf{e}_1 \times \mathbf{e}_3 \tag{3.18d}$$

$$\mathbf{e}_1 \times \mathbf{e}_1 = \mathbf{e}_2 \times \mathbf{e}_2 = \mathbf{e}_3 \times \mathbf{e}_3 = 0 \tag{3.18e}$$

The vector product **a** and **b** can then be expressed as

$$\begin{aligned}\mathbf{a} \times \mathbf{b} &= (a_2 b_3 - a_3 b_2)\,\mathbf{e}_1 - (a_1 b_3 - a_3 b_1)\,\mathbf{e}_2 + (a_1 b_2 - a_2 b_1)\,\mathbf{e}_3 \\ &= -\mathbf{b} \times \mathbf{a}\end{aligned} \tag{3.18f}$$

where a_1, a_2, a_3 and b_1,b_2, b_3 are the projections of the vectors **a** and **b** along $\mathbf{e}_1$, $\mathbf{e}_2$, and $\mathbf{e}_3$ directions, respectively.

3.5 Moment of a Force

An important example of vector multiplication is the concept of moment of a force with respect to a point in space. The moment $\mathbf{M}^o$ is a measure of the capacity of force **F** acting on point P to cause rotation about point O. It is defined as follows:

$$\mathbf{M}^o = \mathbf{r}^{P/o} \times \mathbf{F} \tag{3.19}$$

where $\mathbf{r}^{P/o}$ is the position vector connecting point O to P (Fig. 3.6b). The magnitude of $\mathbf{M}^o$ is the product of the magnitude of the force and the perpendicular distance of the point O from the line of action of the force. Note that in the evaluation of the moment, the position vector from O to P can be replaced with any other that connects point O with a point on the line of action of force **F** (Fig. 3.6b). Note also that, if the distance between point O and the line of action of force **F** is zero, then this force creates no moment with respect to point O.

If two forces are equal in magnitude and opposite in direction, their sum is equal to zero. Nevertheless, these two forces exert a moment with respect to any point in space so long as they do not share the same line of action (Fig. 3.6c). Such a pair of forces form what is called a *force couple*. The moment created by the force couple about a point in the plane of the couple can be shown to be equal to the product of the perpendicular distance d between the line of action of the opposing forces and the magnitude of one of the forces comprising the couple. The moment of a force couple about any point in its plane is the same.

Example 3.5. Abduction of Arms. An athlete whose arms are 66 cm long stands with his hands at the thighs holding 10-kg dumbbells. The athlete contracts his front, middle, and rear deltoids and pulls the weights up directly to the side (Fig. 3.7). He raises his arms to the full-flexed shoulder position with the weights above the elbow joint and higher than the shoulder level. Then he slowly lowers the weight to the starting position, and repeats the exercise. Compute the moment generated by the weight of the dumbbell at the shoulder when the arm makes 0°, 45°, and 90° with the vertical axis.

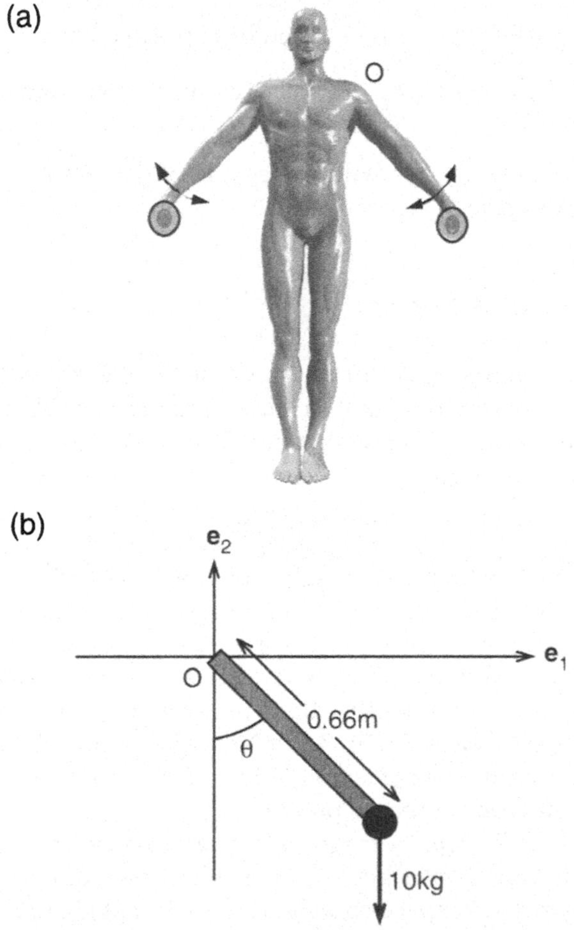

FIGURE 3.7a,b. A workout scheme for the shoulder muscle deltoids incorporates abduction of the arms while carrying free weights **(a)**. The representation of the arm as a slender rod is illustrated in **(b)**.

Solution: Let us draw a coordinate system E whose origin coincides with the center of the shoulder joint. The position of the dumbbell with respect to the origin is then given by the following equation:

$$\mathbf{r}^{P/O} = L \sin\theta\, \mathbf{e}_1 - L \cos\theta\, \mathbf{e}_2 \tag{3.20}$$

where L denotes the length of the arm and θ is the angle the arm makes with the vertical axis as shown in Fig. 3.7. The force exerted by the dumbbell on the athlete equals to the weight of the dumbbell if the exercise is done slowly. The moment this force generates with respect to the center of the shoulder is

$$\mathbf{M} = (L \sin\theta\, \mathbf{e}_1 - L \cos\theta\, \mathbf{e}_2) \times (-M\, g\, \mathbf{e}_2) = -\mathrm{M}\, g\, L \sin\theta\, \mathbf{e}_3 \tag{3.21}$$

Thus, the magnitude of $\mathbf{M}$, $\|\mathbf{M}\| = 0$, 45.8 N-m, and 64.7 N-m for $\theta = 0°$, 45°, and 90°, respectively.

Note that we could have computed the moment $\mathbf{M}$ without going through the vector product. From the definition of vector product, we know that the magnitude of $\mathbf{r} \times \mathbf{F}$ must be equal to the magnitude of $\mathbf{F}$ times the distance d from point O to the line of action of $\mathbf{F}$. As an exercise, identify d for each of the cases considered.

Example 3.6. Bent-Knee Abdominal Crunch. Lie on back on floor, resting lower legs across bench with arms behind base of neck. Slowly curl head and upper torso up off floor in one even-paced movement. Slowly lower until almost touching head and torso and repeat (Fig. 3.8). Compute the moment created by the weight of the upper body on the pelvic joint at the beginning of the crunch where the torso is only slightly off the ground. For the athlete shown in the figure, the distance L between the pelvic joint and the center of mass of the upper body is 34 cm. The weight of the upper body is 25 kg.

Solution: We can find the answer to this question without utilizing vector mathematics. The magnitude of moment $\mathbf{M}$ must be equal to the moment

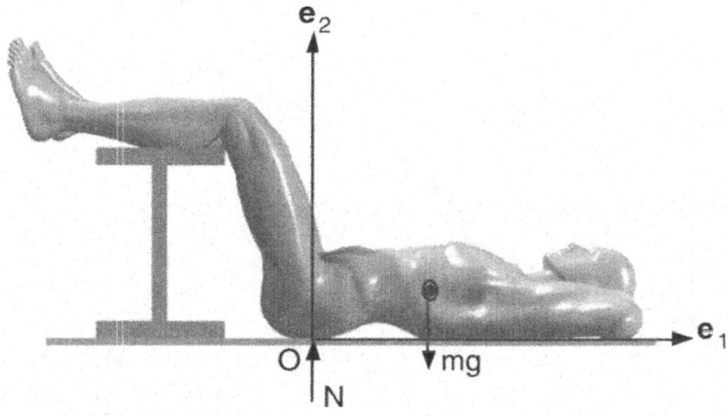

FIGURE 3.8. Schematic diagram of bent-knee abdominal crunches.

arm (0.34 m) times the force (25 kg $\times$ 9.81 m/s^2). Its direction, from the right-hand rule, is clockwise, or into the paper. Thus $\mathbf{M} = -83.4\ \mathbf{e}_3$ (N-m).

3.6 Moment of Momentum About a Stationary Point

Another example of vector product from the field of classical mechanics is a vector called the moment of momentum. Moment of momentum of a particle i about point O, $\mathbf{H}^{\circ}$, is defined as

$$\mathbf{H}^{o} = \mathbf{r}^{i/o} \times m^{i}\, \mathbf{v}^{i} \tag{3.22a}$$

where $\mathbf{r}^{i/o}$ is the position vector connecting point O to i, m^i is the mass of particle i, and $\mathbf{v}^i$ is its velocity, as measured with respect to a coordinate system E that is fixed on earth (Fig. 3.9). The moment of momentum of a particle of mass m tracing a circle of radius r with speed v is

$$\mathbf{H}^{o} = m\, v\, r\, \mathbf{e} \tag{3.22b}$$

where O is the center of the circle and $\mathbf{e}$ is a unit vector perpendicular to the circle. This equation shows that moment of momentum about point O remains constant so long as the particle traverses a circle with constant speed. Let us next consider the moment of momentum about O of a particle of mass m moving in a straight line with constant speed v:

$$\mathbf{H}^{o} = m\, v\, d\, \mathbf{e} \tag{3.22c}$$

where d is the distance from point O to the straight line traversed by the particle and $\mathbf{e}$ is the unit normal to the plane created by point O and the particle path. Equations 3.22b and 3.22c are quite similar, and yet the motion with which each is associated is different (circular versus linear mo-

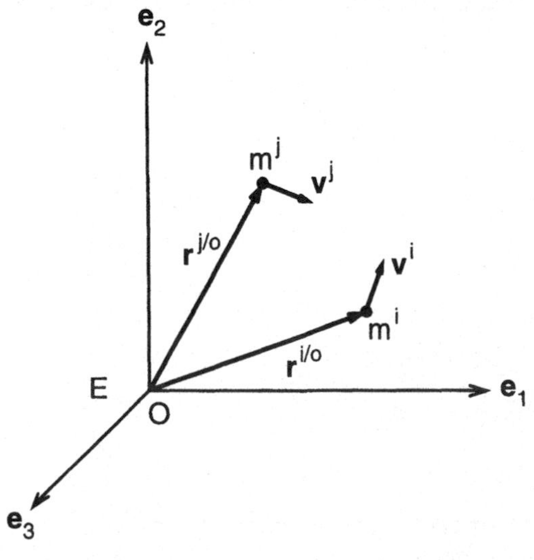

FIGURE 3.9. Moment of momentum of particles about the origin O of the Cartesian reference frame E.

tion). Thus, it is difficult to assign a simple physical meaning to the concept of moment of momentum. Nevertheless, this concept plays a central role in the analysis of motion of solids and multibody systems.

For a system of particles, the vector $\mathbf{H}^o$ is defined as follows:

$$\mathbf{H}^o = \Sigma\, \mathbf{r}^{i/o} \times m^i\, \mathbf{v}^i \tag{3.23}$$

where m^i is the mass and $\mathbf{v}^i$ is the velocity of the particle i, and the summation is over all particles in the system considered ($i = 1, n$).

Let us illustrate this definition with a simple example. Let A (0, 1, 2), B(−3, 1, 0), and C (6, 7, 3) represent the positions of points A, B, and C at time $t = 0$ in a Cartesian reference frame E. Let us assume further that each have the same velocity, $\mathbf{v} = 3\,\mathbf{e}_1$ (m/s). Particle A has a mass of 2 kg, B 4 kg, and C 3 kg. The moment of momentum about point O (0, 0, 0) for the system of particles at time $t = 0$ is

$$\mathbf{H}^o = 2\,(\mathbf{e}_2 + 2\mathbf{e}_3) \times 3\,\mathbf{e}_1 + 4\,(-3\mathbf{e}_1 + \mathbf{e}_2) \times 3\,\mathbf{e}_1 + 3\,(6\,\mathbf{e}_1 + 7\mathbf{e}_2 + 3\mathbf{e}_3) \times 3\,\mathbf{e}_1 = (39\,\mathbf{e}_2 - 81\mathbf{e}_3)\ (\text{kg-m}^2/\text{s})$$

In the following, we derive an equation that relates the time rate of change of moment of momentum of a system of particles to the rotational capacity of the external forces acting on the system. First, we take the time derivative of moment of momentum given by Eqn. 3.23:

$$d\mathbf{H}^o/dt = \Sigma\,(d\mathbf{r}^i/dt) \times m^i\,\mathbf{v}^i + \Sigma\,\mathbf{r}^i \times m^i\,(d\mathbf{v}^i/dt) \tag{3.24}$$

where time derivative of velocity $\mathbf{v}^i$ ($d\mathbf{v}^i/dt$) is the acceleration of particle i. The summation i is over all particles of the system under consideration. The first term on the right-hand side of Eqn. 3.24 is equal to zero because, by definition, $d\mathbf{r}^i/dt$ is equal to the velocity $\mathbf{v}^i$, and the cross product of two identical vectors, by definition of the vector product, is always equal to zero. Thus the rate of change of moment of momentum becomes equal to

$$d\mathbf{H}^o/dt = \Sigma\,\mathbf{r}^i \times m^i\,\mathbf{a}^i \tag{3.25}$$

where $\mathbf{a}^i = (d\mathbf{v}^i/dt)$.

Using Newton's second law, the time rate of change of moment of momentum of a system of particles can be related to the external forces acting on the system:

$$d\mathbf{H}^o/dt = \Sigma\,\mathbf{r}^i \times m^i\,\mathbf{a}^i = \Sigma\,\mathbf{r}^i \times \mathbf{F}^i \tag{3.26}$$

where $\mathbf{F}^i$ is the external force acting on particle i of the system under consideration. According to this equation, the time rate of change of moment of momentum with respect to a point fixed in space is equal to the resultant moment about the same point of external forces and force couples. Thus for a system of particles (bodies), the laws of motion state that:

$$(\Sigma m^i)\,\mathbf{a}^c = d\mathbf{L}/dt = \Sigma\,\mathbf{F}^i \tag{3.27a}$$

$$d\mathbf{H}^o/dt = \mathbf{M}^o \tag{3.27b}$$

where $\mathbf{M}^o$ is the moment generated by external forces about point O plus the sum of force couples. These two equations can be used to determine the translational as well as the average rotational motions of a system of particles. According to Eqn. 3.27b, the moment of momentum remains constant unless the system is acted on by an external moment.

In the following, we discuss examples of simulation of movement and motion based on the method of lumped masses. In this method, the mass of a body part is lumped and positioned at the center of mass of the body part. For example, a body segment such as an upper arm is represented as a weightless rigid rod with a point mass attached to the center of the rod. Lumped-mass models of human body have been used in simulating car crashes. The method is also commonly used in structural analysis. For example, the earthquake analysis of a four-story plane frame could be reduced to determination of the horizontal displacements of four lumped masses, each representing the mass of a level. As we see in the next chapter, this simplification results in modest errors in the estimates of forces acting on a system. Nevertheless, the lumped-mass analysis provides insights as to how inertial terms interact with the forces acting on a system.

Example 3.7. Gymnast Holds onto a Bar. A gymnast holding onto a bar rotates from a horizontal position under the action of gravitational force while keeping her body aligned in a straight line (Fig. 3.10). Determine the equations governing the motion of the gymnast.

Solution: We model the gymnast as a long and slender rod of length L and mass m. We lump the mass of the gymnast at the center of the rod. We assume that the wrists of the gymnast behave as hinge joints as the gymnast rotates around the bar in the clockwise direction under the influence of gravity. Let ϕ be the angle the rod makes with the vertical axis. With respect to the Cartesian coordinate system shown in Fig. 3.10, the center of mass of the rod is given by the following position vector:

$$\mathbf{r} = (L/2) \sin \phi \, \mathbf{e}_1 - (L/2) \cos \phi \, \mathbf{e}_2 \tag{3.28}$$

The velocity and acceleration of the center of mass are determined by taking the time derivative of the position vector given by Eqn. 3.28:

$$\mathbf{v} = (d\phi/dt) \, [(L/2) \cos \phi \, \mathbf{e}_1 + (L/2) \sin \phi \, \mathbf{e}_2] \tag{3.29a}$$

$$\begin{aligned} \mathbf{a} = {} & (d^2\phi/dt^2) \, [(L/2) \cos \phi \, \mathbf{e}_1 + (L/2) \sin \phi \, \mathbf{e}_2] \\ & + (d\phi/dt)^2 \, [-(L/2) \sin \phi \, \mathbf{e}_1 + (L/2) \cos \phi \, \mathbf{e}_2] \end{aligned} \tag{3.29b}$$

Using the free-body diagram shown in Fig. 3.10, we can now write the equation of motion for the mass center as follows:

$$F_1 = m \, [(d^2\phi/dt^2) \, (L/2) \cos \phi - (d\phi/dt)^2 \, (L/2) \sin \phi] \tag{3.30a}$$

$$-mg + F_2 = m \, [(d^2\phi/dt^2) \, (L/2) \sin \phi + (d\phi/dt)^2 \, (L/2) \cos \phi] \tag{3.30b}$$

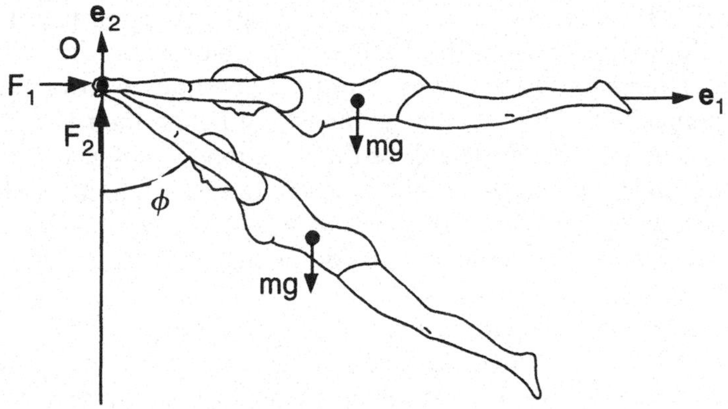

FIGURE 3.10. A gymnast swinging around a horizontal bar. The gymnast is kept at rest at the horizontal configuration and then let go. She rotates clockwise, with angle ϕ indicating the orientation of her body with the vertical direction.

where F_1 and F_2 are the horizontal and the vertical components, respectively, of the force exerted on the gymnast by the bar. These equations contain three unknowns that need to be determined as a function of time during swing: reaction forces F_1 and F_2 and the angle $\phi = \phi(t)$. We need yet another equation to determine these parameters. The moment of momentum for the swinging rod can be expressed as follows:

$$\begin{aligned}\mathbf{H}^o &= \mathbf{r} \times m\,\mathbf{v}\\ &= m\,[(L/2)\sin\phi\,\mathbf{e}_1 - (L/2)\cos\phi\,\mathbf{e}_2]\\ &\quad \times (d\phi/dt)\,[(L/2)\cos\phi\,\mathbf{e}_1 + (L/2)\sin\phi\,\mathbf{e}_2]\\ &= m\,(d\phi/dt)\,[(L/2)^2]\,[\sin^2\phi + \cos^2\phi]\,\mathbf{e}_3\\ &= m\,(d\phi/dt)\,[(L/2)^2]\,\mathbf{e}_3\end{aligned} \tag{3.31}$$

Taking the time derivative of Eqn. 3.31, we arrive at an equation for the rate of change of moment of momentum:

$$(d\mathbf{H}^o/dt) = m\,(d^2\phi/dt^2)\,[(L/2)^2]\,\mathbf{e}_3 \tag{3.32}$$

The external moment acting on the rod must be equal to $(d\mathbf{H}^o/dt)$. The free-body diagram in Fig. 3.10 shows that only gravitational force creates moment with respect to the hinge O. The forces F_1 and F_2 act on the hinge at O and therefore create no moment. Thus, conservation of moment of momentum requires that

$$\begin{aligned}m\,(d^2\phi/dt^2)\,[(L/2)^2]\,\mathbf{e}_3 &= \mathbf{r} \times (-\,mg\,\mathbf{e}_2)\\ &= [(L/2)\sin\phi\,\mathbf{e}_1 - (L/2)\cos\phi\,\mathbf{e}_2] \times (-mg\,\mathbf{e}_2)\end{aligned}$$

This leads to the following differential equation:

$$(d^2\phi/dt^2) = -(2\,g/L)\sin\phi \tag{3.33}$$

Equation 3.33 is identical to the differential equation governing the motion of a simple pendulum. Thus, this simple analysis predicts that the period in which a gymnast rotates from horizontal down to vertical position depends on the height of the gymnast but not on their weight. The taller the gymnast, the longer will be the duration of swing.

Example 3.8. Swinging of Arms. While standing straight a man begins swinging his arms at constant frequency (Fig. 3.11). Compute the moment of momentum about the center of mass of the man in the standard standing configuration.

Solution: Let θ be the angle that one of the arms, arm 1, makes with the vertical axis. Because of the asymmetry of motion, the other arm (arm 2) will make angle $-\theta$ with the vertical axis. The position vector of the center of mass of each arm with respect to the Cartesian coordinates shown in Fig. 3.11 are given by the following equations:

$$\mathbf{r}^1 = h\,\mathbf{e}_2 + d\,\mathbf{e}_1 - (L/2)\cos\theta\,\mathbf{e}_2 - (L/2)\sin\theta\,\mathbf{e}_3 \qquad (3.34a)$$

$$\mathbf{r}^2 = h\,\mathbf{e}_2 - d\,\mathbf{e}_1 - (L/2)\cos\theta\,\mathbf{e}_2 + (L/2)\sin\theta\,\mathbf{e}_3 \qquad (3.34b)$$

in which h is the vertical distance between the center of mass in the standard standing position and the line connecting the shoulders, d is half the length between the shoulders, and L is the length of an upper limb.

The velocity of the center of mass of each arm can be obtained by taking the time derivative of Eqn. 3.34:

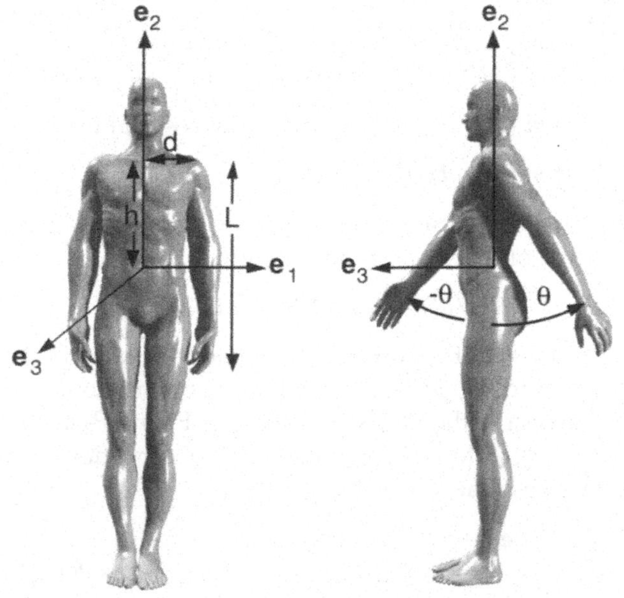

FIGURE 3.11. While standing straight, a man begins swinging his arms at constant frequency.

$$\mathbf{v}^1 = (d\theta/dt)\, L \sin\theta\, \mathbf{e}_2 - (d\theta/dt)\, L \cos\theta\, \mathbf{e}_3 \quad (3.35a)$$

$$\mathbf{v}^2 = (d\theta/dt)\, L \sin\theta\, \mathbf{e}_2 + (d\theta/dt)\, L \cos\theta\, \mathbf{e}_3 \quad (3.35b)$$

In estimating the total moment of momentum $\mathbf{H}^o$, we lump the distributed masses of the arms and assign them as point masses to the center of mass of each arm:

$$\mathbf{H}^o = \mathbf{r}^1 \times m\, \mathbf{v}^1 + \mathbf{r}^2 \times \mathrm{m}\, \mathbf{v}^2$$

When the vector multiplication specified in this equation is carried out, the result can be expressed as

$$\mathbf{H}^o = 2d\, L\, m\, (d\theta/dt) \cos\theta\, \mathbf{e}_2 \quad (3.36)$$

This equation shows that $\mathbf{H}^o$ varies as a function of time. The ground force must create enough moment about point O to compensate for the time variation in moment of momentum. In practice, what happens is that the trunk rotates slightly to the left and right during swinging of the arms in such a way to make $\mathbf{H}^o$ to be equal to zero at all times.

Example 3.9. Push-ups. Determine the reaction forces exerted on the feet and the hands during push-ups.

Solution: The person performing the push-ups is represented by a slender rod of length L with a lumped mass m (body weight) at its center. The arms are assumed to be composed of two weightless rods that are linked to each other. The shoulders are considered to lie 0.2 L from the free end of the rod that represents the body as shown in Fig. 3.12.

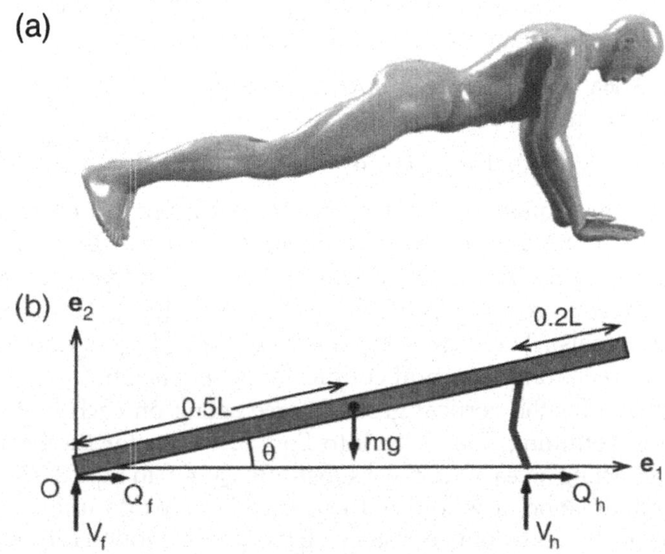

FIGURE 3.12a,b. Representation of a man doing pushups **(a)** as a slender rod with a lumped mass at the center **(b)**.

The position, velocity, and acceleration of the lumped mass m can be expressed as

$$\mathbf{r} = (L/2)\,(\cos\theta\,\mathbf{e}_1 + \sin\theta\,\mathbf{e}_2)$$

$$\mathbf{v} = (L/2)\,(d\theta/dt)\,(-\sin\theta\,\mathbf{e}_1 + \cos\theta\,\mathbf{e}_2)$$

$$\mathbf{a} = (L/2)\,(d^2\theta/dt^2)\,(-\sin\theta\,\mathbf{e}_1 + \cos\theta\,\mathbf{e}_2) - (L/2)\,(d\theta/dt)^2\,(\cos\theta\,\mathbf{e}_1 + \sin\theta\,\mathbf{e}_2)$$

Let us now write the equations of motion of the center of mass:

$$2\,(V_f + V_h) - mg = m\,[(L/2)\,(d^2\theta/dt^2)\cos\theta - (L/2)\,(d\theta/dt)^2 \sin\theta\,\mathbf{e}_2] \quad (3.37a)$$

$$2\,(Q_f + Q_h) = m\,[-(L/2)\,(d^2\theta/dt^2)\sin\theta - (L/2)\,(d\theta/dt)^2\cos\theta] \quad (3.37b)$$

In these equations, the symbol V denotes vertical forces whereas Q is used for horizontal friction forces. The subscripts f and h refer to foot and hand, respectively. Arms are assumed to transmit vertical forces only ($Q_h = 0$).

The rate of rotation of the body during the push-ups is governed by the conservation of moment of momentum. Let us take moment of momentum with respect to point O, which lies between the points at which the feet touch the floor in the midplane of motion.

$$\mathbf{H}^o = \mathbf{r} \times m\,\mathbf{v} = m\,(L/2)^2\,(d\theta/dt)\,\mathbf{e}_3$$

$$d\mathbf{H}^o/dt = m\,(L/2)^2\,(d^2\theta/dt^2)\,\mathbf{e}_3 \quad (3.38a)$$

The time rate of change of moment of momentum must be equal to the sum of the external moment acting on the body:

$$\mathbf{M}^o = (L/2)\,(\cos\theta\,\mathbf{e}_1 + \sin\theta\,\mathbf{e}_2) \times (-mg\,\mathbf{e}_2) + (0.8\,L\cos\theta\,\mathbf{e}_1) \times (2V_h\,\mathbf{e}_2) = [-mg\,(L/2) + 1.6\,LV_h\,]\cos\theta\,\mathbf{e}_3 \quad (3.38b)$$

Equating 3.38a and 3.38b, we obtain an expression for the vertical ground force exerted on each hand:

$$V_h = 0.31\,m\,[g\cos\theta + (L/2)\,(d^2\theta/dt^2)] \quad (3.38c)$$

According to this equation, the ground force carried by each arm is slightly less than one-third of the body weight when $d^2\theta/dt^2 = 0$. This is the force that provides an excellent workout for the triceps muscle. Another major muscle group essential for push-ups is the vertebral column flexors (abdominals). These muscles stabilize the body position along a straight line, preventing vertebral column hyperextension.

An expression for the vertical ground force acting on each foot can be obtained by substituting Eqn. 3.38c into Eqn. 3.37a. As for the horizontal ground forces acting on each foot, we substitute $Q_h = 0$ in Eqn. 3.37b. Without this approximation, it would not be possible to assign unique values for these ground forces by only considering the laws of motion for the whole body. Such force systems are said to be statically indeterminate.

A few more remarks about the mechanical analysis of the push-up are in order. First, the slender rod actually represents the body from head to the ankle. We have not taken into account the fact that the feet remain more or less vertical during push-up rather than aligned with the rest of the body. Second, it is relatively easy to determine the time history of θ using equipment available in most engineering departments. One could hook up a videocamera to a computer and digitize the videoimage of the push-up events. Appropriate software can then be used to assess the values of θ, $d\theta/dt$, and $d^2\theta/dt^2$ as a function of time. In this way, it is possible to study the effect the rate of motion has on the ground forces. Note that the ground reaction forces can also be measured directly, but one would need two force plates to quantify these contact forces.

3.7 Moment of Momentum About the Center of Mass

Moment of momentum of a system of particles with respect to a point O fixed in reference frame E was defined as

$$\mathbf{H}^{o} = \Sigma\, \mathbf{r}^{i/o} \times m^{i}\mathbf{v}^{i}$$

where as usual O denotes the origin of E and i represents particle i in the system of particles. In many instances, particularly when we want to investigate locomotion (walking, running, jumping) or movement in air, it is more convenient to consider moment of momentum about the center of mass. Using parallegram law of vector addition:

$$\mathbf{r}^{i/o} = \mathbf{r}^{c/o} + \mathbf{r}^{i/c}$$

Substituting this expression into the equation for moment of momentum about point O, we find:

$$\mathbf{H}^{o} = \mathbf{r}^{c/o} \times \mathbf{L} + \Sigma\, \mathbf{r}^{i/c} \times m^{i}\mathbf{v}^{i} \tag{3.39a}$$

in which $\mathbf{L}$ is the linear momentum of the system of particles, as defined by Eqn. 3.5:

$$\mathbf{L} = \Sigma m^{i}\mathbf{v}^{i} = (\Sigma m^{i})\, \mathbf{v}^{c}$$

The last term on the right-hand side of Eqn. 3.39a represents the moment of momentum of the system about the center of mass, $\mathbf{H}^{c}$. Thus,

$$\mathbf{H}^{o} = \mathbf{r}^{c/o} \times (\Sigma m^{i})\, \mathbf{v}^{c} + \mathbf{H}^{c} = \mathbf{r}^{c/o} \times \mathbf{L} + \mathbf{H}^{c} \tag{3.39b}$$

Let us next take the time derivative of Eqn. 3.39b:

$$d\mathbf{H}^{o}/dt = \mathbf{v}^{c/o} \times \mathbf{L} + \mathbf{r}^{c/o} \times \Sigma\mathbf{F}^{i} + d\mathbf{H}^{c}/dt$$

The first term on the right-hand side of this equation is equal to zero because $\mathbf{v}^{c/o}$ and $\mathbf{L} = (\Sigma m^{i})\, \mathbf{v}^{c/o}$ are in the same direction and hence their vector product must be equal to zero. Because the time derivative of the moment of momentum about point O is equal to the resultant moment acting on the system with respect to point O, we find:

$$\mathbf{r}^{c/o} \times \Sigma\mathbf{F}^{i} + d\mathbf{H}^{c}/dt = \Sigma\mathbf{r}^{i/o} \times \mathbf{F}^{i} = \Sigma(\mathbf{r}^{c/o} + \mathbf{r}^{i/c}) \times \mathbf{F}^{i}$$

$$d\mathbf{H}^{c}/dt = \Sigma\, \mathbf{r}^{i/c} \times \mathbf{F}^{i} \tag{3.40}$$

According to this equation, the time rate of change of moment of momentum about the center of mass is equal to the sum of the moments of external forces and force couples with respect to the center of mass.

Example 3.10. Arm Swinging During Walking. Determine how rigorously one has to swing the arms to avoid twisting during walking.

Solution: We have previously shown (in Example 3.8) that the moment of momentum from arm swinging is given by the following expression (see Eqn. 3.36):

$$\mathbf{H} = 2d\, L\, m\, (d\theta/dt) \cos\theta\, \mathbf{e}_2$$

in which d is half the length between the shoulders, and L is the length of an upper limb. The moment of momentum for the legs is given by a similar expression. Thus, the total moment of momentum with respect to the center of mass can be written as

$$\mathbf{H}^{c} = 2d_{a}\, L_{a}\, m_{a}\, (d\theta_{a}/dt) \cos\theta_{a}\, \mathbf{e}_2 - 2d_{l}\, L_{l}\, m_{l}\, (d\theta_{l}/dt) \cos\theta_{l}\, \mathbf{e}_2$$

in which the subscripts a and l refer to the upper and lower limbs, respectively. Several observations can be made with regard to this equation. First, the left arm must rotate in the same direction as the right leg and vice versa to cancel the contributions of arm and leg swing to the moment of momentum of the human body. Second, because the lower limbs are longer and heavier than the upper limbs, the arms must rotate faster to cancel out the twisting effect of the legs. As the arms and the legs have the same period of swing, the way to achieve zero moment of momentum from swing is to increase the amplitude of swing of the arms while also increasing their rate of rotation.

3.8 Summary

Any object, whether living or nonliving, can be considered as being composed of a large number of small particles, and for each particle with mass m and a vanishingly small volume dV, Newton's laws of motion hold. Linear momentum of a system of particles is defined as the sum of the products of the mass of each particle with its velocity:

$$\mathbf{L} = \Sigma(m^{i}\, \mathbf{v}^{i})$$

in which m^i is the mass of particle i and $\mathbf{v}^i$ is its velocity. Newton characterized linear momentum as the quantity of motion. Writing Newton's second and third law for each particle in a system of particles and summing over all particles in the system, one can show that

$$d\mathbf{L}/dt = \Sigma\mathbf{F}^{i}$$

where $\Sigma\mathbf{F}^i$ denotes the sum of all external forces acting on the system of particles. According to this equation, the time rate of change of linear momentum is equal to the sum of external forces. It is known as the equation for conservation of linear momentum. Gravitational force and the forces that arise as the result of contact of particles in the system with the particles outside the system are external forces. Newton's third law requires that forces that act between particles in the system under study do not contribute to the change of linear momentum. Such forces are called internal forces.

The center of mass of a system of particles is defined by the relation:

$$(\Sigma m^i)\,\mathbf{r}^c = \Sigma(m^i\,\mathbf{r}^i)$$

in which $\mathbf{r}^c$ denotes the position of center of mass with respect to a Cartesian reference frame and $\mathbf{r}^i$ is the position vector of particle i. The center of mass is not necessarily occupied by any particle in the system of particles. Using this definition in the equation for the conservation of linear momentum, one obtains an equation governing the position of the center of mass as a function of time:

$$\Sigma\mathbf{F}^i = (\Sigma m^i)\,\mathbf{a}^c$$

in which $\mathbf{a}^c$ is the acceleration of the center of mass.

The term moment of momentum about a point fixed on earth is defined by the following equation:

$$\mathbf{H}^o = \Sigma\,\mathbf{r}^{i/o} \times m^i\mathbf{v}^i$$

in which $\mathbf{r}^{i/o}$ denotes the position vector from the stationary point O to the particle i. The conservation of moment of momentum dictates that

$$d\mathbf{H}^o/dt = \Sigma\mathbf{r}^{i/o} \times \mathbf{F}^i$$

Again, in this equation, $\mathbf{F}^i$ represents the external force i acting on the ith particle of the system. Conservation of moment of momentum about the center of mass is governed by an equation of the same form:

$$d\mathbf{H}^c/dt = \Sigma\mathbf{r}^{i/c} \times \mathbf{F}^i$$

where

$$\mathbf{H}^c = \Sigma\,\mathbf{r}^{i/c} \times m^i\mathbf{v}^i$$

and $\mathbf{r}^{i/c}$ is the position vector from the center of mass to the particle i.

3.9 Problems

Problem 3.1. To hit a ball, a female volleyball player jumps in the vertical direction. If her center of mass rises 0.4 m during the airborne portion of the jump, determine the time duration $2t^*$ during which the

player is airborne. Determine the velocity of her center of mass at the instant of takeoff.
Answer: $2t^* = 0.57$ s, $v_o = 2.8$ m/s.

Problem 3.2. During a triple axel, a figure skater covers a horizontal distance of 3 m and reaches a maximum height of 0.6 m above the ice. Determine the initial velocity of the skater as she lifts off the ice.
Answer: $\mathbf{v}_o = 4.28\,\mathbf{e}_2 + 3.43\,\mathbf{e}_2$ (m/s), where $\mathbf{e}_2$ and $\mathbf{e}_2$ are unit vectors in horizontal and vertical directions, respectively.

Problem 3.3. In speed skating, the skater pushes off against the ice as the skate is gliding forward (Fig. P.3.3). The direction of push-off is perpendicular to the gliding direction of the skate. This action results in a sinusoidal trajectory of the center of mass of the skater when skating along the straightaways. A sideways push off the right leg causes a leftward movement of the center of mass and vice versa. During the sideways push-off of the right leg of a 70-kg skater, projection of the push-off force onto the horizontal plane was measured to be 3600 N. The duration of the push-off was 0.32 s. The velocity of the center of mass before the pushoff was in the direction of gliding and the speed was equal to 15 m/s. Determine the speed of the center of mass immediately after the completion of the push-off. (Take $\mathbf{e}_1$ be the direction of the velocity before the push-off [direction of gliding] and let $\mathbf{e}_2$ be the

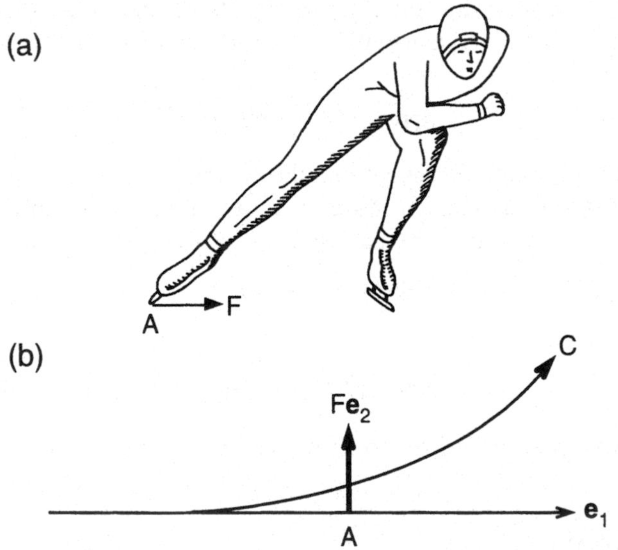

FIGURE P.3.3a,b. Schematic diagram of a speed skater during a sideways push **(a)**. The *curved line* in **(b)** that is identified with symbol C represents the path of the center of mass of the skater whereas the straight line along the $\mathbf{e}_1$ axis represents the path of the skate A.

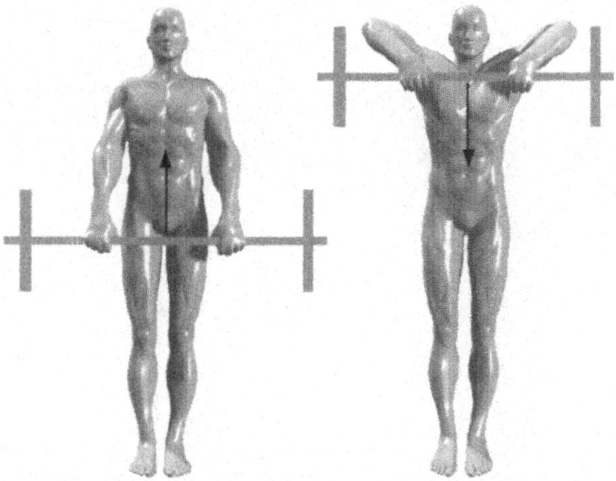

FIGURE P.3.4. A man performing upright row to strengthen his back muscles.

direction of the push-off. Note that $\mathbf{e}_1$ and $\mathbf{e}_2$ make a right-handed coordinate system in the horizontal plane.)
Answer: $\mathbf{v} = 15.0\ \mathbf{e}_1 + 16.4\ \mathbf{e}_2$ (m/s).

Problem 3.4. Using a palms-down grip and hands close together, an athlete holds a barbell at the thighs at 0.7 m above the floor. Then he pulls the barbell up to his chin with an acceleration of 6 m/s^2 upward (Fig. P.3.4). Compute the force the ground exerts on the athlete at the initiation of motion. The athlete is 75 kg and the barbell weighs 20 kg. Assume that the center of mass of the athlete remains at 0.8 m above the floor during the exercise. Note that this exercise is called upright row. It is designed to work the back muscle trapezius.
Answer: F = 1051 N.

Problem 3.5. A man does chin-ups by pulling himself up toward a bar fixed on a side wall (Fig. P.3.5). Let θ and ϕ be the angles the forearm and the upper arm make with the horizontal axis $\mathbf{e}_1$, as shown in Fig. P.3.5. Using the fact that the horizontal distance between the man's two hands (D) does not change during the chin-ups, develop an equation that relates θ to ϕ. This equation is called a constraint equation.
Answer: $D - d = 2\ (L \cos\theta + L \cos\phi)$

Problem 3.6. A man does chin-ups (Fig. P.3.5). Let θ and ϕ be the angles the forearm and the upper arm make with the horizontal axis $\mathbf{e}_1$. At time $t = 0$, $\theta = \phi = 45°$ and $d\theta/dt = d\phi/dt = 0$. Assuming also that $d^2\phi/dt^2 = 2$ rad/s^2 at $t = 0$, determine the force (F) exerted by the holding bar on the fists of the man at the initiation of motion. The man weighs 72 kg. Both the upper arm and the lower arm are 37 cm long; the shoulder width $d = 54$ cm.

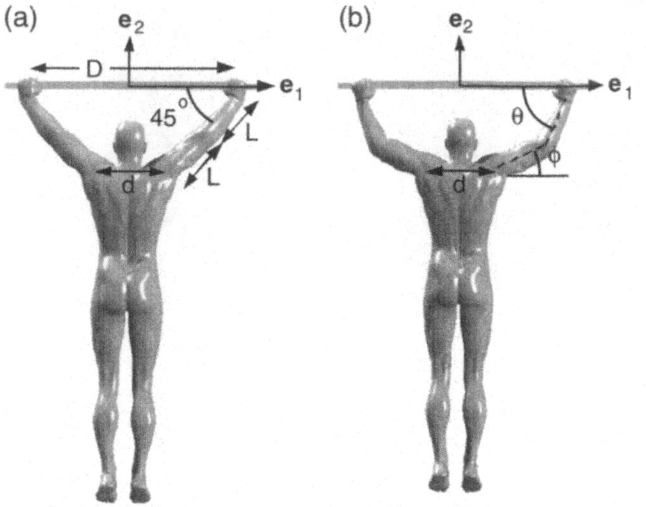

FIGURE P.3.5a,b. Schematic diagram of a man doing chin-ups.

Answer: The trunk of the man moves up and down as a single body. Thus the acceleration of the center of mass of the man in the vertical direction is obtained by taking succesive time derviatives of the vertical distance between the shoulders and the bar:

$$y = -L\,(\sin\theta + \sin\phi),\ v = dy/dt = -L\,[(d\theta/dt)\cos\theta + (d\phi/dt)\cos\phi]$$

$$a = d^2y/dt^2 = -L\,[(d^2\theta/dt^2)\cos\theta + (d^2\phi/dt^2)\cos\phi] + L\,[(d\theta/dt)^2\sin\theta + (d\phi/dt)^2\sin\phi]$$

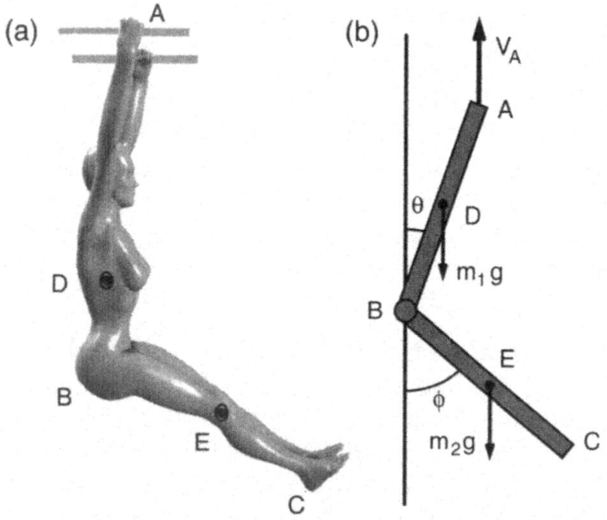

FIGURE P.3.7a,b. A woman performing leg lifts while holding onto bars.

At $t = 0$, $(d\theta/dt) = (d\phi/dt) = 0$ and $(d^2\phi/dt^2) = 2$ rad/s. Compute $d^2\theta/dt^2$ by taking the succesive time derivatives of the constraint equation given as the answer of Problem 3.7: $d^2\theta/dt^2 = -(d^2\phi/dt^2)\sin\phi/\sin\theta = -2$ rad/s. Thus $a = 0$ and $F = 72\text{ kg}\cdot 9.81\text{ m/s}^2$ at $t = 0$.

Problem 3.7. While holding onto bars, an athlete raises her legs upward (Fig. P.3.7) such that the angle ϕ her legs make with the vertical axis varies with time in accordance with the following equation:

$$\phi = 50t^2$$

What is the differential equation governing the angle of inclination θ of her upper body with the vertical axis? The woman weighs 50 kg and has a height of 1.9 m. One can represent her as composed of two slender rods of equal length L (L = 0.95 m). The mass of her body is then divided into two (m_1, m_2) and lumped equally at the center of each rod ($m_1 = m_2 = 25$ kg). Hint: This is a mathematically tedious problem. The following equations provide a road map for the solution:

$$\mathbf{r}^{D/A} = -(L/2)(\sin\theta\,\mathbf{e}_1 + \cos\theta\,\mathbf{e}_2)$$

$$\mathbf{v}^{D/A} = d(\mathbf{r}^{D/A})/dt$$

$$\mathbf{r}^{E/A} = \mathbf{r}^{B/A} + \mathbf{r}^{E/B} = -L(\sin\theta\,\mathbf{e}_1 + \cos\theta\,\mathbf{e}_2) + (L/2)(\sin\phi\,\mathbf{e}_1 - \cos\phi\,\mathbf{e}_2)$$

$$\mathbf{v}^{E/A} = d(\mathbf{r}^{E/A})/dt$$

$$\mathbf{H}^{A} = \mathbf{r}^{D/A} \times m_1\mathbf{v}^{D/A} + \mathbf{r}^{E/A} \times {}_2\mathbf{v}^{E/A}$$

$$d\mathbf{H}^{A}/dt = \mathbf{M}^{A}$$

Problem 3.8. A seesaw is released from rest at the position shown in Fig. P.3.8. The weight at B is half the weight at D. Using the conservation of moment of momentum with respect to the fulcrum point A, derive a differential equation that governs the time course of the orientation of the seesaw until D hits the ground. Assume that the seesaw itself is weightless. Masses m and $2m$ are fixed to the seesaw.
Answer: $d^2\theta/dt^2 = -(g/3L)\cos\theta$

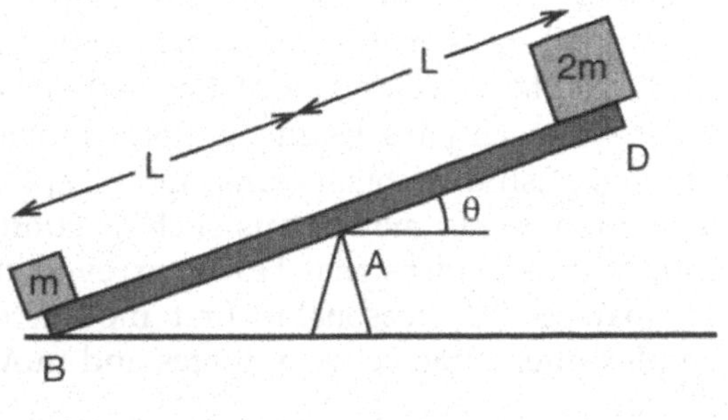

FIGURE P.3.8. A seesaw held at rest is released at time $t = 0$ with two weights acting on opposite sides. The seesaw itself is assumed weightless.

4

Bodies in Planar Motion: Jumping, Diving, Push-Ups, Back Curls

4.1 Introduction

In the previous chapter, we analyzed the forces involved in athletic movements and human motion by lumping the mass of each body segment at its geometric center. In reality, the mass of a body is distributed more or less uniformly throughout the body rather than being concentrated at a few points. In this chapter, we will lift the assumption of lumped masses. The objects that are under consideration are the ones that behave as rigid under the application of external forces. The distance between any two points in a rigid body does not change with time as the body undergoes motion. Therefore, the angle between any two line segments in a rigid object, and hence the shape of the object, remains constant. In reality, even the stiffest materials like steel or bone undergo subtle shape changes in response to applied forces. However, the extent of deformation is typically small so that the distance between any two particles in the object is hardly affected by the external forces acting on the object.

What is the significance of rigid body analysis to human body dynamics? First, the human body can be reasonably well represented by an interconnected chain of rigid links in the analysis of upper and lower limb movement. Typically, movement is a result of rotations performed at joints between the body segments. Then, some modes of human motion such as diving can be considered as a series of rapid shape changes followed by longer durations of constant shape motion. Additionally, the analysis presented here provides an estimate of the errors involved in using lumped-mass analysis of human movement and motion.

This chapter is focused on planar motion. We call the motion of a body planar when all particles in the body move in parallel planes. Vertical jumping, push-ups, somersaults, and biceps curls are all examples of planar motion. Other movements such as running or long jumping are essentially three dimensional. However, even in these modes of motion, planar analysis is a reasonable first model to provide insights into the complex interaction between forces and movement.

4.2 Planar Motion of a Slender Rod

In this section, we consider the planar motion of a thin and slender rod with uniformly distributed mass. We explore the motion of its center of mass as well as its rotation. The mechanical analysis of motion of a rod allows us to introduce in elementary form all the basic principles of rigid body mechanics. Also, in many applications, either the human body or various long bones of the extremities can be considered as rigid rods; thus, the geometry chosen has significance to human body dynamics.

Consider the swinging motion of a rod with uniformly distributed mass in a vertical plane as shown in Fig. 4.1. The moment of momentum of the particles in the rod with respect to the hinge O can be written as an integral summation over small mass elements of the body:

$$\mathbf{H}^{o} = \int \mathbf{r} \times \mathbf{v}\, dm \tag{4.1}$$

where **r** is the position vector from point O to the center of mass of the mass element *dm*, and **v** is the velocity of the center of mass of *dm* in the reference frame E. The integration is over all the small mass elements of the slender rod (Fig. 4.1). From the geometry of the pendulum movement, we deduce the following equations for the position **r** and velocity **v** of *dm*, which is located at a distance *s* away from point O:

$$\mathbf{r} = s\,[\sin\phi\ \mathbf{e}_1 - \cos\phi\ \mathbf{e}_2] \tag{4.2a}$$

$$\mathbf{v} = (d\phi/dt)\, s\,[\cos\phi\ \mathbf{e}_1 + \sin\phi\ \mathbf{e}_2] \tag{4.2b}$$

in which ϕ is the angle of the rod with the $\mathbf{e}_2$ axis as shown in Fig. 4.1. The time derivative of angle ϕ, $(d\phi/dt)$, is the rate of rotation of the rod. The rod rotates counterclockwise when $(d\phi/dt)$ is positive and rotates clockwise when $(d\phi/dt)$ is negative.

The angular velocity ω of the rod undergoing planar motion is defined in the vectorial form as follows:

$$\boldsymbol{\omega} = (d\phi/dt)\ \mathbf{e}_3 = \omega\ \mathbf{e}_3 \tag{4.3}$$

where $\mathbf{e}_3$ is the unit vector perpendicular to the plane of motion (Fig.4.1a). Because the angles have no dimension, the unit of angular velocity is inverse time. It is expressed as radians per second (rad/s).

Substituting the position vector and velocity expressions given by Eqns. 4.2a and 4.2b into Eqn. 4.1 and noting that $dm = (m/L)\, ds$, we obtain the following relation:

$$\mathbf{H}^{o} = \int\{s\,[\sin\phi\ \mathbf{e}_1 - \cos\phi\ \mathbf{e}_2]\} \times \{(d\phi/dt)\, s\,[\cos\phi\ \mathbf{e}_1 + \sin\phi\ \mathbf{e}_2]\}\ (m/L)\, ds$$

After performing the vector multiplications in the equation written above, the moment of momentum expression reduces to

$$\mathbf{H}^{o} = (m/L)\ (d\phi/dt)\ \mathbf{e}_3 \int[\sin^2\phi + \cos^2\phi]\ s^2 ds$$

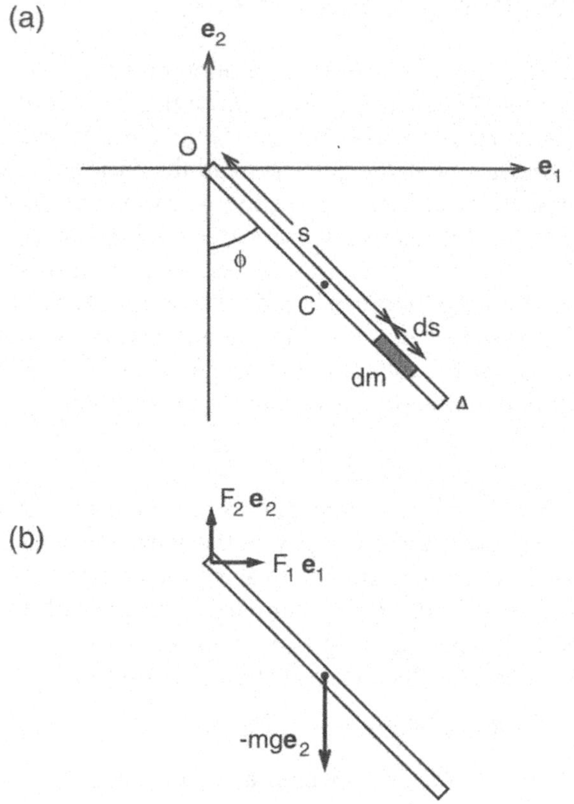

FIGURE 4.1a,b. Oscillation of a pendulum about the vertical axis. The motion takes place in the ($\mathbf{e}_1$, $\mathbf{e}_2$) vertical plane. The slender rod OA has uniform mass density. A small mass element of the rod is shown in **(a)**. The free-body diagram of the rod showing all the external forces acting on it is given in **(b)**.

The term $(d\phi/dt)\ \mathbf{e}_3$ is equal to $\boldsymbol{\omega} = \omega\ \mathbf{e}_3$ and that $[\sin^2\phi + \cos^2\phi] = 1$. After integrating this equation from 0 to L, we obtain:

$$\mathbf{H}^{\mathrm{o}} = (mL^2/3)\ (d\phi/dt)\ \mathbf{e}_3 = (mL^2/3)\ \omega\ \mathbf{e}_3 \tag{4.4}$$

In the previous chapter we had found for a pendulum of the same geometry but with mass m concentrated at the midway point:

$$\mathbf{H}^{\mathrm{o}} = (mL^2/4)\ (d\phi/dt)\ \mathbf{e}_3 = (mL^2/4)\ \omega\ \mathbf{e}_3 \tag{4.5}$$

Thus, the assumption that the mass of the rod is localized at the center of the rod rather than being uniformly distributed throughout results in the underestimation of the moment of momentum by about one-third.

As we have seen before, the conservation of momentum for a system of particles dictates that

$$(d\mathbf{H}^o/dt) = \mathbf{M}^o \tag{4.6}$$

where $\mathbf{M}^o$ denotes the moments of external forces with respect to point O. A free-body diagram of the pendulum is shown in Fig. 4.1b. Because the reaction force exerted by the pin passes through point O, its lever arm with respect to this point is zero and hence the pin force contributes no moment with respect to point O. The only external moment results from the gravitational force acting at the center of the rod. Thus, the conservation of moment of momentum yields the following equation:

$$(mL^2/3)\,(d^2\phi/dt^2)\,\mathbf{e}_3 = (L/2)\,(\sin\phi\,\mathbf{e}_1 - \cos\phi\,\mathbf{e}_2) \times (-mg\,\mathbf{e}_2)$$

from which we obtain

$$(2L/3)\,(d^2\phi/dt^2) = -g\sin\phi$$

$$(d^2\phi/dt^2) = -(3g/2L)\sin\phi \tag{4.7a}$$

In Chapter 2, we have shown that for a pendulum composed of a slender rod and a lump mass m positioned at one end, the corresponding equation was Eqn. 2.17b:

$$(d^2\phi/dt^2) = -(g/L)\sin\phi \tag{4.7b}$$

Comparison of Eqns. 4.7a and 4.7b indicates that resistance to angular acceleration decreases when the mass is distributed over the length rather than concentrated at the end of the rod.

When the oscillations of the rod around the vertical axis is small ($\sin\phi \cong \phi$) the solution for Eqn. 4.7a can be approximated by the following equation:

$$\phi = A\sin[(3g/2L)^{1/2}\,t] + B\cos[(3g/2L)^{1/2}\,t]$$

in which A and B are arbitrary constants determined by the intial conditions. If we let

$$\phi = \pi/6 \text{ and } (d\phi/dt) = 0 \text{ at } t = 0$$

we obtain

$$\phi = (\pi/6)\cos[(3g/2L)^{1/2}\,t] \text{ for } t \geq 0 \tag{4.8}$$

This equation shows that the rod with distributed mass swings around the vertical axis with the period of $2\pi(2L/3g)^{1/2}$. For a simple pendulum composed of a slender rod of length L and a bob of mass m, the period of oscillation is equal to $2\pi(L/g)^{1/2}$. Therefore, the rod with uniform mass distribution rotates around point O much like a classical pendulum with effective length equal to $2L/3$. Thus, in using the lumped-mass approach, we would have achieved an exact solution if we had placed the lumped mass at a distance $2L/3$ from the fixed point O.

Next, let us turn our attention to the forces exerted by the hinge on the rod at point O. The free-body diagram in Fig. 4.1b shows all the external forces acting on the rod. These forces are the gravitational force $-mg\ \mathbf{e}_2$ acting at the center of the rod and the force $(F_1\ \mathbf{e}_1 + F_2\ \mathbf{e}_2)$ exerted by the pin at point O. According to the equation of motion of the center of mass, the net resultant force acting on an object must be equal to the mass of the object times the acceleration of the center of mass. The position, velocity, and acceleration of the center of mass are given by the following expressions:

$$\mathbf{r} = (L/2)\ [\sin\phi\ \mathbf{e}_1 - \cos\phi\ \mathbf{e}_2] \tag{4.9a}$$

$$\mathbf{v} = (d\phi/dt)\ (L/2)\ [\cos\phi\ \mathbf{e}_1 + \sin\phi\ \mathbf{e}_2] \tag{4.9b}$$

$$\begin{aligned}\mathbf{a} = {} & (d^2\phi/dt^2)\ (L/2)\ [\cos\phi\ \mathbf{e}_1 + \sin\phi\ \mathbf{e}_2] \\ & + (d\phi/dt)^2\ (L/2)\ [-\sin\phi\ \mathbf{e}_1 + \cos\phi\ \mathbf{e}_2]\end{aligned} \tag{4.9c}$$

Then, the components of the equation of motion give us

$$F_1 = m\ (L/2)\ [(d^2\phi/dt^2)\cos\phi - (d\phi/dt)^2 \sin\phi] \tag{4.10a}$$

$$F_2 = -mg + m\ (L/2)\ [(d^2\phi/dt^2)\sin\phi + (d\phi/dt)^2\cos\phi] \tag{4.10b}$$

Once we compute the angle ϕ as a function of time by using Eqn. 4.8, the reaction forces at the hinge O can be determined with the use of Eqns. 4.10a and 4.10b. Note that when the pendulum is at rest in the vertical position we have $F_1 = 0$ and $F_2 = -mg$.

4.3 Angular Velocity

The mass elements of an arbitrarily shaped object are not aligned along a straight line as was the case for a rod, and therefore the estimation of moment of momentum requires the evaluation of a difficult integral of the vector product of velocity of a small mass element and its position vector over the volume of the object. However, because the distance between any two points in a rigid body remains constant, this integral can be reduced to a simple algebraic form. Let O denote the origin of the coordinate system E that is fixed on earth. The point C is the center of mass, and P and Q are two arbitrary points of the rigid body B undergoing planar motion parallel to the $(\mathbf{e}_1, \mathbf{e}_2)$ plane (Fig. 4.2a). Using the vector addition property, it is clear that

$$\mathbf{r}^{\mathrm{Q}} = \mathbf{r}^{\mathrm{P}} + \mathbf{r}^{\mathrm{Q/P}} \tag{4.11}$$

where $\mathbf{r}^{\mathrm{P}}$ and $\mathbf{r}^{\mathrm{Q}}$ are the position vectors of points P and Q and $\mathbf{r}^{\mathrm{Q/P}}$ is the vector connecting point P to point Q. As can be seen in Fig. 4.2b, this latter vector can be written as

$$\mathbf{r}^{\mathrm{Q/P}} = r^{\mathrm{Q/P}}\ [\sin\theta\ (\cos\phi\ \mathbf{e}_1 + \sin\phi\ \mathbf{e}_2) + \cos\theta\ \mathbf{e}_3] \tag{4.12}$$

where $r^{\mathrm{Q/P}}$ denotes the length of $\mathbf{r}^{\mathrm{Q/P}}$, θ is the angle between $\mathbf{e}_3$ and $\mathbf{r}^{\mathrm{Q/P}}$, and ϕ is defined as the angle between $\mathbf{e}_1$ and the projection of $\mathbf{r}^{\mathrm{Q/P}}$ into the $(\mathbf{e}_1, \mathbf{e}_2)$ plane.

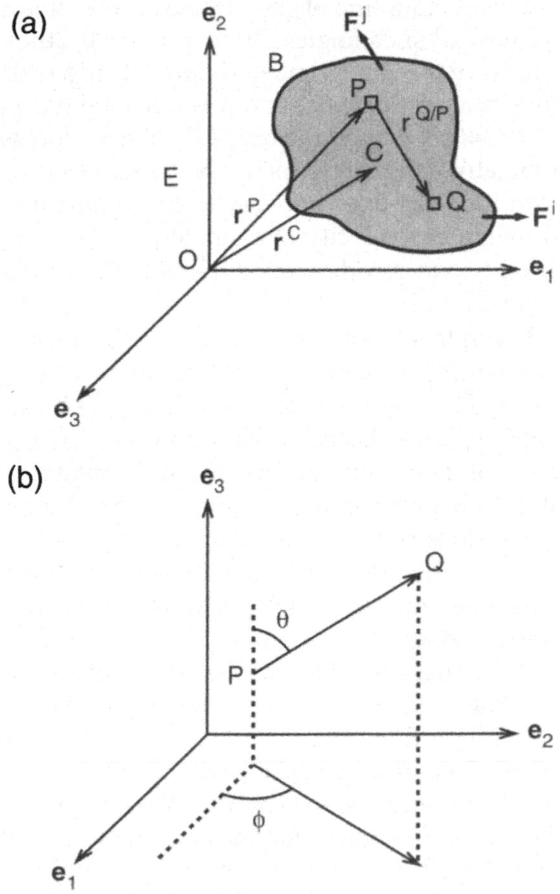

FIGURE 4.2a,b. General plane motion of a rigid object B parallel to the ($\mathbf{e}_1$, $\mathbf{e}_2$) plane **(a)**. The reference frame E is fixed on earth. The symbol C denotes the center of mass. The angles θ and ϕ shown in **(b)** are defined in the text.

Let us substitute Eqn. 4.12 into Eqn. 4.11 and take the time derivative of the result:

$$\mathbf{v}^{Q} = \mathbf{v}^{P} + (d\phi/dt)\, r^{Q/P} \sin\theta\, [-\sin\phi\, \mathbf{e}_1 + \cos\phi\, \mathbf{e}_2] \qquad (4.13)$$

In deriving this equation we have used the fact that the distance between any two points in a rigid body remains constant, and therefore we set $dr^{Q/P}/dt = 0$. Also, because the motion occurs parallel to the ($\mathbf{e}_1$,$\mathbf{e}_2$) plane, the angle θ remains constant; therefore, $d\theta/dt = 0$.

Using the definition of vector multiplication of two vectors, Eqn. 4.13 can be shown to be equivalent to the following vector equation:

$$\mathbf{v}^{Q} = \mathbf{v}^{P} + d\mathbf{r}^{Q/P}/dt = \mathbf{v}^{P} + \boldsymbol{\omega} \times \mathbf{r}^{Q/P} \qquad (4.14)$$

$$\boldsymbol{\omega} = (d\phi/dt)\, \mathbf{e}_3 \qquad (4.15)$$

where $\boldsymbol{\omega}$ is the angular velocity of the rigid body B with respect to the coordinate system E. Note that in a plane parallel to the ($\mathbf{e}_1$, $\mathbf{e}_2$) plane, angle ϕ can be any angle, taken counterclockwise from a line element fixed in E to a line element fixed in the object B (Fig. 4.3). This is true because all such angles differ from each other by a constant and therefore have the same time derivative. If $(d\phi/dt) > 0$, then the object B rotates counterclockwise and it is said to have a positive angular velocity. If $(d\phi/dt) < 0$, then the object B rotates clockwise and it is said to have a negative angular velocity. Note also that angular velocity may vary with time but does not vary from point to point in a rigid body. Thus, knowing the velocity of a single point in a rigid body and its angular velocity, we can determine the velocity of any other point in the rigid body.

Example 4.1. Vertical Jumping. In this example we seek to understand the contributions of segmental rotations of body parts to the vertical velocity of the body's mass center during vertical jumping. As shown in Fig 4.4, the jumper keeps his hands on his hips and jumps as high as he can. We will determine the velocity of the center of mass of the athlete 0.2 s after he begins the preparatory countermovement phase of the jump. The dimensions of the athlete are given as follows: L_f (length of the foot) = 27 cm, L_l (length from ankle to knee) = 48 cm, L_t (length from knee to hip) = 50 cm, and L_c (length from hip to center of mass) = 28 cm. He weighs 68 kg.

We model the athlete as composed of four segments as shown in Fig. 4.4. Angles between body segments and the horizontal in the fixed reference frame E are indicated by ϕ^f, ϕ^l, ϕ^t, and ϕ^c. As usual, the segment orientation angle is positive when taken counterclockwise from horizontal. The counter movement begins at $t = 0$ when the body segments make the following angles with the horizontal: $\phi^f = 0$, $\phi^l = 66°$, $\phi^t = -43°$, and $\phi^c = 23°$. Thus, at this instant, the feet are flat on the ground, the knee bent, and the upper body bent forward. Employing the inverse dynamics approach and using a videocamera and a computer, these body segment angles were measured as a function of time for $t > 0$:

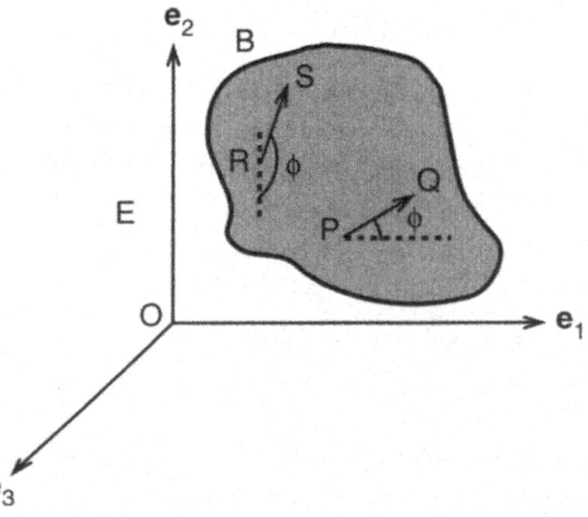

FIGURE 4.3. The rate of rotation of a rigid body in planar motion is equal to the time rate of change of angle ϕ. The angle ϕ need not be uniquely defined because the rate of rotation is the time derivative of angle ϕ. The figure shows two angles whose time derivatives give the same value for the angular velocity.

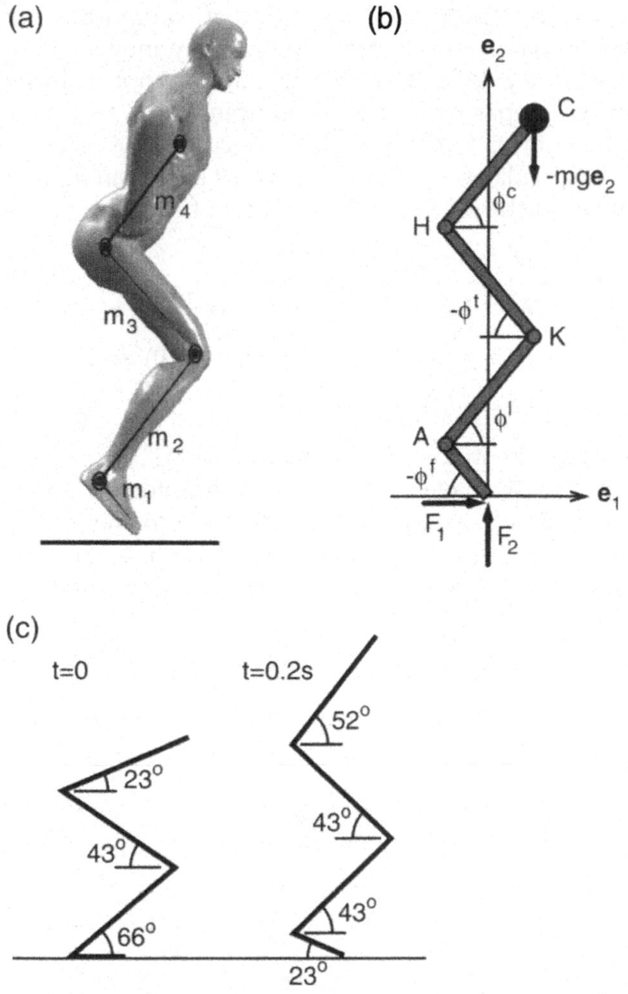

FIGURE 4.4a–c. The schematic diagram of a professional vertical jumper **(a)**. The jumper is represented by a four-link model **(b)**. The symbols F, A, K, H, and C denote the tip of the foot, ankle, knee, hip, and the center of mass of the athlete, respectively. The angles these links make with the horizontal axis are shown in **(c)**.

$$\phi^{f} = -2\,t \tag{4.16a}$$

$$\phi^{l} = 20\,(t - 0.15)^2 + 0.7 \tag{4.16b}$$

$$\phi^{t} = -15\,(t - 0.10)^2 - 0.6 \tag{4.16c}$$

$$\phi^{c} = 2.5\,t + 0.4 \tag{4.16d}$$

In these expressions, time t is in seconds and the segment angles of rotation are in radians. Note that although the jump begins from the standing upright position, the jumper is not in that configuration at time $t = 0$. The center of mass of the jumper has already moved downward toward the ground at $t = 0$ so that this instant does not correspond to the beginning of the preparatory phase of jumping.

Solution: Our aim is to compute the velocity of the center of mass of the athlete.The angular velocities of segments are obtained by taking the time derivatives of the corresponding angles of rotation:

$$\begin{aligned}
\boldsymbol{\omega}^{f} &= (d\phi^{f}/dt)\,\mathbf{e}_3 = -2\,\mathbf{e}_3 \\
\boldsymbol{\omega}^{l} &= (d\phi^{l}/dt)\,\mathbf{e}_3 = 40\,(t - 0.15)\,\mathbf{e}_3 \\
\boldsymbol{\omega}^{t} &= (d\phi^{t}/dt)\,\mathbf{e}_3 = -30\,(t - 0.10)\,\mathbf{e}_3 \\
\boldsymbol{\omega}^{c} &= (d\phi^{c}/dt)\,\mathbf{e}_3 = 2.5\,\mathbf{e}_3
\end{aligned} \tag{4.17}$$

Angular velocities are in the units of rad/s.

At time $t = 0.2$ s, we have the following values for the rotation angles: $\phi^{f} = -23°$, $\phi^{l} = 43°$, $\phi^{t} = -43°$, and $\phi^{c} = 52°$. We converted the angles in radians to angles in degrees by using the relation π (rad) = 180°.

The angular velocities at $t = 0.2$ s are computed by inserting the value of $t = 0.2$ s in Eqn. 4.17:

$$\boldsymbol{\omega}^{f} = -2\,\mathbf{e}_3;\ \boldsymbol{\omega}^{l} = 2\,\mathbf{e}_3;\ \boldsymbol{\omega}^{t} = -3\,\mathbf{e}_3;\ \boldsymbol{\omega}^{c} = 2.5\,\mathbf{e}_3$$

Substituting these values successively into Eqn. 4.14a, we obtain the following values for the velocities of joints and the mass center:

$$\mathbf{v}^{A} = 0 + (-2\,\mathbf{e}_3) \times [(0.27)\,(-\cos 23°\,\mathbf{e}_1 + \sin 23°\,\mathbf{e}_2)] = [0.2\,\mathbf{e}_1 + 0.5\,\mathbf{e}_2]$$

$$\begin{aligned}
\mathbf{v}^{K} &= [0.2\,\mathbf{e}_1 + 0.5\,\mathbf{e}_2] + (2\,\mathbf{e}_3) \times [(0.48)\,(\cos 43°\,\mathbf{e}_1 + \sin 43°\,\mathbf{e}_2)] \\
&= [-0.5\,\mathbf{e}_1 + 1.2\,\mathbf{e}_2]
\end{aligned}$$

$$\begin{aligned}
\mathbf{v}^{H} &= [-0.5\,\mathbf{e}_1 + 1.2\,\mathbf{e}_2] + (-3\,\mathbf{e}_3) \times [(0.50)\,(-\cos 43°\,\mathbf{e}_1 + \sin 43°\,\mathbf{e}_2)] \\
&= [0.6\,\mathbf{e}_1 + 2.3\,\mathbf{e}_2]
\end{aligned}$$

$$\begin{aligned}
\mathbf{v}^{C} &= [0.6\,\mathbf{e}_1 + 2.3\,\mathbf{e}_2] + (2.5\,\mathbf{e}_3) \times [(0.28)\,(\cos 52°\,\mathbf{e}_1 + \sin 52°\,\mathbf{e}_2)] \\
&= +2.7\,\mathbf{e}_2
\end{aligned}$$

where all velocities are in m/s. Note that vertical velocities increase steadily as one goes up from ankle to knee, to the hips, and finally to the center of mass of the trunk. Bobbert and van Ingen Schenau (1988) found that for 10 skilled jumpers the average velocity of the center of mass increased from 0 to approximately 3.5 m/s during the preparatory phase of the jump.

Rate of Change of Angle Between Body Segments

In some instances, when we consider a series of rigid bodies linked to each other, a physically meaningful quantity is the time rate of change of the angle between adjoining links. After all, muscles act on a joint to affect the angle between the articulating bones. The time rate of change of angle between adjoining body segments, when multiplied by the unit vector perpendicular to the plane of motion, is the angular velocity of one link relative to another in a multibody system. The angular velocity of body B (say the lower arm) with respect to body D (say the upper arm) is denoted by the symbol $\boldsymbol{\omega}^{B/D}$. Take a straight line fixed in D and another fixed in B as shown in Fig. 4.5. Let $\phi^{B/D}$ be the angle taken in counterclockwise direction from the straight line fixed in D to the straight line fixed in B; then $\boldsymbol{\omega}^{B/D}$ is given by the following equation:

$$\boldsymbol{\omega}^{B/D} = (d\phi^{B/D}/dt)\,\mathbf{e}_3 \tag{4.18}$$

For bodies B1, B2, and B3 that are serially linked to each other, by substracting or adding angles, it can be shown that

$$\boldsymbol{\omega}^{B3/E} = \boldsymbol{\omega}^{B3/B2} + \boldsymbol{\omega}^{B2/B1} + \boldsymbol{\omega}^{B1/E} \tag{4.19}$$

This equation is called the principle of superposition of angular velocity. We illustrate its use by considering the relative angular velocities of various body segments during vertical jumping.

Example 4.2. Body Segment Motions During Jumping. In Example 4.1 we found the following values for the angular velocities of the foot (f), leg (l), thigh (t), and trunk (c) at 0.2 s after the beginning of the prepatory phase of the vertical jumping:

$$\boldsymbol{\omega}^{f} = -2\,\mathbf{e}_3;\ \boldsymbol{\omega}^{l} = 2\,\mathbf{e}_3;\ \boldsymbol{\omega}^{t} = -3\,\mathbf{e}_3;\ \boldsymbol{\omega}^{c} = 2.5\,\mathbf{e}_3$$

Using Eqn. 4.19, we can show that the angular velocity of the leg with respect to the foot is given by the following equation:

$$\boldsymbol{\omega}^{l/f} = \boldsymbol{\omega}^{l} - \boldsymbol{\omega}^{f} = 2\,\mathbf{e}_3 - (-2\,\mathbf{e}_3) = 4\,\mathbf{e}_3$$

This result indicates that the angle between the leg and the foot at the ankle is increasing at $t = 0.2$ s. This rate of rotation is principally caused by the contraction of the calf muscle.

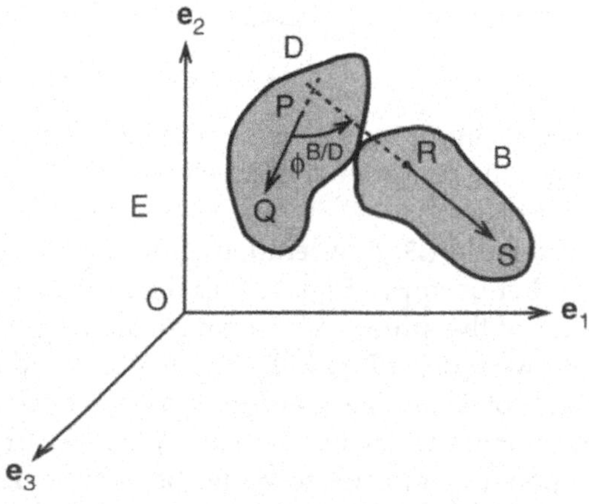

FIGURE 4.5. Relative rotation of rigid object B with respect to another rigid object D in planar motion parallel to the ($\mathbf{e}_1$, $\mathbf{e}_2$) plane. The angle $\phi^{B/D}$ is drawn counterclockwise from a straight line fixed in D to another straight line that rotates with B. Both lines are parallel to the ($\mathbf{e}_1$, $\mathbf{e}_2$) plane.

Let us next compute the angular velocity of the thigh with respect to the leg:

$$\boldsymbol{\omega}^{t/l} = \boldsymbol{\omega}^t - \boldsymbol{\omega}^l = -3\ \mathbf{e}_3 - (2\ \mathbf{e}_3) = -5\ \mathbf{e}_3$$

This value corresponds to the extension of the leg, and the primary muscle group that actuates leg extension are the quads.

The angular velocity of the trunk relative to the thighs is given by the following equation:

$$\boldsymbol{\omega}^{c/t} = \boldsymbol{\omega}^c - \boldsymbol{\omega}^t = 2.5\ \mathbf{e}_3 - (-3\ \mathbf{e}_3) = 5.5\ \mathbf{e}_3$$

This equation points to extension at the hip joint, which is the primary action of the hamstrings. Thus, at the instant of vertical jumping, calf muscles, quads, and hamstrings appear as actuators of relative motion of various body segments. Antagonists to these muscle groups are also activated to bring accuracy and control to the movement. Bobbert and van Ingen Schenau (1988) found that the sequence that is realized by the pattern of muscle activation is upper body, upper legs, lower legs, and feet.

4.4 Angular Acceleration

How do we determine the acceleration of a small volume element in a rigid object? We do this by taking the time derivative of the equation that relates the velocities of two points in a rigid object:

$$\mathbf{v}^Q = \mathbf{v}^P + \boldsymbol{\omega} \times \mathbf{r}^{Q/P} \qquad (4.20a)$$

$$\mathbf{a}^Q = d\mathbf{v}^Q/dt = \mathbf{a}^P + \mathbf{a}^{Q/P}$$

$$\mathbf{a}^Q = \mathbf{a}^P + \boldsymbol{\alpha} \times \mathbf{r}^{Q/P} + \boldsymbol{\omega} \times (\boldsymbol{\omega} \times \mathbf{r}^{Q/P}) \qquad (4.20b)$$

$$\boldsymbol{\alpha} = d\boldsymbol{\omega}/dt = (d^2\phi/dt^2)\ \mathbf{e}_3 \qquad (4.20c)$$

in which $\boldsymbol{\alpha}$ is defined as the angular acceleration. In the planar motion, the rules concerning the sequential multiplication of three vectors lead to the following simplification for the third term in Eqn. 4.20b:

$$\boldsymbol{\omega} \times (\boldsymbol{\omega} \times \mathbf{r}^{Q/P}) = -\omega^2\ \mathbf{r}^{Q/P}$$

This equation indicates that the higher the rate of rotation of a limb, the larger the component of acceleration that is directed toward the center of rotation.

Example 4.3. Acceleration of the Center of Mass During Vertical Jumping. Let us apply Eqn. 4.20 to determine the acceleration of the center of mass of the athlete who is preparing for vertical jump. The details of this case were described in Example 4.1. The lengths of the various body segments of the athlete were given as L_f (length of the foot) = 27 cm, L_l (length from ankle to knee) = 48 cm, L_t (length from knee to hip) = 50 cm, and L_c (length from hip to center of mass) = 28 cm. The athlete weighed 68 kg. The angles that various body segments made with the horizontal axis were presented as follows:

$$\phi^{\mathrm{f}} = -2\,t \tag{4.16a}$$

$$\phi^{\mathrm{l}} = 20\,(t - 0.15)^2 + 0.7 \tag{4.16b}$$

$$\phi^{\mathrm{t}} = -15\,(t - 0.10)^2 - 0.6 \tag{4.16c}$$

$$\phi^{\mathrm{c}} = 2.5\,t + 0.4 \tag{4.16d}$$

The angular velocity and angular acceleration of the various body segments can be deduced from these equations by taking the first and the second time derivative, respectively:

$$\boldsymbol{\omega}^{\mathrm{f}} = -2\,\mathbf{e}_3;\ \boldsymbol{\alpha}^{\mathrm{f}} = 0$$

$$\boldsymbol{\omega}^{\mathrm{l}} = 40\,(t - 0.15)\,\mathbf{e}_3;\ \boldsymbol{\alpha}^{\mathrm{l}} = 40\,\mathbf{e}_3$$

$$\boldsymbol{\omega}^{\mathrm{t}} = -30\,(t - 0.10)\,\mathbf{e}_3;\ \boldsymbol{\alpha}^{\mathrm{t}} = -30\,\mathbf{e}_3$$

$$\boldsymbol{\omega}^{\mathrm{c}} = 2.5\,\mathbf{e}_3;\ \boldsymbol{\alpha}^{\mathrm{c}} = 0$$

At time $t = 0.2$, the angles of rotatation, angular velocity, and angular acceleration are found to be

$$\phi^{\mathrm{f}} = -23°;\ \boldsymbol{\omega}^{\mathrm{f}} = -2\,\mathbf{e}_3;\ \boldsymbol{\alpha}^{\mathrm{f}} = 0$$

$$\phi^{\mathrm{l}} = 43°;\ \boldsymbol{\omega}^{\mathrm{l}} = 2\,\mathbf{e}_3;\ \boldsymbol{\alpha}^{\mathrm{l}} = 40\,\mathbf{e}_3$$

$$\phi^{\mathrm{t}} = -43°;\ \boldsymbol{\omega}^{\mathrm{t}} = -3\,\mathbf{e}_3;\ \boldsymbol{\alpha}^{\mathrm{t}} = -30\,\mathbf{e}_3$$

$$\phi^{\mathrm{c}} = 52°;\ \boldsymbol{\omega}^{\mathrm{c}} = 2.5\,\mathbf{e}_3;\ \boldsymbol{\alpha}^{\mathrm{c}} = 0$$

We next use these values in Eqn. 4.20b to determine the acceleration of the ankle, knee, hip, and finally the center of mass:

$$\begin{aligned}\mathbf{a}^{\mathrm{A}} &= \mathbf{a}^{\mathrm{O}} + \boldsymbol{\alpha} \times \mathbf{r}^{\mathrm{A/O}} + \boldsymbol{\omega}^{\mathrm{f}} \times (\boldsymbol{\omega}^{\mathrm{f}} \times \mathbf{r}^{\mathrm{A/O}}) \\ &= -2\,\mathbf{e}_3 \times [-2\,\mathbf{e}_3 \times 0.27\,(-\cos 23°\,\mathbf{e}_1 + \sin 23°\,\mathbf{e}_2)] = 1\,\mathbf{e}_1 - 0.4\,\mathbf{e}_2\end{aligned}$$

$$\begin{aligned}\mathbf{a}^{\mathrm{K}} = {} & (1.0\,\mathbf{e}_1 - 0.4\,\mathbf{e}_2) + 40\,\mathbf{e}_3 \times 0.48\,(\cos 43°\,\mathbf{e}_1 + \sin 43°\,\mathbf{e}_2) + 2\,\mathbf{e}_3 \\ & \times [2\,\mathbf{e}_3 \times 0.48\,(\cos 43°\,\mathbf{e}_1 + \sin 43°\,\mathbf{e}_2)] = -13.4\,\mathbf{e}_1 + 12.3\,\mathbf{e}_2\end{aligned}$$

$$\begin{aligned}\mathbf{a}^{\mathrm{H}} = {} & (-13.4\,\mathbf{e}_1 + 12.3\,\mathbf{e}_2) - 30\,\mathbf{e}_3 \times 0.50\,(-\cos 43°\,\mathbf{e}_1 + \sin 43°\,\mathbf{e}_2) \\ & -3\,\mathbf{e}_3 \times [-3\,\mathbf{e}_3 \times 0.50\,(-\cos 43°\,\mathbf{e}_1 + \sin 43°\,\mathbf{e}_2)] = 0.1\,\mathbf{e}_1 + 20.2\,\mathbf{e}_2\end{aligned}$$

$$\begin{aligned}\mathbf{a}^{\mathrm{C}} = {} & (0.1\,\mathbf{e}_1 + 20.2\,\mathbf{e}_2) + 2.5\,\mathbf{e}_3 \times [2.5\,\mathbf{e}_3 \\ & \times 0.28\,(\cos 52°\,\mathbf{e}_1 + \sin 52°\,\mathbf{e}_2)] = -1.0\,\mathbf{e}_1 + 18.8\,\mathbf{e}_2\end{aligned}$$

Let us next compute the ground force exerted on the athlete at time $t =$ 0.2 s. According to the laws of motion, the mass of the athlete times the acceleration of the center of mass must be equal to the resultant force acting on the athlete:

$$F_1 = 68\,(-1) = -68\ \mathrm{N}$$

$$F_2 = +68\,(9.81) + 68\,(18.8) = 1945.5\ \mathrm{N}$$

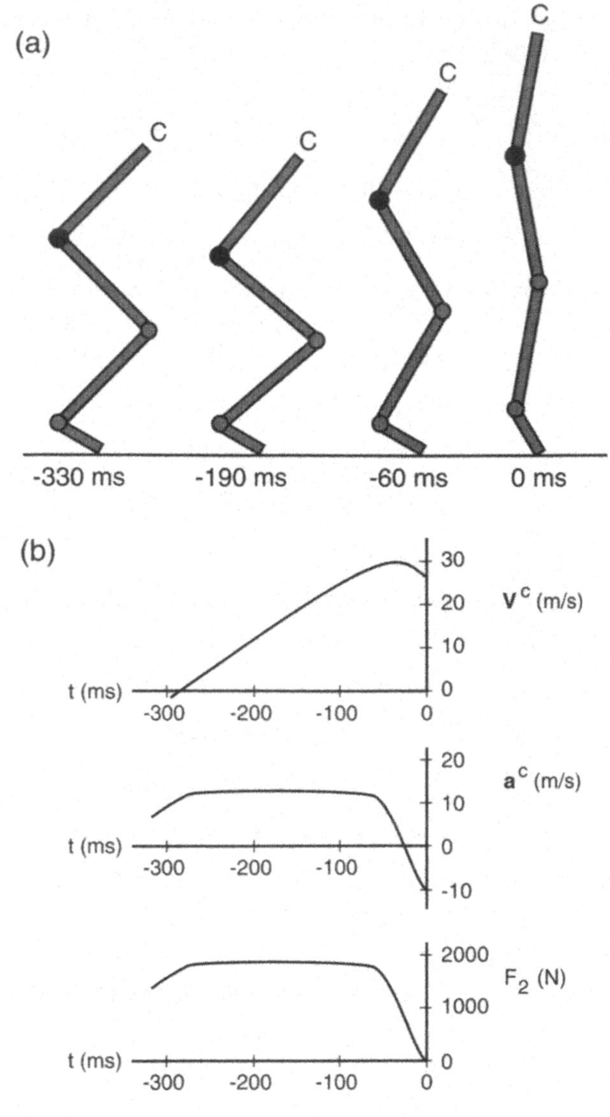

FIGURE 4.6a,b. Approximate positions of the body segments of a vertical jumper 330, 190, 60 and 0 ms before toe-off **(a)**. The mean velocity and the mean acceleration of the center of mass of the athlete and the ground force acting on him as a function of time before the takeoff **(b)**.

The vertical ground force acting on the athlete is approximately three times his weight at 0.2 s into the progression of the jump.

Bobbert and van Ingren Schenau (1988) measured the time course of the ground force, electrical activity of leg muscles, and the angle of rotation of body segments during the course of repeated vertical jumps for a group of 10 proficient male volleyball players. Figure 4.6 reproduces some of the experimental data obtained by these authors. The results indicate that during the period before the takeoff the vertical ground force is at first smaller then the weight of the subject, then larger than the weight by about twofold, and finally decreases toward zero. At the takeoff, the ground force must be equal to zero.

4.5 Angular Momentum

The moment of momentum of a rigid object B with respect to a point P that belongs to the object can be written as

$$\mathbf{H} = \int \mathbf{r}^{Q/P} \times \mathbf{v}^{Q}\, dm \tag{4.21}$$

in which dm is a small mass element, point Q is at the geometric center of the mass element, $\mathbf{r}^{Q/P}$ is the position vector from P to Q, $\mathbf{v}^{Q}$ is the velocity of Q, and the integration is over all small mass elements of the object B (Fig. 4.7). The relation that was earlier presented in the chapter about the velocity of two points in a rigid object can be used to reduce the moment of momentum expression into the following form:

$$\mathbf{H} = \int \mathbf{r}^{Q/P} \times (\mathbf{v}^{P} + \boldsymbol{\omega} \times \mathbf{r}^{Q/P})\, dm$$

Note that both $\mathbf{v}^{P}$ and $\boldsymbol{\omega}$, although they may vary with time, are constants with respect to the integration over the mass elements of object B. Thus we can take these terms outside the integral sign.

$$\mathbf{H} = \left(\int \mathbf{r}^{Q/P}\, dm\right) \times \mathbf{v}^{P} + \int \mathbf{r}^{Q/P} \times (\boldsymbol{\omega} \times \mathbf{r}^{Q/P})\, dm \tag{4.22}$$

Note that the first term on the right-hand side of Eqn. 4.22 is always equal to zero for two cases: if point P is fixed in space then $\mathbf{v}^{P} = 0$, or if point

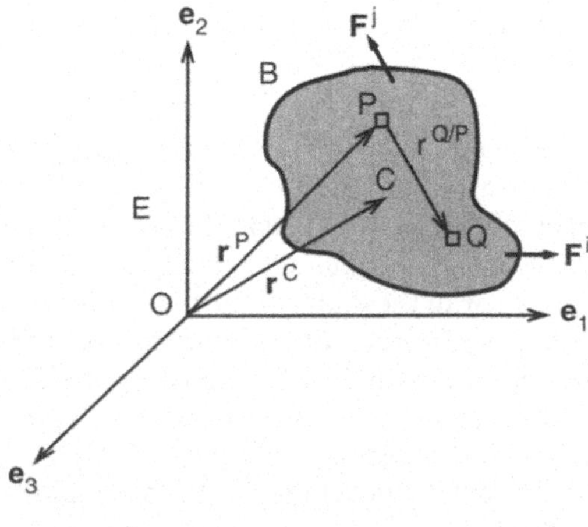

FIGURE 4.7. General plane motion of a rigid object B parallel to the ($\mathbf{e}_1$, $\mathbf{e}_2$) plane. The reference frame E is fixed on earth. The symbol C denotes the center of mass.

P coincides with the center of mass then the integral of position vector with respect to the center of mass must be equal to zero. Under one of these conditions, the moment of momentum expression reduces to

$$\mathbf{H} = \int \mathbf{r}^{Q/P} \times (\boldsymbol{\omega} \times \mathbf{r}^{Q/P})\, dm \tag{4.23}$$

Let us next make the following substitutions into Eqn. 4.23:

$$\mathbf{r}^{Q/P} = x_1\, \mathbf{e}_1 + x_2\, \mathbf{e}_2 + x_3\, \mathbf{e}_3$$

$$\boldsymbol{\omega} = \omega\, \mathbf{e}_3$$

in which $\mathbf{e}_3$ is the unit vector of the reference frame E that is perpendicular to the plane of motion. Note that these unit vectors can be taken outside the integral because they do not depend on the mass elements of the rigid object B.

The resultant expression for moment of momentum becomes

$$\begin{aligned} \mathbf{H} &= \int (x_1\mathbf{e}_1 + x_2\mathbf{e}_2 + x_3\mathbf{e}_3) \times (\boldsymbol{\omega}\, \mathbf{e}_3) \times (x_1\mathbf{e}_1 + x_2\mathbf{e}_2 + x_3\mathbf{e}_3)\, dm \\ &= \omega \int (-x_1\, x_3)\, dm\, \mathbf{e}_1 + \boldsymbol{\omega} \int (-x_2\, x_3)\, dm\, \mathbf{e}_2 + \omega \int (x_1{}^2 + x_2{}^2)\, dm\, \mathbf{e}_3 \\ &= \omega\, (I_{13}\, \mathbf{e}_1 + I_{23}\, \mathbf{e}_2 + I_{33}\, \mathbf{e}_3) \end{aligned} \tag{4.24}$$

in which the mass moment of inertia components are defined as

$$I_{13} = \int (-x_1\, x_3)\, dm \tag{4.25a}$$

$$I_{23} = \int (-x_2\, x_3)\, dm \tag{4.25b}$$

$$I_{33} = \int (x_1{}^2 + x_2{}^2)\, dm \tag{4.25c}$$

Mass moment of inertia depends strictly on the geometry and the distribution of mass of the rigid object B. Note that if the plane of the motion coincides with a plane of symmetry of the object B, then $I_{13} = I_{23} = 0$. Figure 4.8 shows two objects where the plane of motion is also a plane of symmetry (a) and two others when the plane of motion is not a plane of symmetry (b). We focus on the former case. Thus, for the rigid bodies undergoing planar motion in the plane of symmetry we have

$$\mathbf{H} = \omega \int (x_1{}^2 + x_2{}^2)\, dm\, \mathbf{e}_3 = I_{33}\, \omega\, \mathbf{e}_3 \tag{4.26}$$

The parameter I_{33} depends on the geometry and mass distribution characteristics of the rigid object as well as the position of the point with respect to which it is calculated. The parameter I_{33} is denoted as I^c when it denotes the mass moment of inertia with respect to an axis that is perpendicular to the plane of motion and passes through the center of mass. Moments of inertia of rigid objects of different shapes have been tabulated in Appendix 2 of this book. For a slender circular arc of angle θ and radius r, the mass moment of inertia $I^c = mr^2$, where m is the mass of the arc and r is its radius. For a rectangular solid, $I^c = (m/12)\,(a^2 + b^2)$, where a and b denote the lengths of the sides of the solid in the plane of motion.

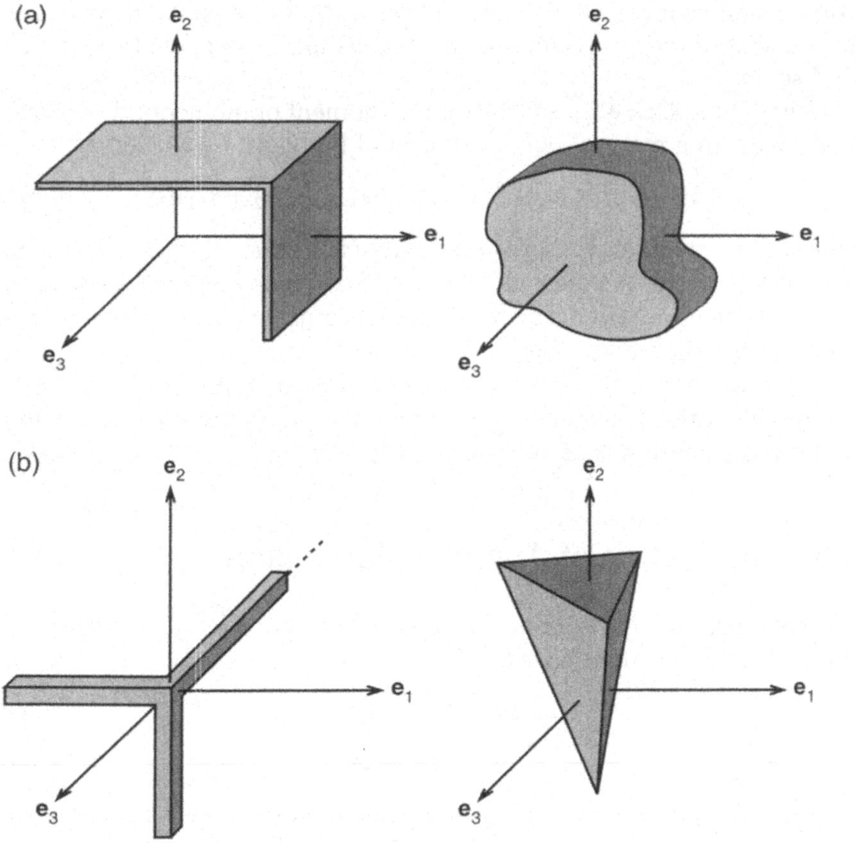

FIGURE 4.8a,b. Examples of rigid objects that have planar symmetry **(a)** and those that do not have any plane of symmetry **(b)**.

Mass moment of inertia with respect to any other point, say point O, can be found by using the geometric relation:

$$I^o = (x_1{}^2 + x_2{}^2)\ dm = I^c + mr^2 \tag{4.27a}$$

where m is the mass of the object and r is the distance between O and C. This relation is called the parallel axis theorem. Earlier in the chapter, we found that the mass moment of inertia of a slender rod with respect to one of its ends was given by the following relation:

$$I_{33} = I^o = mL^2/3$$

Using Eqn. 4.27, we see that the mass moment of inertia with respect to the mass center $I^c = mL^2/12$ for the slender rod considered here.

Mass moment of inertia is an entity that readily applies to three-dimensional geometries, as we shall see later in the book. In three dimensions, mass moment of inertia with respect to center of mass has nine components that can be organized into a 3×3 symmetrical matrix. That means that of the nine components only six are independent; $I_{23} = -I_{32}$, and so on.

Using Eqns. 4.22, 4.24, and 4.27a, the moment of momentum of a rigid body B about a material point of the rigid body can be written as

$$\mathbf{H}^{O} = \mathbf{r}^{C/O} \times m\, \mathbf{v}^{O} + \omega\, (I_{13}\, \mathbf{e}_1 + I_{23}\, \mathbf{e}_2 + I_{33}\, \mathbf{e}_3) \tag{4.27b}$$

where O is a point in the rigid body and m is its mass. If O is a point that is fixed in a reference frame which is stationary with respect to earth, then $\mathbf{v}^{O} = 0$. Furthermore, if the plane of motion coincides with a plane of symmetry of the rigid body, then $I_{13} = I_{23} = 0$.

One final comment about the moment of momentum of a rigid body: because the rate of rotation of all points in a rigid body is the same, moment of momentum is called angular momentum.

4.6 Conservation of Angular Momentum

We have previously seen that, for a system of particles, conservation of moment of momentum dictated that

$$(d\mathbf{H}^{o}/dt) = \mathbf{M}^{o} \tag{4.28a}$$

$$(d\mathbf{H}^{c}/dt) = \mathbf{M}^{c} \tag{4.28b}$$

in which $\mathbf{H}^{o}$ and $\mathbf{H}^{c}$ refer to the moment of momentum with respect to fixed point O and center of mass, respectively. The symbols $\mathbf{M}^{o}$ and $\mathbf{M}^{c}$ represent, respectively, the resultant moment of external forces about point O and the center of mass.

When a rigid object undergoes planar motion, we have seen that its moment of momentum (angular momentum) can be expressed by the following simple equations:

$$\mathbf{H}^{o} = I^{o}\, \omega\, \mathbf{e}_3 \tag{4.29a}$$

$$\mathbf{H}^{c} = I^{c}\, \omega\, \mathbf{e}_3 \tag{4.29b}$$

When one substitutes these expressions into Eqn. 4.28 depicting the conservation of angular momentum, we arrive at the following equation:

$$I^{o}\, \alpha\, \mathbf{e}_3 = \mathbf{M}^{o} \tag{4.30a}$$

$$I^{c}\, \alpha\, \mathbf{e}_3 = \mathbf{M}^{c} \tag{4.30b}$$

As we noted before, α is the angular acceleration of the rigid body B and I^{o}, the mass moment of inertia with respect to point O that is fixed in E, is related to I^{c} by the following equation:

$$I^{o} = mr^2 + I^{c}$$

in which r is the distance between center of mass of the object and point O.

Equation 4.30 is the fundamental equation governing the rate of rotation of a rigid object. It is similar in structure to the equation governing the motion of center of mass, recapitulated below:

$$m\ \mathbf{a}^c = \mathbf{F} \tag{4.31}$$

where m is the mass of the rigid body, $\mathbf{a}^c$ is the acceleration of the center of mass, and $\mathbf{F}$ is the resultant external force acting on the rigid body. It is clear from Eqns. 4.30 and 4.31 that mass is the measure of an object's resistance to uniform acceleration, and mass moment of inertia is the measure of a rigid body's resistance to changes in the rate of rotation. The resultant external force determines the path of the center of mass during motion, whereas the resultant external moment is responsible for a change in the angular velocity of a rigid body undergoing planar motion. Equations 4.30 and 4.31 comprise a complete set of equations of motion for a rigid object. In the following, we present a number of examples that illustrate the use of these equations.

Example 4.4. Swinging of a Disk Around a Fixed Axis. A disk of radius r and mass m is hinged at point O. It is otherwise free to swing under the application of an external force (Fig. 4.9). Let ϕ be the angle that the radial line of the disk passing through point O makes with the vertical axis. Derive the equations of motion of the disk.

Solution: The free-body diagram shown in Fig 4.9 illustrates the forces acting on the disk. These forces are the gravitational force $-mg\ \mathbf{e}_2$ acting at the center of mass of the disk and the reaction force $(F_1\ \mathbf{e}_1 + F_2\ \mathbf{e}_2)$ exerted by the pin at point O on the disk B. The only moment acting on the disk is the moment of the gravitational force, and through vector multiplication it can be shown to be given by the following equation:

$$\mathbf{M} = -mgr \sin\phi\ \mathbf{e}_3 \tag{4.32}$$

The moment of inertia of the disk with respect to its center of mass is $I^c = mr^2/2$ (see Appendix 2). Substituting this value into Eqn. 4.27 we find $I^o = 3mr^2/2$. Conservation of angular momentum with respect to point O yields:

$$(3mr^2/2)\ \alpha = (3mr^2/2)\ (d^2\phi/dt^2) = -mgr \sin\phi$$

$$(d^2\phi/dt^2) = -[2g/(3r)] \sin\phi \tag{4.33}$$

In the case of small oscillations, $\sin\phi$ can be replaced by ϕ. The small angle swinging solution for Eqn. 4.33 is then equal to

$$\phi = A \sin\,[(2g/3r)^{1/2}\ t] + B \cos\,[(2g/3r)^{1/2}\ t] \tag{4.34}$$

where A and B are arbitrary constants to be determined by the initial conditions. For the initial conditions $\phi = \pi/6$ and $(d\phi/dt) = 0$ at $t = 0$, we find that $A = 0$ and $B = \pi/6$. This is equivalent to the frequency of a pen-

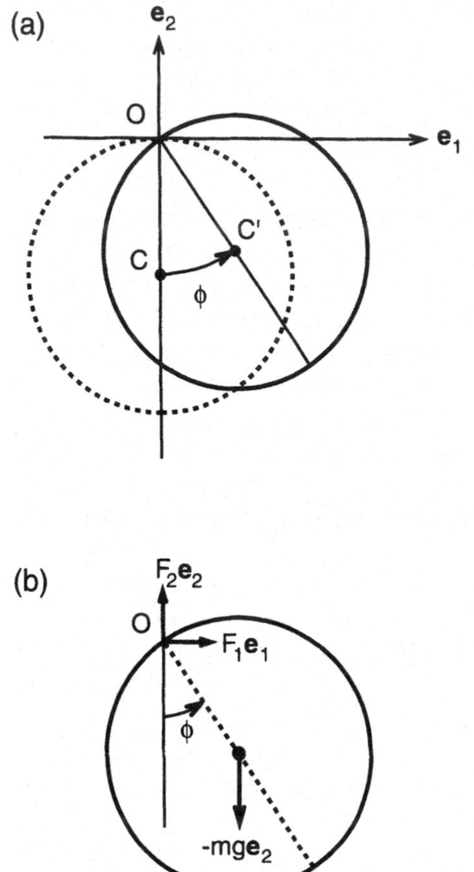

FIGURE 4.9a,b. A pendulum composed of a disk with uniform distribution of mass **(a)**. Free-body diagram of the disk showing external forces acting on it is also illustrated **(b)**.

dulum (weightless rod and bob with mass m) with length equal to $r_g = 3r/2$. Reaction forces acting on the disk at the pin joint O can be computed by using the law of motion for the center of mass, as was illustrated for the swinging of a rod in the previous example.

Example 4.5. Rotation of an Airborne Disk. Let us consider the rotation of a disk as it moves in air. It is given an initial angular velocity of $\omega\, \mathbf{e}_3$ before it becomes airborne. How does its angular velocity change as it traverses a parabolic path in air?

Solution: Because the resultant gravitational force acts on the center of mass, it creates no moment with respect to the center of mass. The angular momentum with respect to the center of mass must remain constant. According to Eqn. 4.30b, the rotational velocity of an object moving in air is constant and this constant is equal to the initial angular velocity at the time the object becomes airborne.

Example 4.6. An Airborne Rod Breaks into Two. A rod of length L and mass m rotating counterclockwise in air with angular velocity $\omega\, \mathbf{e}_3$ suddenly breaks into two parts. Determine the angular velocity of the two pieces of rod immediately after the breakup. Assume the rods to be of equal length ($L/2$).

Solution: Before the rod breaks into two, its angular momentum with respect to its center of mass is given by the following equation:

$$\mathbf{H}^c = I^c\, \omega\, \mathbf{e}_3 = (mL^2/12)\, \omega\, \mathbf{e}_3$$

in which the term on the right in parentheses is the mass moment of inertia of the rod with respect to its center of mass. Because the only external force that acts on the object before, during, and after the break is the force of gravity and this force creates no moment with respect to the center of mass, angular momentum $\mathbf{H}^c$ must be conserved. Immediately after the breakup, the original rod is replaced by a system composed of two rigid bodies. Using Eqn. 4.22 for each of the two rods and summing the two equations, the moment of momentum about point C immediately after the break can be shown to be

$$\mathbf{H}^c = 2\, [(m/2)(L/2)^2/3]\, \omega^f\, \mathbf{e}_3$$

in which $\omega^f\, \mathbf{e}_3$ is the angular velocity of the two rods after the break, and the term in the brackets represents the moment of inertia of one-half of the original rod with respect to its end that coincides with the center of mass of the system. Equating these equations we find that

$$\omega^f = \omega$$

Thus, immediately after the breakup, the two rods rotate with the same angular velocity as before the breakup.

4.7 Applications to Human Body Dynamics

The most fundamental step in the analysis of movement and motion is drawing a free-body diagram showing all external forces acting on an object. If the free body under consideration is part of a body segment, then external forces to be shown include those forces that the rest of the body apply on the body segment under consideration. Although such forces are treated as internal forces in the overall motion of the entire body, they need to be considered as external forces when the movement of an individual body segment is studied.

Some of the external forces acting on an object (part of an object) may be known in both in direction and magnitude. An example of such a force is the gravitational force. In other cases, we might know something about the direction of an external force. When the friction between objects can be neglected, direction of contact force becomes perpendicular to the surface of contact. Frictional force is always in the direction that opposes relative sliding on the surface of contact. There are also propulsive forces that actuate movement and motion. As in the case of bicycling, the propulsive force is in the direction of motion. In speed skating, the propulsive force lies in a plane that is at right angles to the gliding direction. In the following, we present examples concerning the planar motion of humans.

Example 4.7. Diving. Competitive dives in swimming involve several turns before the diver enters the water with as little splash as possible. In one such dive, the diver began the dive with hands at his sides and at the end of the board with his back toward the water. He quickly adopted the layout position where his arms extended in the line of the body, and then assumed the tuck position in which the thighs and the lower legs are pulled in toward the trunk (Fig. 4.10). The angular velocity of the athlete

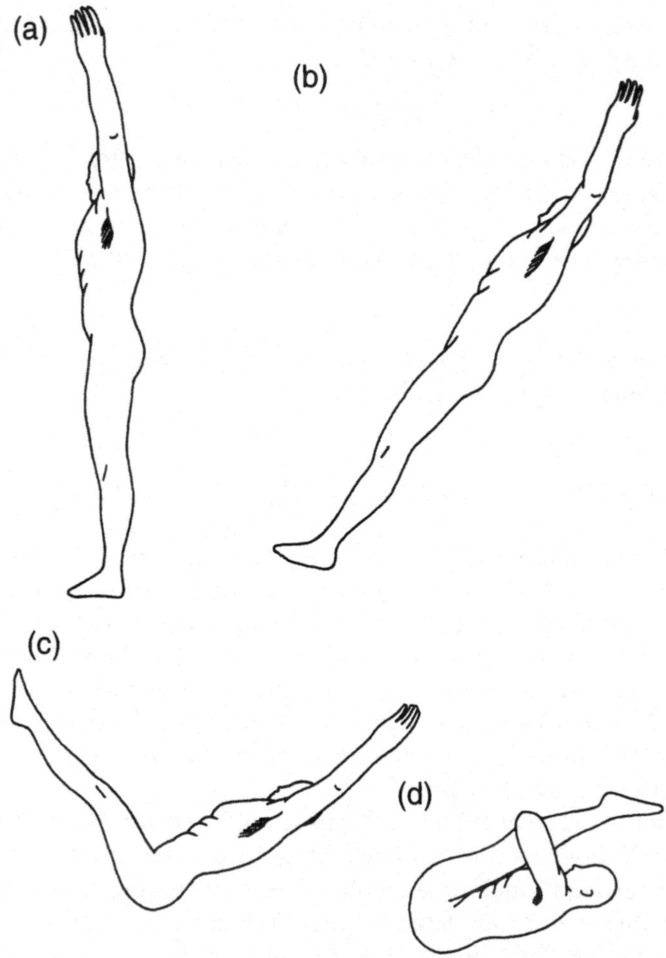

FIGURE 4.10a–d. Planar diving in which the diver changes shape from fully extended **(a)** to tucked configuration **(d)** through intermediate shapes **(b)** and **(c)**.

in the layout position was $\omega = -2\pi$ (s^{-1}). Determine his angular velocity in the tuck position. The moment of inertia I^c of the athlete in the layout position was 134 lb-in-s^2 and in the tuck position 34 lb-in-s^2.

Solution: Once the swimmer is airborne the only force acting on him is the gravitational force passing through his center of mass, and as there is no external moment acting on him while he is airborne, his angular momentum with respect to his center of mass must be constant:

$$\begin{aligned} \mathbf{H}^c &= I^c \omega \mathbf{e}_3 \\ &= 134 \text{ lb-in-s}^2 \times -2\pi\ (\text{s}^{-1})\ \mathbf{e}_3 \\ &= 34 \text{ lb-in-s}^2 \times \omega\ \mathbf{e}_3 \\ \omega &= -24.75 \text{ rad/s} \end{aligned}$$

According to this finding, the rate of rotation of the athlete increases approximately fourfold when he assumes the tuck position. In diving, the angular momentum is determined at the time of the takeoff. Once the angular momentum is set, then it remains constant until the diver encounters the water. Note that during the fraction of a second when the swimmer switches from the layout position to the tucked position, his body is changing shape and therefore cannot be idealized as a rigid body. In that brief time period, $\mathbf{H}^c = I^c \omega \mathbf{e}_3$ will not hold, because it was derived under the assumption that all the various segments of the body rotated with the same angular velocity. This is clearly not the case when the thighs and lower legs are moving toward the trunk as the diver assumes the tuck position.

Example 4.8. Gymnast on Rings. The feet of a gymnast of mass $2m$ and height $2L$ are attached to two rings as shown in Fig. 4.11. The gymnast is let go from rest in the horizontal position as indicated in the figure. The gymnast swings down keeping her body aligned in a straight line. To assess the loads carried by her abdominal and back muscles during the swing, let us model the gymnast with two rods (OA and AB) connected by a hinge at point A. The point O represents the feet, and we assume it to be stationary in the reference frame E. When these two rods let go from rest in the horizontal configuration, will they begin to rotate as one solid body? Does the gymnast contract back (or abdominal) muscles to remain straight? What are the angular accelerations of rods OA and AB?

Solution: Free-body diagrams of rods OA and AB shown in Fig. 4.11 indicate that there are six unknowns in this problem: the horizontal and vertical reaction forces exerted by the hinge (rings) on the rod OA, the horizontal and vertical forces OA exerts on AB, and the angular accelerations OA and AB. Newton's third law dictates that the resultant force AB exerts on OA must be equal in magnitude but opposite in direction of the force OA exerts on AB. We use the equations of motion for the center of mass of each rod (points D and E in Fig. 4.11c, lower diagrams) and the conservation of angular momentum to evaluate these unknowns.

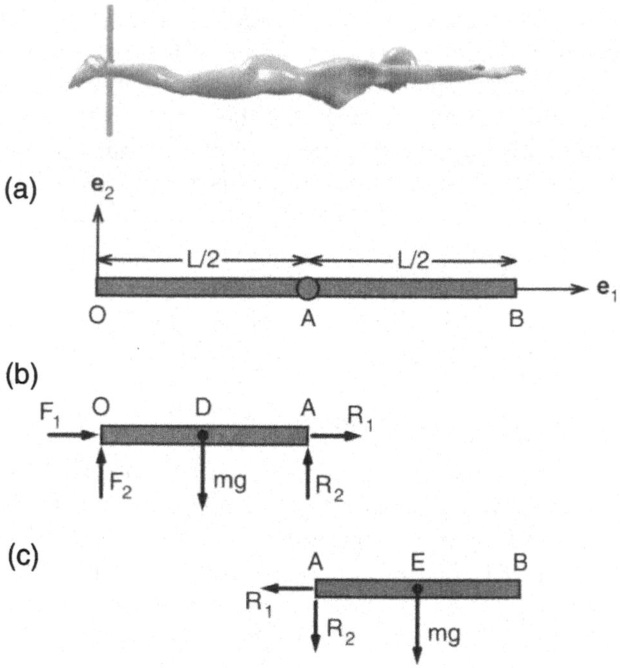

FIGURE 4.11a–c. A gymnast rotating downward from a straight horizontal configuration while the positioning of the feet remains constant **(a)**. The gymnast is modeled as two rods linked to each other **(b)**. The free-body diagrams of each of the rods are shown in **(c)**.

First, let us express the accelerations of points D and E in terms of the angular accelerations of rods OA and AB. We begin with the determination of the acceleration of D by using Eqn. 4.20b for points O and D. The acceleration at point O is zero because this point is fixed in space. Furthermore, the rods under consideration are released from rest, and hence their angular velocities are zero. Therefore, Eqn. 4.20b dictates that

$$\begin{aligned}\mathbf{a}^{\mathrm{D}} &= \alpha_1\, \mathbf{e}_3 \times (L/2)\, \mathbf{e}_1 \\ &= \alpha_1\, (L/2)\, \mathbf{e}_2\end{aligned}$$

in which $\alpha_1\, \mathbf{e}_3$ is the angular acceleration of rod AB. Let us use the same equation to determine the acceleration of E:

$$\begin{aligned}\mathbf{a}^{\mathrm{E}} &= \mathbf{a}^{\mathrm{A}} + \alpha_2\, \mathbf{e}_3 \times (L/2)\, \mathbf{e}_1 \\ &= \alpha_1\, L\, \mathbf{e}_2 + \alpha_2\, \mathbf{e}_3 \times (L/2)\, \mathbf{e}_1 \\ &= [\alpha_1\, L + \alpha_2\, (L/2)]\, \mathbf{e}_2\end{aligned}$$

in which $\alpha_2\, \mathbf{e}_3$ is the angular acceleration of rod AB.

Let us next consider the equations of motion for the center of mass of each of the rods. These equations can be written in the $\mathbf{e}_1$ direction as follows:

$$\text{For OA: } F_1 + R_1 = m\,(0) \Rightarrow F_1 = -R_1$$

$$\text{For AB: } -R_1 = m\,(0) \Rightarrow R_1 = 0 \Rightarrow R_1 = -F_1 = 0$$

Thus, the horizontal reaction forces acting on joints O and A are equal to zero. The equations of motion for the rods OA and AB in the $\mathbf{e}_2$ direction are

$$F_2 + R_2 - mg = m\,\alpha_1\,(L/2) \qquad (4.35a)$$

$$-R_2 - mg = m\,[\alpha_1\,L + \alpha_2\,(L/2)] \qquad (4.35b)$$

Note that the reacting force $-R_2$ creates a counterclockwise (positive) moment with respect to point E.

These two equations involve four unknowns: F_2, R_2, α_1, and α_2. We need two more equations to solve for them. The principle of conservation of angular momentum provides two scalar equations. First, the conservation of angular momentum of bar OA about the fixed point O:

$$-mg(L/2) + R_2\,L = (mL^2/3)\,\alpha_1 \qquad (4.36a)$$

Note that $I^o = (mL^2/3)$ is the moment of inertia of OA about point O. The conservation of angular momentum of the bar AB about its mass center E requires that

$$R_2\,(L/2) = (mL^2/12)\,\alpha_2 \qquad (4.36b)$$

We can eliminate R_2 from Eqns. 4.36a and 4.36b:

$$-3g + L\,\alpha_2 = 2L\,\alpha_1 \qquad (4.36c)$$

We can similarly eliminate R_2 from Eqns. 4.35b and 4.36a:

$$-9g = 8L\,\alpha_1 + 3L\,\alpha_2 \qquad (4.36d)$$

Solving these last two equations for α_1 and α_2 we find that:

$$\alpha_1 = -(9g/7L) \text{ and } \alpha_2 = (3g/7L)$$

Thus, bar OA (legs) will begin rotating clockwise whereas the bar AB (trunk) will begin rotating counterclockwise. The gymnast will have to use abdominal muscles to remain aligned along a straight line.

Example 4.9. Body Curls. In body curls, one uses a specially designed bench to support the heels, upper thighs, and pelvis on padded supports (Fig. 4.12a). Lower the upper torso slowly down to vertical position. Then slowly pull up to the horizontal position and repeat the cycle. Assuming that the back muscle, the erector spinae, is the only muscle involved in this exercise, develop a procedure to evaluate muscle tension as a function of the angle the body trunk makes with the vertical axis.

Solution: Figure 4.12b shows a free-body diagram for the conceptual analysis of the back extension exercise. The upper body is represented by

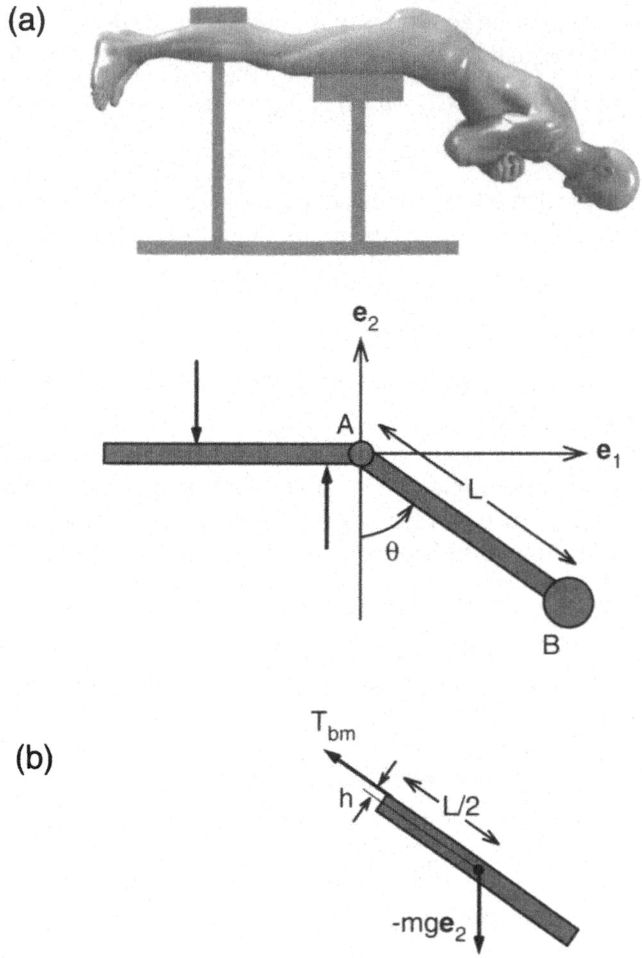

FIGURE 4.12a,b. A man performing back curls to strengthen the back muscle, erector spinae **(a)**. The free-body diagram of the trunk of the man during a back curl **(b)**.

rod AB of length L and mass m. The B end of the rod represents the head. It is free to move in the plane of the motion. The A end represents the pelvic girdle, which remains stationary during the course of the exercise. Gravity pulls the upper trunk down, tending to turn it in the clockwise direction. On the other hand, the back muscle generates a moment in the counterclockwise direction.

The angular momentum of the upper trunk with respect to the fixed point A and its time derivative can be written as

$$\mathbf{H}^{\mathrm{A}} = (mL^2/3)\,(d\theta/dt)\,\mathbf{e}_3 \tag{4.37a}$$

$$(d\mathbf{H}^{\mathrm{A}}/dt) = (mL^2/3)\,(d^2\theta/dt^2)\,\mathbf{e}_3 \tag{4.37b}$$

in which θ is the angle the trunk makes with the vertical axis, and the symbol t denotes the time. As usual, $\mathbf{e}_3$ is the unit vector perpendicular to the plane of the motion.

The conservation of moment of momentum dictates that $(d\mathbf{H}^A/dt)$ is equal to the total external moment acting on the upper body with respect to point A. Thus follows the next equation:

$$(mL^2/3)\,(d^2\theta/dt^2) = -mg\,(L/2)\sin\theta + T_{bm}\,h \tag{4.38}$$

where g denotes the gravitational acceleration, T_{bm} is the force generated by the back muscle, and h denotes the distance from point A to the line of action of the back muscle force. Using Eqn. 4.38, T_{bm} can be expressed as

$$T_{bm} = m\,[g(L/2)\sin\theta + (L^2/3)(d^2\theta/dt^2)]/h \tag{4.39}$$

Equation 4.39 shows that the force exerted by the erector spinae depends on the angle of inclination of the upper body with the vertical axis as well as the angular acceleration of the trunk. Assuming $m = 34$ kg, $L = 0.9$ m, $h = 0.06$ m, $\theta = 45°$, and $d^2\theta/dt^2 = 2$ rad/s, we find for the force exerted by the erector muscle $T = 2074$ N. Thus, the force exerted by the erector spinae is much greater than the weight of the athlete (667 N). About one-quarter of the force results from the upward acceleration of the trunk. However, the dominant reason why erector spinae has to exert such a high level of force during the back curl is because its moment arm with respect to the center of rotation at the hip is small, 0.06 m, as opposed to the moment arm of the weight of the upper body, 0.32 m.

4.8 Instantaneous Center of Rotation

The human knee is a multiple joint in which the femur of the thigh interacts with the tibia of the lower leg and the patella, the kneecap (Fig. 4.13). The geometry of the opposing surfaces in the femorotibial joint is complex, with the radius of curvature changing signs within the region of interaction. Where exactly is the center of rotation of this joint?

The answer to this question lies in the definition of instantaneous center of rotation. In planar motion, continuous motion of a rigid body may be considered as a succession of infinitely small displacements. Each small displacement can be brought about by rotation about some fixed axis. Thus, the motion can be regarded as a series of rotations about a moving axis. The axis of rotation at any instant is called the instantaneous axis of rotation. The point where the instantaneous axis meets the plane of motion is called the instantaneous center of rotation. How do we determine the position of the instantaneous center of rotation?

For locating this center, one needs to know the velocities of two points of a rigid body at an instant of time. Consider a rigid object B undergoing planar rotation as shown in Fig. 4.13a. Let P and Q be two points of

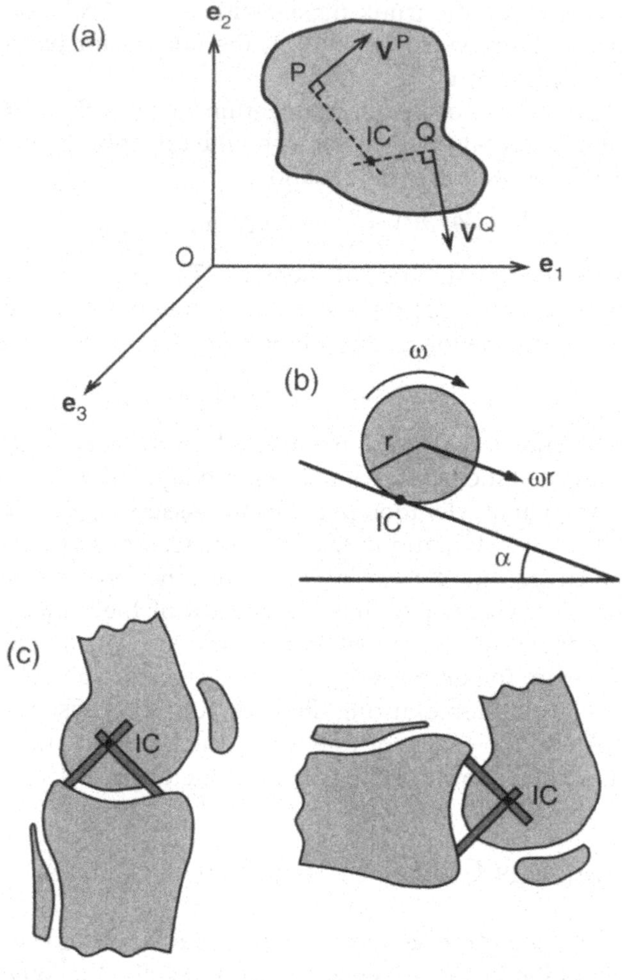

FIGURE 4.13a–c. The instantaneous center of rotation of an arbitrary rigid body in planar motion **(a)**. The instantaneous center is marked with the symbol *IC*. The instantaneous center of a sphere rolling down an inclined plane is the point of contact with the plane **(b)**. The instantaneous center of the knee is illustrated in **(c)**.

the object on a plane parallel to the plane of the motion. The velocities of P and Q are represented by $\mathbf{v}^P$ and $\mathbf{v}^Q$, respectively. The center of rotation is then the intersection of the line that passes through P and is perpendicular to $\mathbf{v}^P$ with the corresponding line passing through point Q (Fig. 4.13a). The velocity of the instantaneous center is zero. The velocities of points in the rigid body are proportional to their distances from the instantaneous center. The instantaneous center of a sphere rolling down an inclined plane without slip is the point of the sphere in contact with the plane at time t (Fig. 4.13b).

In the human knee, the instantaneous center coincides with the intersection of the cruciate ligaments (Fig. 4.13c). Because these ligaments are very stiff, they hardly change length, and therefore the velocities of points of origin and insertion of these ligaments remain perpendicular to the ligament itself. The instantaneous center, thus defined for the knee, may not correspond to a material point belonging to the head of the femur. The arcs corresponding to the successive instant centers of rotation of the tibia rotating in relation to the femur and that of the femur moving in relation to the tibia are called *polodes*. These two curves give a picture of the movement between the femur and tibia projected on the sagittal plane. In the unstable knee, the instantaneous center may vary significantly from time to time, and doctors have used polodes to detect instability associated with the knee. In the normal knee, polodes trace a compact curve around a point. The length scale of this curve is small in comparison to the lengths of the interacting bones. The center of the polodes is for practical purposes the center of rotation of the joint. For more information on the geometry of articulating surfaces of human joints and their instantaneous centers of rotation, the reader is referred to the article by Kento R. Kaufman and Kai-Nan An, *Joint-Articulating Surface Motion*, that appeared in Bronzino (1995).

4.9 Summary

A rigid body is a solid object such that the distance between any two of its material points remains constant during resting state or in motion. Various body segments of the human body such as head, thighs, and forearms can be reasonably assumed as rigid in the analysis of movement and motion. In planar motion parallel to the ($\mathbf{e}_1$, $\mathbf{e}_2$) plane, the angular velocity ω of a rigid object B with respect to reference frame E is defined as the time rate of change of angle between a straight line fixed in E and another straight line in the rotating body B in the ($\mathbf{e}_1$, $\mathbf{e}_2$) plane, taken counterclockwise.

$$\omega = d\theta/dt\ \mathbf{e}_3$$

in which θ is the aforementioned angle and $\mathbf{e}_3$ is the unit vector that is perpendicular to the plane of motion.

Angular acceleration α is defined by the following relation:

$$\alpha = (d^2\theta/dt^2)\ \mathbf{e}_3 = \alpha\ \mathbf{e}_3$$

When angular acceleration is in the positive $\mathbf{e}_3$ direction, then the rate of rotation increases in the counterclockwise direction.

Velocity vectors of any two points in a rigid object are related by the following equation:

$$\mathbf{v}^{Q} = \mathbf{v}^{P} + \boldsymbol{\omega} \times \mathbf{r}^{Q/P}$$

in which $\mathbf{v}^Q$ and $\mathbf{v}^P$ denote the velocities of points Q and P, and $\mathbf{r}^{Q/P}$ is the position vector connecting point P to point Q.

Acceleration vectors of any two points in a rigid body obey the following relation:

$$\mathbf{a}^Q = d\mathbf{v}^Q/dt = \mathbf{a}^P + \boldsymbol{\alpha} \times \mathbf{r}^{Q/P} + \boldsymbol{\omega} \times (\boldsymbol{\omega} \times \mathbf{r}^{Q/P})$$

in which $\mathbf{a}^Q$ and $\mathbf{a}^P$ denote the acceleration vectors of Q and P, respectively.

The moment of momentum of a rigid object is called angular momentum. For rigid objects that are undergoing planar motion in a plane of symmetry of the object, angular momentum with respect to the center of mass is given as

$$\mathbf{H}^c = I^c \, \alpha \, \mathbf{e}_3$$

in which $\mathbf{H}^c$ denotes the angular momentum with respect to the center of mass and I^c is the mass moment of inertia of the object with respect to the center of mass. It is a measure of resistance of the object to the changes in rate of rotation.

If a point of the object, say point O, is fixed on earth and the object rotates around O, the angular momentum with respect to point O is given by the relation

$$\mathbf{H}^o = I^o \, \alpha \, \mathbf{e}_3$$

The parameter I^o, the mass moment of inertia with respect to point O that is fixed in E, is related to I^c by the following equation:

$$I^o = mr^2 + I^c$$

in which r is the distance between the center of mass of the object and point O. The conservation of angular momentum dicates that

$$I^o \, \alpha \, \mathbf{e}_3 = \mathbf{M}^o$$

$$I^c \, \alpha \, \mathbf{e}_3 = \mathbf{M}^c$$

The right-hand side of these equations refers to the resultant external moment acting on the object with respect to the fixed point O and the center of mass, respectively. The principle of conservation of angular momentum relates the changes in rate of rotation to the resultant moment acting on an object.

4.10 Problems

Problem 4.1. Provide an estimate of the mass moment of inertia of your forearm about the three principal axes that pass through its center of mass. Assume the forearm to be a circular cylinder. Measure its length and girth. Assume that the mass density ρ is equal to 1.0 g/cm^3.

Problem 4.2. The moment of inertia of an athlete with respect to his center of mass along an axis from posterior to anterior was experimentally determined to be equal to $I^c = 13.6$ kg-m². The height of the man is 1.83 m and his weight is 78.6 kg. Represent the athlete as a slender rigid rod and determine an approximate value (I^*) for his moment of inertia. Does the slender rod assumption overestimate the moment of inertia? What would be the effective length h* of the rod that would correctly predict the moment of inertia of the athlete?
Answer: $I^* = 21.2$ kg-m²; $h^* = 1.44$ m.

Problem 4.3. Using the oscillating table method, Matsuo et al. (1995) proposed the following equations for determining the mass moment of inertia of adolescent boys between the ages of 13 and 18:

$$I_{33} = 3.44\ H^2 + 0.144\ W - 8.04$$

$$I_{11} = 3.52\ H^2 + 0.125\ W - 7.78$$

in which I_{33} and I_{11} are the mass moments of inertia with respect to the center of mass around the posterior-to-anterior and right-to-left axis, respectively (see Fig. P.4.3). The parameter H is the height of the adolescent, measured in meters, and W is his mass, measured in kilograms. To check whether this formula could also be applicable to adult men, a group of Air Force researchers measured the mass moment inertia of a select group of Air Force men. Following are the data obtained for three men in the group:

Age	Height (m)	Mass (kg)	I_{33} (kg-m²)	I_{11} (kg-m²)
29	1.83	78	14.9	13.4
22	1.71	66	10.8	9.2
20	1.75	63	11.4	10.3

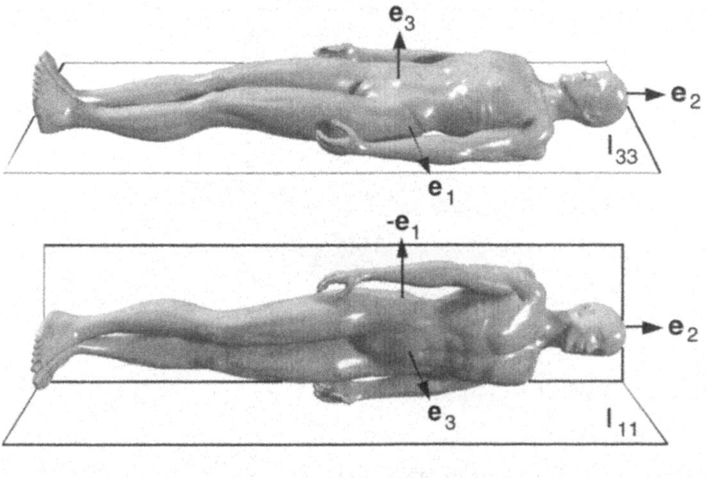

FIGURE P.4.3. Coordinate system attached to an individual in horizontal position.

How applicable are the formulas of Matsuo et al. (1995) to the data on the three men? Compare the model predictions with the data. How far off would be the predictions of these mass moment of inertia components if one represented each individual with a slender rod whose length and mass are equal to that of the individual?

Problem 4.4. Investigate the applicability of the equations developed by Matsuo et al. (1995) for other population groups such as adolescent girls, adult women, elderly men, and elderly women. Conduct a literature search and find relevant data. Determine if there are phenomenological equations already developed for these subpopulations. If not, how would you go about coming up with your own set of empirical equations?

Problem 4.5. Discuss the various ways of determining the mass moment of inertia of body parts. Provide examples from the literature.

Problem 4.6. Provide an estimate of the spatial location of the center of mass C of the dancer leaping in air as shown in Fig. P.4.6. Compute the location of C using the data given below. Specify in detail any additional assumptions you had to make to arrive at your results. Note that you need to establish a reference frame to compute and specify the location of the center of mass. Assume that the reference frame is centered at the top of the head.

Body segment	Weight (kg)	Length (cm)
Trunk	24	51
Head	3.4	27.5
Thigh	4.	46
Shank	2.9	43
Foot	0.8	22
Upper arm	1.7	27
Forearm	1.3	24
Hand	0.4	19

FIGURE P.4.6. Dancer leaping on stage.

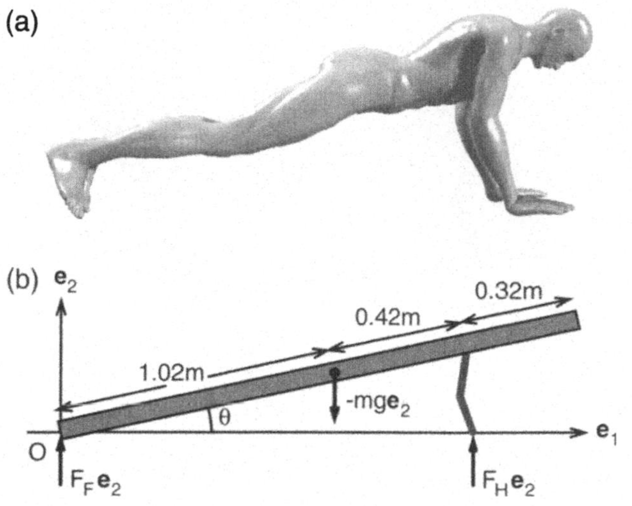

FIGURE P.4.7a,b. Schematic diagram of an individual doing push-ups **(a)**. In this exercise, the man is represented as a rod with uniform distribution of mass **(b)**. The arms are modeled as weightless links **(b)**.

Problem 4.7. Determine the vertical ground forces acting on a man at the feet (F_F $\mathbf{e}_2$) and hands (F_H $\mathbf{e}_2$) while performing push-ups as shown in Fig. P.4.7. The man weighs 71 kg and has a height of 1.76 m. At the instant considered ($t = 0$), the angle his body makes with the horizontal plane (θ) is 20°. Again at $t = 0, d\theta/dt = 4$ rad/s and $(d^2\theta/dt^2) = 7$ rad/s^2. The distance from the bottom of his feet to his center of gravity = 1.02 m. The distance between his head and shoulders is 0.32 m. The body is aligned straight and rotates around the fixed point O as shown in the figure. Assume that the arms can be represented by weightless rods hinged at the middle. Arms transmit vertical forces only.
Answer: $F_H = 872$ N, $F_F = -95.5$ N. The fact that F_F is negative implies that somebody must have been pressing at the ankles of the man doing the push-ups.

Problem 4.8. The rods B1 and B2 shown in Fig. P.4.8 each have mass m and length L. They are hinged together and in the resting position are aligned on a straight line. The rods are released while making 30° with the vertical axis. The rod B1 slides on the smooth, frictionless surface of the floor and the center of mass of the system moves parallel to the floor. Determine the reaction force F_2 and the angular accelerations of B1 and B2 right after the release. Take $m = 28$ kg, and $L = 0.78$ m. Note that this two-rod system might capture some of the essential features of sideway falls. Among the elderly population, a sideway fall is a most frequent cause of hip fracture. The answer to this problem may provide information about the nature of shape change during such a fall

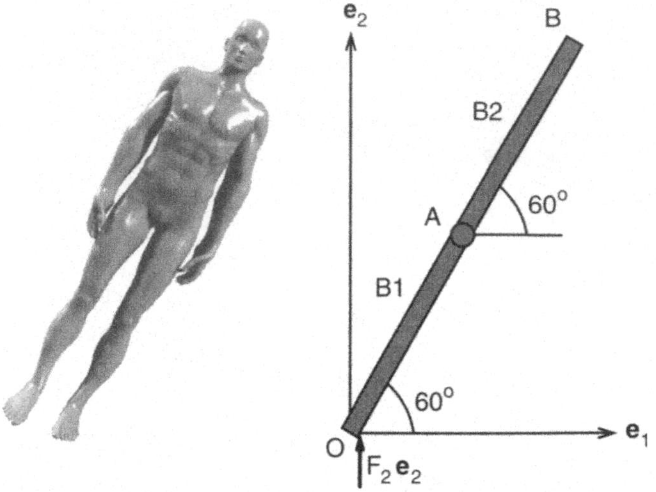

FIGURE P.4.8. Sideway fall of a person onto a floor (*left*) and its representation using a two-link model (*right*).

in the absence of a strong reactive contraction of the back and abdominal muscles.
Answer: $F_2 = 550$ N, $d^2\theta_{B1}/dt^2 = 0$, $d^2\theta_{B2}/dt^2 = -75.0$ rad/s^2.

Problem 4.9. A diver is airborne in full extended position rotating with clockwise angular velocity $\omega = -5$ rad/s. At time $t = 0$ he begins to pull his legs toward his chest at a rate of $-\pi$ rad/s (Fig. P.4.9). Determine the angular velocity of his trunk and that of the lower extremities. Assume that the diver is composed of two slender rods each weighing 32 kg and 0.86 m long. Hint: Use Eqn. 4.27b and show $\omega_{\text{trunk}} = -3.43$ rad/s and $\omega_{\text{leg}} = -6.57$ rad/s.

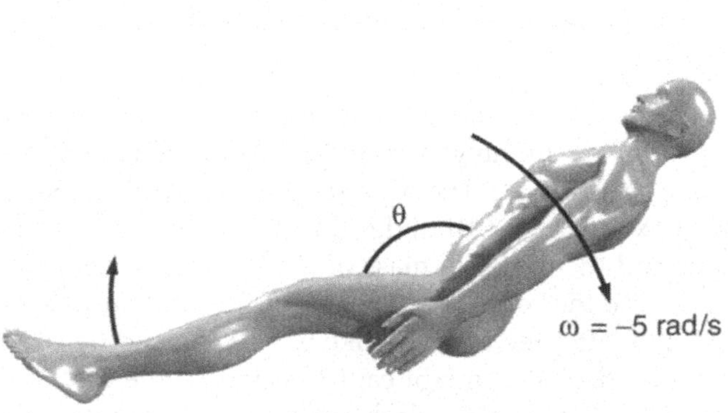

FIGURE P.4.9. A diver pulling his legs toward his body while airborne.

5

Statics: Tug-of-War, Weight Lifting, Trusses, Cables, Beams

5.1 Introduction

An object that either moves with constant velocity or remains at rest is said to be in a state of static equilibrium. A building is in static equilibrium because its weight is balanced vertically by the upward ground force exerted on it. A ballerina keeps a delicate balance by positioning her center of mass on a vertical line that passes through the tip of her feet in contact with the floor. What are the conditions of static equilibrium? How do we use the equations of static equilibrium to determine some of the unknown forces acting on an object? These questions are addressed in this chapter.

5.2 Equations of Static Equilibrium

We have previously shown that the motion of a body B is governed by the following equations:

$$\Sigma \mathbf{F} = m\mathbf{a}^{c} \tag{5.1}$$

$$\Sigma \mathbf{M}^{o} = d\mathbf{H}^{o}/dt \tag{5.2}$$

in which $\Sigma \mathbf{F}$ is the resultant external force acting on B, m is its mass, and $\mathbf{a}^{c}$ is the acceleration of its center of mass. The symbol $\Sigma \mathbf{M}^{o}$ denotes the resultant moment acting on B with respect to point O, and $\mathbf{H}^{o}$ is the moment of momentum of B with respect to the same point. Equation 5.1 is called the equation of motion of the center of mass. According to this equation, the resultant force acting on an object must be equal to the mass of the object times the acceleration of its center of mass. Equation 5.2 is the mathematical expression of the principle of conservation of moment of momentum with respect to a fixed point O:

$$\mathbf{H}^{o} = \int \mathbf{r} \times \mathbf{v}\, dm \tag{5.3}$$

in which **r** and **v** denote, respectively, the position and the velocity of the mass element *dm*, and the integration is over the mass of body B.

The acceleration of the center of mass a body is equal to zero when the body is at rest or in constant motion. Hence, the resultant force acting on the body must be equal to zero:

$$\Sigma \mathbf{F} = 0 \tag{5.4}$$

This equation is called the condition of force balance. The balance of external forces acting on an object may in certain cases be sufficient to ensure static equilibrium. For example, the weight of an elevator hanging from a cable is supported by the pull of the cable. Thus, the tension in the cable must be equal to the weight of the elevator. Another example is the force balance in a game of tug-of-war. If two boys can pull on a rope with the same intensity, they and the rope remain in static equilibrium.

An object, however, is not necessarily in static equilibrium even if the resultant force acting on the object is equal to zero. That is because if there is a resultant moment acting on the object, it will, in accordance with Eqn. 5.2, alter the moment of momentum of the object, resulting in rotational motion. Moment of momentum of an object with respect to a fixed point O must be equal to zero as long as the object remains at rest. To assure static equilibrium we must also satisfy the following condition:

$$\Sigma \mathbf{M}^{o} = 0 \tag{5.5}$$

The term on the left-hand side of Eqn. 5.5 is the resultant moment with respect to any point fixed on earth. Thus, if the moment of forces acting on an object can be shown to be zero with respect to one point, then it is zero for any other point fixed in an inertial reference frame.

Equations 5.4 and 5.5 comprise the equations of static equilibrium. They are valid for living systems as well as nonliving objects. When the forces and the moments acting on an object are three-dimensional, the vector equations, Eqns. 5.4 and 5.5, correspond to six independent scalar equations. If the forces acting on an object all lie in a plane, then Eqn. 5.4 reduces to two independent scalar equations. Because the moment of forces in a plane is perpendicular to that plane, Eqn 5.5 yields a single scalar equation in this case. Thus, there are three independent equations in planar statics.

According to the equations of static equilibrium, if only two forces act on an object, not only that the forces must be equal in magnitude and opposite in direction but also they must have the same line of action (Fig. 5.1). The condition of force balance is satisfied when the two forces are equal in magnitude but in opposite direction. The condition of balance of moment of momentum requires that they must have the same line of action. Otherwise, the moment of one of these forces with respect to the

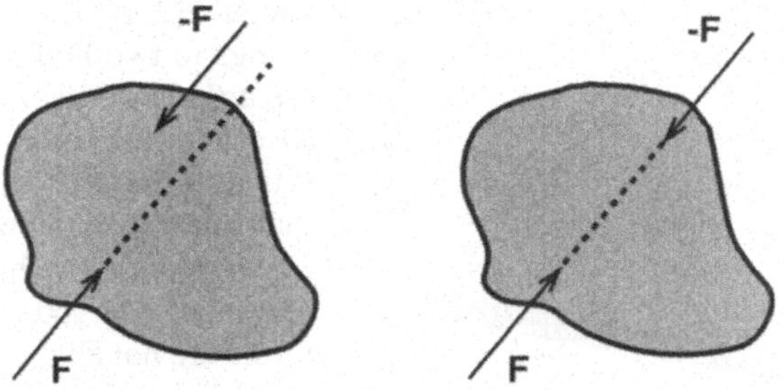

FIGURE 5.1. An object subjected to two point forces. For this object to be in equilibrium, forces involved must be of same magnitude, opposite sense of direction, and must also share the same line of action.

point of application of the other force will not be equal to zero. This result has implications on the forces carried by the long bones of the human body. The weight of a bone acts at its center of mass but may be negligible in comparison with the forces acting on the joints. This is illustrated in Fig. 5.2, in which a rod is shown to be positioned in space using strings tied to its ends. When the tensions in the two strings are large enough, the rod aligns along the tension line as if it were weightless and all the forces acting on it were at its endpoints. Note, however, that in the case of a long bone, muscle forces do not act right at the ends of the bone but have small lever arms. Thus, the assumption that the bone transmits force along the direction of its long axis may not be reasonable under a variety of loading conditions.

Another simple result directly derived from the equations of static equilibrium concerns objects under the application of three forces (Fig. 5.3). If the lines of action of two of these forces intersect each other at some point in space, then the line of action of the third force must also pass through that point. This follows from the condition of balance of external moments. The only way the resulting moment on the object will be equal to zero is if the line of action of the third force also passes through this point.

Investigation of static equilibrium requires computations of moments created by external forces. As we have already demonstrated, the moment of a force with respect to point A is defined as

$$\mathbf{M}^{A} = \mathbf{r}^{P/A} \times \mathbf{F} \tag{5.6a}$$

in which $\mathbf{r}^{P/A}$ is the vector connecting the point of application of force $\mathbf{F}$ to the point A (Fig. 5.4). We could determine moment $\mathbf{M}^{A}$ by going through the formal procedures of vector multiplication. However, sometimes it is easier to adopt an approach that employs only scalar algebra. From the definition of the vector product we know that the direction of

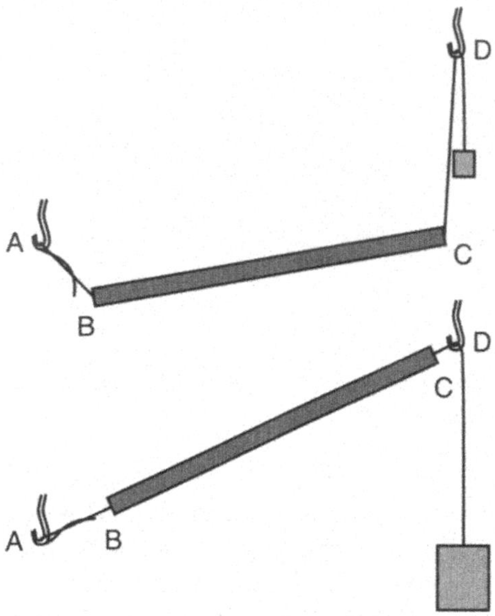

FIGURE 5.2. A rod connected by strings to two hooks that are fixed on earth. The direction of the strings specifies the directions of the forces acting at the ends of the rod. Note that when the tensions in the strings are comparable to the weight of the rod, then the rod and the string forces are not aligned. However, as the string connecting the rod to the hook D is pulled with increasing force, the rod acts as if it is under the influence of just two forces. The weight of the rod can then be neglected in the condition of force balance.

the moment must be perpendicular to the plane created by $\mathbf{r}^{P/A}$ and $\mathbf{F}$. Also, the magnitude of $\mathbf{M}^A$ (M^A) is given by the following equation:

$$M^A = r^{P/A} \cdot F \cdot \sin\theta \tag{5.6b}$$

in which θ is the angle between $\mathbf{r}^{P/A}$ and $\mathbf{F}$. The sense of direction of the moment is such that counterclockwise rotation is considered positive whereas clockwise rotation is considered negative (Fig. 5.4).

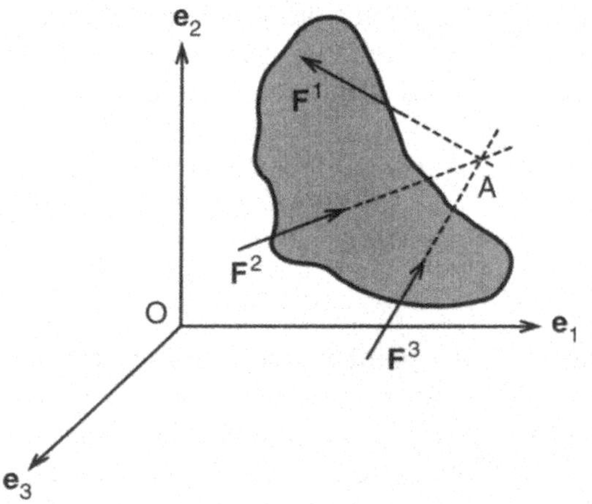

FIGURE 5.3. An object under the influence of three force vectors. If two of these vectors intersect at a point in space, then the third force vector must also pass through the same point to ensure static equilibrium.

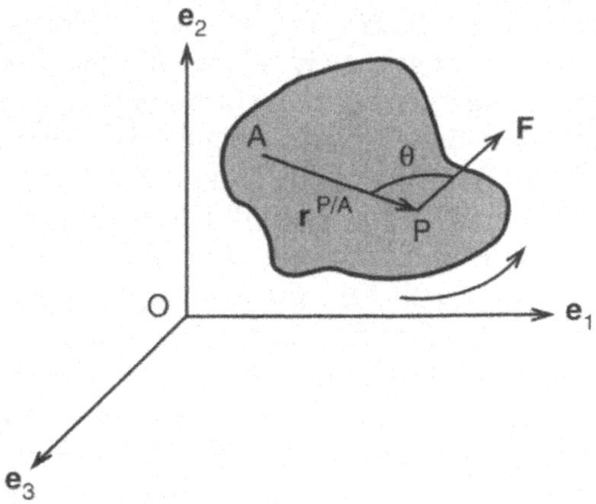

FIGURE 5.4. The moment produced by force **F** about point A. The magnitude of the moment is equal to the products of magnitude of the force, the length between A to the point of application of the force, and the sine of the angle θ the force makes with the position vector connecting A to the point of application of the force.

5.3 Contact Forces in Static Equilibrium

Humans and animals spend a good proportion of their times at static equilibrium, sitting, sleeping, resting. Even when they are in motion, sometimes the inertial terms (mass times acceleration) are small compared to the magnitude of some of the forces acting on them. In such situations, the equilibrium analysis serves as a valid first approximation of the contact forces involved in the movement.

Example 5.1. Stretching of the Achilles Tendon. Consider an athlete who is pushing against a wall to stretch his Achilles tendons during a warm-up before an athletic event (Fig. 5.5). The athlete has a mass $m = 80$ kg and a height L of 183 cm. The distance DB between the top of his head and his

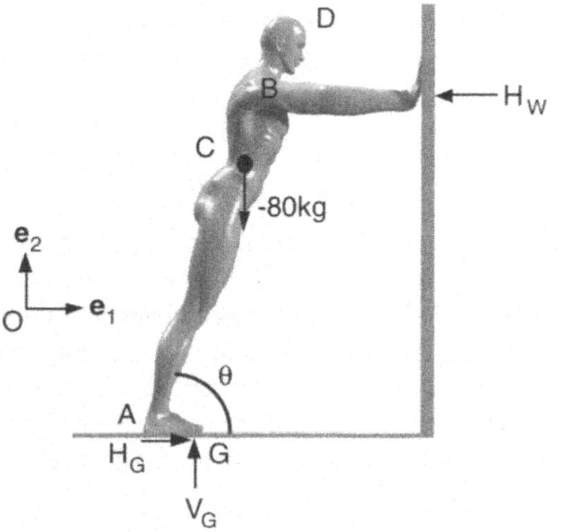

FIGURE 5.5. An athlete leaning on a wall to stretch the Achilles tendon before an athletic event. The external forces acting on the athlete are marked in the figure.

shoulders is 30 cm. The distance AC between the bottom of his foot and his center of gravity is 101 cm, and the distance AG between the heel and the toes is 25 cm. Assume further that the friction between the hands of the athlete and the wall is negligible. Determine the forces exerted on the athlete by the wall and the ground as a function of the angle of inclination θ.

Solution: This is a two-dimensional problem in which all the forces acting on the individual lie in a plane. The free-body diagram shown in Fig. 5.5 indicates that the forces acting on the individual are the force of gravity (the weight of the individual), the horizontal force exerted by the frictionless wall, and the vertical and horizontal forces exerted by the ground. Notice that because the person is stretching the Achilles tendon, the heels barely touch the ground, if they touch at all. Thus the point of application of the ground force is positioned somewhere on the toes.

The condition of static force balance (Eqn. 5.4) yields two scalar equations: the resultant force in both the horizontal ($\mathbf{e}_1$) and vertical ($\mathbf{e}_2$) directions must be equal to zero:

$$H_G - H_W = 0 \Rightarrow H_G = H_W \tag{5.7a}$$

$$V_G - mg = 0 \Rightarrow V_G = 80\ \text{kg}\cdot 9.81\ \text{m/s}^2 = 784.8\ \text{N} \tag{5.7b}$$

Thus the condition of force balance enables us to compute the vertical force exerted by the ground on the athlete. To determine the horizontal contact forces acting on the athlete, we need to set the resultant external moment equal to zero. For simplicity in the computations, we choose to evaluate the moments about point A. Thus, the moment created by the horizontal ground force H_G will be equal to zero. The condition of balance of external moments about A then leads to the following expression:

$$-m\,g \cdot AC \cdot \cos\theta + V_G \cdot AG + H_W \cdot AB \cdot \sin\theta = 0 \tag{5.8}$$

in which θ denotes the angle between the long axis of the athlete's body and the horizontal $\mathbf{e}_1$ direction. Note that all moments acting on the athlete are in either positive or negative $\mathbf{e}_3$ direction.

Equation 5.8 can now be used to compute the unknown force $H_G = H_W$ as a function of θ. Using a hand calculator, one can show that $H_G = H_W =$ 641 N, 336 N, and 151 N, respectively, when $\theta = 30°$, 45°, and 60°. Notice that the horizontal force exerted by the ground and by the wall on the athlete increases sharply with decreasing θ. If at some low value of θ the ground and the athlete's shoes cannot generate enough friction as required by equilibrium, then the athlete will slip, rotate clockwise, and fall, reducing the vertical distance from the center of gravity to the ground.

The magnitude of the resultant force exerted by the ground on the individual can be found by using the equation that relates the magnitude of a vector to its projections along the coordinate axes:

$$F_G = (H_G{}^2 + V_G{}^2)^{1/2}$$

$F_G = 1013$ N, 854 N, and 799 N, respectively, for $\theta = 30°$, 45°, and 60°.

The direction at which the ground force acts can be found by using the following equation:

$$\tan \Phi = (H_G/V_G)$$

We find $\tan \Phi = 0.816, 0.428$, and 0.194, respectively, for $\theta = 30°, 45°$, and $60°$. To determine the force exerted on each foot (or arm), we simply have to divide the contact forces thus computed by the factor 2.

Example 5.2. Toppling a Chair. A child weighing 40 kg is seated on a chair (Fig. 5.6). The legs of the chair are 0.4 m apart, and the back of the chair is 1.2 m high. Assuming that the frictional forces at the front legs are large enough to prevent slipping, what is the maximum horizontal force one could exert on the top of the back of the chair without lifting the back legs?

Solution: External forces acting on the child and the chair are either in the $\mathbf{e}_1$ or $\mathbf{e}_2$ direction. We assume that all these forces lie in the $(\mathbf{e}_1, \mathbf{e}_2)$ plane that passes through the center of mass of the child. We then compute the moment of external forces with respect to point B, which is marked in the figure:

$$-40 \cdot 9.81 \cdot 0.20 + F \cdot 1.2 = 0 \Rightarrow F = 65.4 \text{ N}$$

Note that at the beginning of impending rotation of the chair, the legs in the back will lose contact with the floor and, therefore, the ground forces at the back legs of the chair must be equal to zero at that instant. Thus, they contribute nothing to the external moment.

This example shows that a weight of 6.7 kg acting at an elevation of 1.2 m can lift 40 kg of weight. This is the essence of the principle of the lever. Archimedes articulated this principle in Greece in 250 B.C.: "Give me a point of support and a lever long enough and I will lift the earth with my right hand."

Example 5.3. A Woman on Crutches. A woman with a knee injury is using crutches for walking and standing. Determine the contact forces

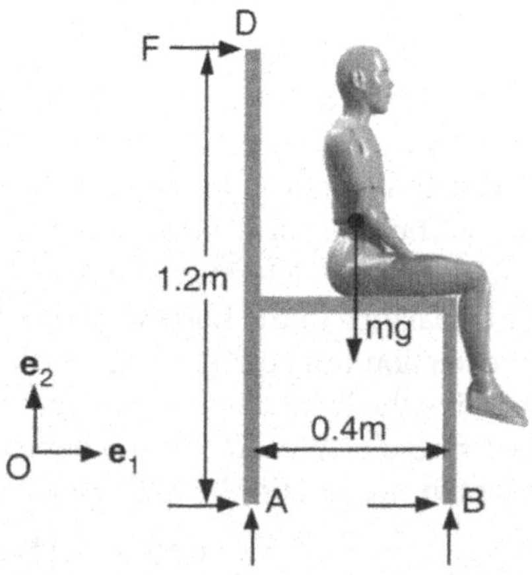

FIGURE 5.6. A child sitting on a chair. The figure depicts the free-body diagram of the child and the chair.

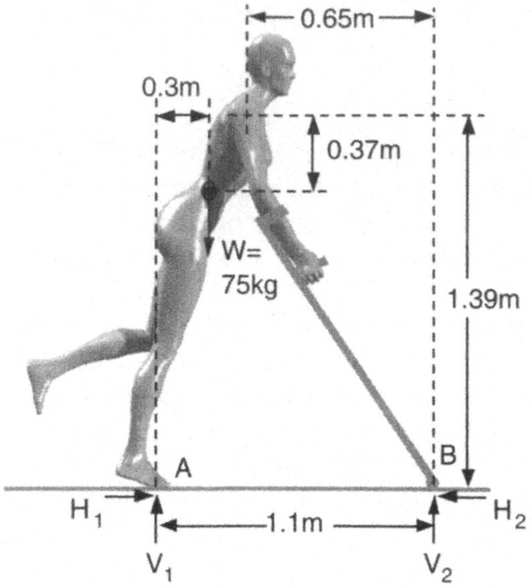

FIGURE 5.7. Contact forces acting on a woman on crutches.

acting on the crutches and on her foot at the standing configuration shown in Fig. 5.7.

Solution: The forces acting on the woman standing with the help of crutches are shown in the figure. There are four contact forces of unknown magnitude acting on the woman and crutches. The equations of statics relating these forces are

$$H_1 - H_2 = 0$$

$$V_1 + V_2 = 75\ \text{kg} \cdot 9.81\ \text{m/s}^2 = 736\ \text{N}$$

$$V_2 \cdot 1.1 = 736 \cdot 0.30 \Rightarrow V_2 = 201\ \text{N}$$

Using the force balance in the vertical direction, we find the ground force acting on the foot as

$$V_1 = -V_2 + 736\ \text{N} = 535\ \text{N}$$

Thus, the crutches used in the form shown in the figure reduce the vertical load on the feet by about 27%. If the woman had kept the crutches closer to her body, the force carried by the crutches would increase and the vertical ground force acting on the foot would decrease.

How do we determine the horizontal forces H_1 and H_2? The condition of force balance in the horizontal direction states that they should be opposite in direction but equal in magnitude. We need additional information to determine the horizontal contact forces uniquely. If one assumes that the contact forces at B create no moment with respect to the shoulder joint of the woman, we obtain the following relationship between H_2 and V_2:

$$V_2 \cdot 0.65\ \text{m} = H_2 \cdot 1.39\ \text{m} \Rightarrow H_2 = 94\ \text{N}$$

Admittedly, this assumption is ad hoc. However, one could defend it by arguing that a finite moment at the shoulder for long durations would result in the excessive use of shoulder muscles, and thus the woman would position the crutch to prevent aching of the shoulder muscles.

Example 5.4. Human Cable. A youth weighing W kg lies on the floor and two other students, each weighing W_s kg, pick him up by the hands and the feet. The arms of the supporting students in combination with the body of the hanging student form a parabola-like curve, which is in tension. Let D denote the span (the horizontal distance) between the shoulders of the supporting students and sag d be the distance from a line between their shoulders to the bottom of the hanging youth (Fig. 5.8).

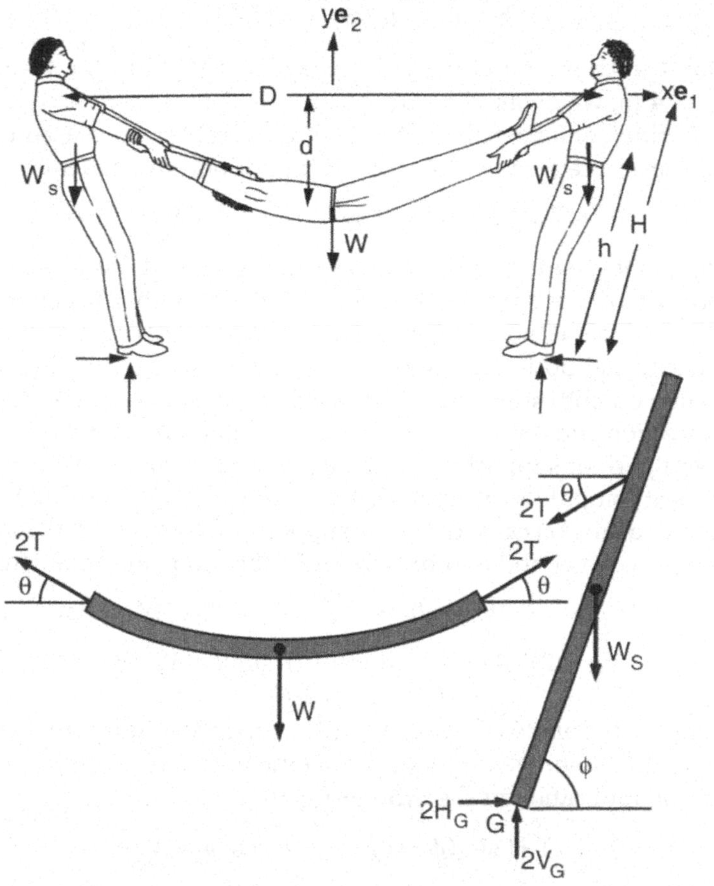

FIGURE 5.8. Two students holding a third at the feet and hands. The free-body diagrams show the forces acting on the supporting students as well as the student who is being lifted. The three students create a structural system that is reminiscent of a cable connecting two poles.

The supporting students lean backward to be able to keep the youth off the ground. Determine the angle of inclination of the supporting students and the force exerted on them by the ground.

Solution: First, we determine the force exerted on the arms of the supporting students. Then we evaluate the ground force. The curve created by the arms of the supporting students and the body of the hanging student is assumed to be given by the parabolic expression:

$$y = d\,[(2x/D)^2 - 1] \tag{5.9a}$$

where x and y denote horizontal and vertical distances along the $\mathbf{e}_1$ and $\mathbf{e}_2$ directions. The angle θ that the human cable makes with the $\mathbf{e}_1$ axis at the shoulders of the supporting students is found by taking the derivative of y with respect to x at $x = D/2$:

$$\tan\theta = (dy/dx) = 4\,(d/D) \tag{5.9b}$$

Thus, for $d = 0.5$ m and $D = 2$ m, we have $\theta = 45°$, and for $d = 0.25$ m and $D = 2.4$ m, we obtain $\theta = 23°$.

The condition of force balance for the human cable can be used to compute the tension carried by the arms of the supporting students:

$$4\mathrm{T}\sin\theta - \mathrm{W} = 0 \Rightarrow \mathrm{T} = \mathrm{W}/(4\sin\theta) \tag{5.10}$$

According to this equation, the tension T increases with decreasing θ and therefore with decreasing sag. When $\theta = 0$, the tension T becomes infinity, indicating that it would be impossible to hold the hanging student in a fully horizontal position. Notice also that the tension T or thrust must be the same on both supporters because there is equilibrium in the horizontal direction and there are only two horizontal forces on the cable: the pull from the right supporter and the equal pull from the left supporter.

Let us next compute horizontal ($\mathrm{H_G}$) and vertical ($\mathrm{V_G}$) components of the ground force acting on each foot of the supporting students. The conditions of force balance in $\mathbf{e}_1$ and $\mathbf{e}_2$ directions yield the following equations:

$$-2\mathrm{T}\cos\theta + 2\mathrm{H_G} = 0 \Rightarrow \mathrm{H_G} = \mathrm{T}\cos\theta = (\mathrm{W}/4)\cot\theta \tag{5.11a}$$

$$\begin{aligned}-\mathrm{W_s} + 2\mathrm{V_G} - 2\mathrm{T}\sin\theta = 0 \Rightarrow \mathrm{V_G} &= \mathrm{T}\sin\theta + \mathrm{W_s}/2\\ &= \mathrm{W}/4 + \mathrm{W_s}/2\end{aligned} \tag{5.11b}$$

To compute the angle ϕ between the body axis of a supporting student and the $\mathbf{e}_1$ direction, we consider the moment of forces with respect to the point of application of the ground forces:

$$2\mathrm{T}\,H\sin(\phi - \theta) - \mathrm{W_s}\,h\cos\phi = 0 \tag{5.12}$$

where H is the length between the shoulder and the sole of the foot and h is the length between the center of gravity of the supporting student and the sole of his feet.

An interesting note about cables, chains, and strings: like living creatures and unlike other solids, they drastically change shape in response

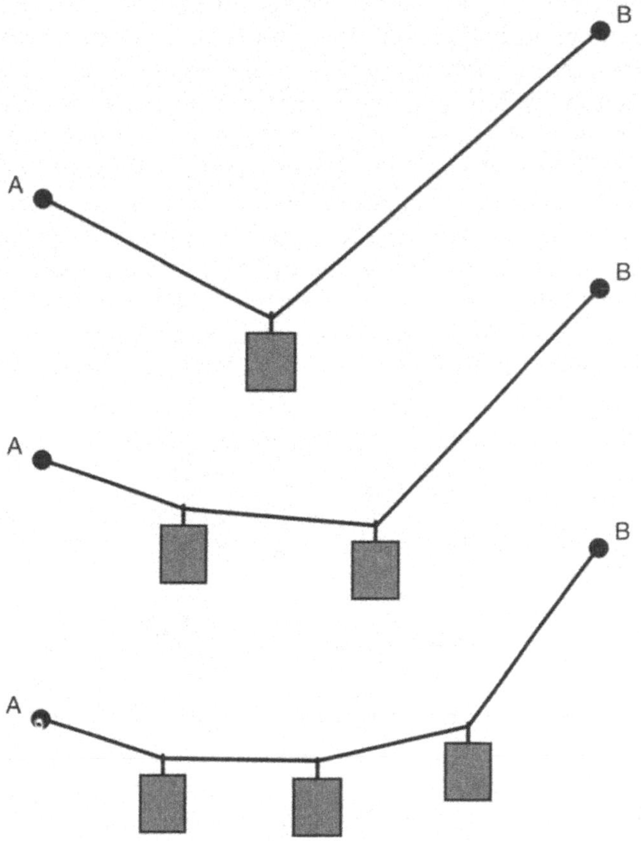

FIGURE 5.9. The shape of a cable changes in response to weights attached to it.

to an external force. Consider, for example, a weightless string that is stretched between two poles. With each addition of weight, the string changes shape in such a way that the portions of the string remains straight between the weights, and between the weights and the supporters (Fig. 5.9). When human-made solid structures undergo significant changes in shape to carry a varying load, they are considered unstable.

5.4 Structural Stability and Redundance

A body is said to be in *stable equilibrium* if a small displacement from the equilibrium position does not result in larger displacements. In other words, when a body is slightly displaced from a stable equilibrium position, the forces brought into play move the body back toward the equilibrium position. On the other hand, if these perturbation forces move the body further from the equilibrium position, the equilibrium is said to be *unstable*. If the perturbation from the equilibrium position brings about no new forces, the equilibrium is *neutral*. These three cases of equilibrium are illustrated in Fig. 5.10a. As shown in the figure, a rod standing on its end with its center of gravity vertically above the point of support is in unstable equilibrium. A small disturbance force will result in an unbalanced moment, taking the rod further away from the equilibrium configuration. The circular cylinder shown in the figure is in *neutral* equilibrium because disturbance from equilibrium position in the horizontal plane does not bring out new forces. If on the other hand the elliptical cylinder is tilted a little, the new contact force that arises will rotate it back to its equilibrium configuration. The elliptical cylinder is therefore in stable equilibrium.

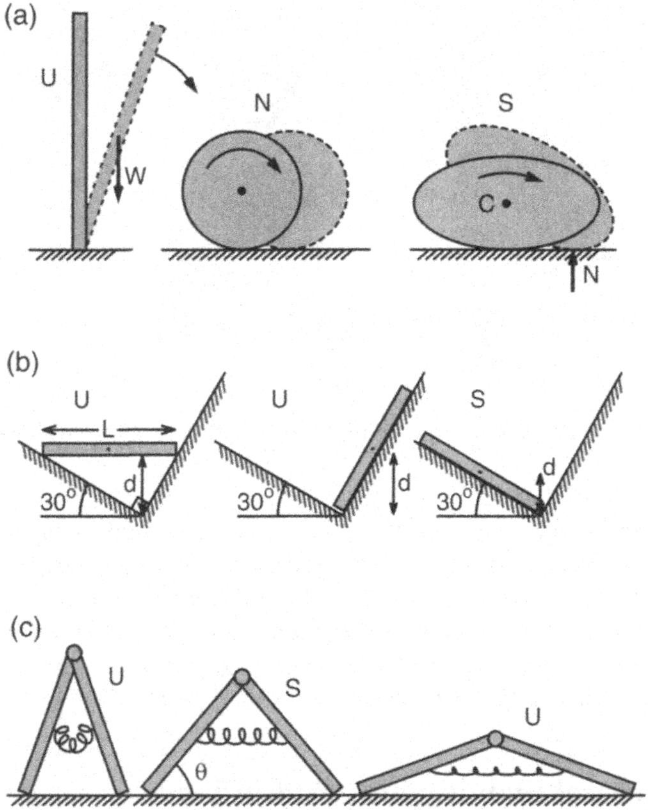

FIGURE 5.10a–c. Classification of static equilibrium of objects. The *top row* **(a)** presents examples of *unstable* (*U*), *neutral* (*N*), and *stable* (*S*) *equilibrium* of rigid objects. The *middle row* **(b)** classifies the various configurations of a rod under static equilibrium. The *bottom row* **(c)** illustrates that for nonrigid structures static equilibrium does not neccessarily correspond to the configuration where the center of gravity is at a minimum height.

A rigid body is in stable equilibrium when its center of gravity is at a minimum height (Fig. 5.10b). All three configurations of the rod shown in the figure are in equilibrium. The height d of the center of mass of the rod for the configurations from left to right are given by the equations:

$$d = L \sin 30° \cos 30°$$

$$d = (L/2) \sin 60°$$

$$d = (L/2) \sin 30°$$

The configuration on the far right corresponds to the smallest value of d and is, therefore, in stable equilibrium. For structures other than a rigid body, stable equilibrium does not necessarily correspond to the configuration that brings the center of gravity to minimum height. Consider, for example, the structure composed of two flat plates and a spring (Fig. 5.10c). The plates are hinged at one edge and are also connected by a spring at midlength. The natural length of the spring is $(L/2)$ where L is the width of the plates. The free ends of the plates are free to slide on the flat surface shown in the figure. If the gravitational force and the reacting contact force were the only external loads acting on the structure, stable equilibrium would have corresponded to the configuration where the two plates lay flat on the horizontal plane. In that configuration, however, the spring would have been stretched drastically and therefore could snap with the smallest of the disturbance. How do we determine which configuration corresponds to static equilibrium in this case? The answer to the question is that the potential energy of the structure is minimum at stable equilibrium. The potential energy V for this structure is given by the relation:

$$V = mg\,(L/2) \sin \theta + mg\,(L/2) \sin \theta + (k/2)\,\delta^2 \qquad (5.13)$$

The first two terms correspond to the gravitational potential energy of the two plates. The last term is the energy stored into the spring as a result of its stretch, with δ denoting the extension of the length of the spring. It can be shown that

$$\delta = L\,(\cos \theta - 0.5) \qquad (5.14)$$

To compute the minimum potential energy, we take the derivative of V with respect to θ and set it equal to zero:

$$kL/(mg) = \cos \theta / [\sin \theta\,(\cos \theta - 0.5)] \qquad (5.15)$$

The term on the left is a dimensionless parameter. It is a measure of the strength of the spring force relative to the force of gravity. Solution of this algebraic equation corresponds to minimum potential energy when the second derivative of V with respect to θ ($d^2\,V/d^2\theta$) is positive. When $kL/(mg) = 9.5$, $\theta = 55°$ corresponds to the minimum potential energy and therefore to stable equilibrium. Note that for very stiff springs the angle θ that corresponds to stable equilibrium will be slightly less than 60°. As the spring stiffness is decreased toward zero, the structure will flatten at static equilibrium, with θ reducing toward zero. Although the structure discussed here does not look anything like the human body or any part of the body, there are resemblances. Muscle–tendon complexes of the human body store energy like the spring of the two-rod structure. When a calf muscle goes into contraction, the stable equilibrium of the leg will be much different than when the muscle is relaxed and therefore has much less stiffness. The reader might have experienced a muscle spasm and how it can distort the resting configuration of a leg.

Example 5.5. Stability of the Human Shoulder. In the human shoulder, the glenoid fossa region of the scapula supports the humerus of the upper arm much like the nose of a seal balancing a ball (Fig. 5.11). The equilibrium of a uniform rod in vertical configuration is an unstable one. Because the humerus is not uniform, it is much more difficult to keep it balanced. How does a seal balance a ball on its nose? What are the implications for the shoulder joint?

Solution: Consider a uniform rod of length L and mass m that is in unstable equilibrium (Fig. 5.11b). Let us apply a small perturbation to the bar in the form of a horizontal force δf. Because the rod will tend to move in the direction of the unbalanced force, the rough substrate on which the rod is resting will exert a frictional force in the direction opposite to δf. Both the perturbation force δf and the frictional force f will produce counterclockwise moment with respect to the center of the rod. The rod will gain angular acceleration of the magnitude given by the equation:

$$(\delta \mathrm{f} + \mathrm{f})\ (L/2) = (mL^2/12)\ \alpha \Rightarrow \alpha = 6\ (\delta \mathrm{f} + \mathrm{f})/(mL)$$

Thus, the rod would begin to rotate in the counterclockwise direction. If, however, the surface on which the rod rests was given a horizontal acceleration a in the direction of δf, the rotation of the rod can be prevented. First, the rough plane moving in the direction of δf will pull the rod with a frictional force in the same direction. Therefore, an imposed acceleration on the surface could alter the direction of the frictional force. Second, if the acceleration is chosen such that

$$a = 2\delta \mathrm{f}/m$$

the resultant couple with respect to the mass center will be equal to zero, and the rod will translate in the direction of the force of perturbation δf. The nose of the seal is certainly capable of imposing lateral movement on a ball it is balancing. Similarly, the scapula is a highly mobile shoulder bone and therefore the glenoid fossa can be laterally displaced through coordinated muscle action. This example illustrates how skeletal muscles can transform an unstable equilibrium into a stable equilibrium.

The role of supporting structures in joint stability can be studied further by considering the two-link system shown in Fig. 5.12a. The two links articulate at a ball-and-socket joint. An ecccentric load will cause rotation

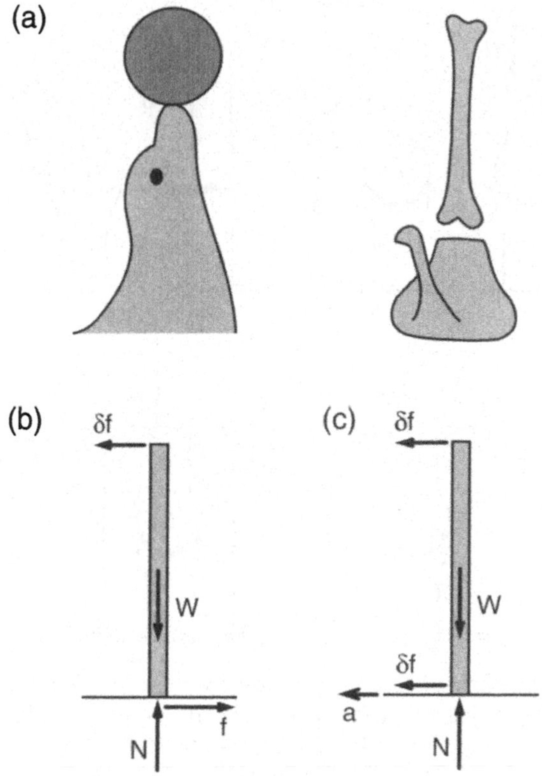

FIGURE 5.11a–c. The stable equilibrium of the humerus on top of the scapula. The arm, positioned vertically above the head, appears to defy the laws of gravity much like a ball standing on the nose of a seal **(a)**. When a small lateral force is applied to a rod standing on one of its ends on a rough horizontal plane, the rod will begin to rotate and ultimately to lie flat on the plane **(b)**. If, on the other hand, a translational acceleration is imposed on the supporting plane in the direction of the perturbation force, the resultant moment acting on the rod would be equal to zero. The rod would have a small displacement but not a rotation in the direction of the applied perturbation force **(c)**.

of one member relative to the other. If the socket was not deep enough, as in the shoulder joint (or the articulating surface had varying curvature, as in the knee joint), an eccentric load could result in the disruption of the joint. The schematic diagrams shown in Fig. 5.12b indicate how muscle–tendon systems could prevent large displacements at the joint by adjusting the tendon tension accordingly. The glenohumeral joint articulating the humerus of the upper arm with the scapula is rather shallow, and therefore an eccentric loading could lead to instability. The joint is made stable by its capsular tissue, ligaments, and musculature, which provide the scapula the necessary movements to balance forces of perturbation.

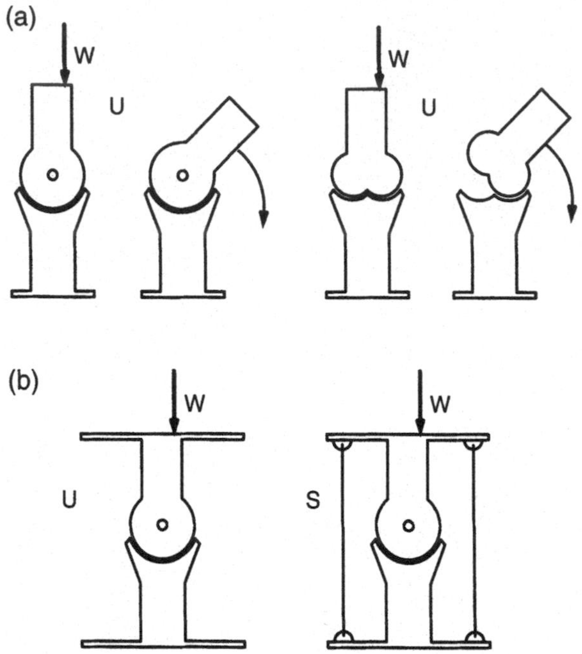

FIGURE 5.12a,b. The effect of eccentric loading on the relative motion of two rigid members articulating at a ball-and-socket joint **(a)**. The joint can be stabilized and the relative motion coordinated by adding two tension-carrying cords (muscle–tendon complexes) on each side of the joint **(b)**.

Appropriately positioned contact forces are needed to ensure the static balance of an object under the action of gravity. In some cases, the number of unknown contact forces is greater than the number of equations of static equilibrium, and such systems are *statically indeterminate*. An example of a statically indeterminate system is shown in Fig. 5.13a, where a ballerina is depicted as leaning on her partner. The couple forms contact with the dance floor through all four feet. Because the dance floor supports frictional forces, four vertical and four horizontal contact forces need to be evaluated. Because there are only three independent equations of equilibrium for a planar case, it is not possible to determine uniquely the numerical values of the unknowns using only the conditions of equilibrium. Static human postures such as sitting on a chair with feet grounded and back leaning against the chair constitute examples of statically indeterminate situations.

Next let us consider Fig 5.13b, where a male dancer is carrying the entire weight of the woman dancer. We could evaluate the location of the center of mass of the couple from their measurements and arbitrarily assume that the ground force acts at the center of the sole of the feet. Using the equations of static equilibrium, we can then compute the vertical reaction forces acting at the male dancer's feet uniquely. But what about

FIGURE 5.13a–e. Snapshots of static equilibrium positions in classical ballet. The couple shown in **(a)** has contact with the floor at four locations. Their bodies together compose a statically indeterminate structure. The vertical ground forces acting on the couple shown in **(b)** can be computed using the equations of static equilibrium. The support forces acting on the ballerina in **(c)** can also be determined with the equations of static equilibrium. The ballerina shown in **(d)** balances her weight by aligning it with the vertical ground force acting on the toes of her foot. The number of static equations in this case exceeds the number of unknown contact forces. The ballerina is unstable in the sense that a perturbation from her equilibrium position will require finite movement at her ankle and her arms to restore the resting configuration. The dancer **(e)** is free-falling under the action of gravity. His pose and facial expressions, however, are intended to convince us that he is able to defy gravity by hanging in air.

the horizontal reaction forces? Static equilibrium requires that the sum of the horizontal forces must be equal to zero, and so we have two unknown horizontal force components and one equation. The system is statically indeterminate in the horizontal direction. The couple is in a statically stable position because a small alteration in the posture will not lead to larger alterations; all it will do is to change the magnitudes of the reaction forces acting on each foot of the male dancer. If a horizontal force is exerted on this couple, frictional forces at the male dancer's feet could balance the applied force and the couple would remain in static equilibrium.

The ballerina in Fig. 5.13c is supported by vertical and horizontal ground forces and also by her dance partner. If we assume that the force exerted by the partner on the ballerina acts along the direction of the ballerina's arm, then we have a statically determinate situation. There are three unknowns and three equations of equilibrium. Furthermore, the ballerina is in a statically stable pose. The force exerted by her partner could be either tension or compression and thus would prevent her rotating clockwise or counterclockwise.

The next case is the figure (Fig. 5.13d) is a ballerina *en pointe* where she stands on top of the toes on one leg, with weight being taken primarily through the first and second toes. This is just one of the several poses in classical ballet in which the ballerina strikes a delicate balance. In another posture, called an *attitude*, the body is supported on one leg with the other lifted to the front, side, or back with the knee flexed. In an *arabesque*, the body is supported on one leg while the other is fully extended behind the dancer. The contact forces acting on a ballerina in such poses are the vertical and horizontal ground forces. There are three equations of equilibrium to be satisfied with two unknown forces. Unless the ballerina can keep her center of mass right on top of the ground force acting on it, the resultant external moment acting on her would not be equal to zero. Keeping in equilibrium in these positions means the accurate positioning of the center of gravity of the body right on the vertical line crossing the point of application of the contact force on the ground. In response to a perturbation, she could realign her center of mass by moving her arms slightly or by bringing her heels down to the floor momentarily. Thus, a small perturbation will lead to movement and even to artistic catastrophe and mean reviews, but she is not helpless in preventing a fall. Even an untrained person can stand on the toes of one foot by rocking on the ankle to keep balance or by slightly bending the knee. The human body is a robust structure that provides various pathways to accomplish a given physical task.

The last picture, in Fig. 5.13e, is that of an airborne ballet dancer. There are no contact forces acting on him. He is free-falling under the action of gravity. His pose and facial expressions, however, create the illusion that he can actually hang in air; this show of apparent defiance of gravity is certainly part of the art of ballet.

The concept of stability of static equilibrium has been the subject of extensive investigation in structural engineering. The requirement of stability in structures is concerned with the danger of having unacceptable motions. When a tree is acted upon by a hurricane wind, and is not properly rooted in the ground or balanced by its own weight, it may topple over without disintegrating. The tree is then said to be unstable in rotation. A regular pendulum is in stable equilibrium when the rod connecting the bob to the point of anchor is vertical. Small perturbations of displacements from that configuration will not lead to larger displacements of the pendulum. After a series of oscillations, the pendulum will again reach equilibrium.

5.5 Structures and Internal Forces

In this section we present several examples on the static equilibrium of human-made structures. We discuss a method of computing the internal forces in a structure that arise in response to external loading. Even though internal forces do not contribute to the course of motion of an object, they could lead to material failure, fractures, tearing, or extensive distortion. As such, they are instrumental in the proper functioning of human-made structures. The first example of this section concerns a truss, a structure extensively used in bridges, cranes, roofs, and so on. The function of a truss is to transmit loads applied to it to its points of anchors to the earth.

Example 5.6. Mechanical Analysis of a Truss. Figure 5.14 shows a typical truss that is subject to external forces in the horizontal and vertical direction. Our first aim is to determine the reaction forces at A and E. At point A, the truss is anchored to the ground. The reaction forces acting on the hinge joint A are denoted as A_1 and A_2, as shown in the figure. The point E may undergo small lateral displacements but is fixed in the vertical direction. Such a joint is called a roller joint. The vertical support force acting on it is denoted as E_2. There are three unknown forces in this case and three independent equations of equilibrium. We will use those equations to determine the reaction forces acting on the truss.

The condition of force balance in the horizontal direction leads to

$$5{,}000 \text{ N} + A_1 = 0 \Rightarrow A_1 = -5{,}000 \text{ N} \tag{5.16a}$$

Let us next take the moment of all external forces with respect to point A and set it equal to zero:

$$-5{,}000 \cdot 0.8 - 20{,}000 \cdot 1.8 + E_2 \cdot 2.4 = 0 \Rightarrow E_2 = 16{,}667 \text{ N} \tag{5.16b}$$

The sum of all forces in the vertical direction must also be equal to zero

$$\begin{aligned} A_2 + E_2 - 20{,}000 &= 0 \\ A_2 + 16{,}667 - 20{,}000 &= 0 \\ A_2 &= 3{,}333 \text{ N} \end{aligned} \tag{5.16c}$$

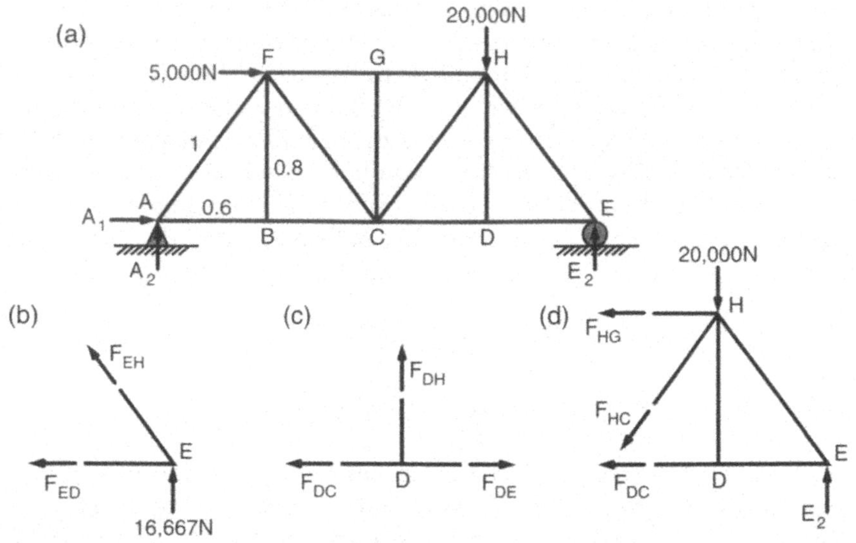

FIGURE 5.14a–d. A truss loaded with a vertical and horizontal force **(a)**. Also shown in the figure are the free-body diagrams of joints E **(b)** and D **(c)** and the free-body diagram of the section of the truss consisting of members that connect at joints D, E, and H **(d)**. All horizontal elements of the truss have equal lengths.

Thus, we were able to compute the reaction forces uniquely. The choice of the particular equations of static equilbrium was somewhat arbitrary. For example, instead of using Eqn. 5.16c, the condition of force balance in the vertical direction, we could have determined the resultant external moment with respect to point E and set it equal to zero:

$$-5{,}000 \cdot 0.8 + 20{,}000 \cdot 0.6 - A_2 \cdot 2.4 = 0 \Rightarrow A_2 = 3{,}333\ \text{N}$$

Thus, one could determine the resultant moment with respect to two distinct points and set them equal to zero. Together with the condition of force balance in the horizontal direction, this set of equations is sufficient to ensure planar static equilibrium.

How much force does each element of the truss bear in transmitting the external loads to the ground at points A and E? This we can also evaluate using the equations of static equilibrium. In a truss, elements are pinned to each other. Thus, if we neglect their own weight in comparison with the external forces acting on the truss, a truss element is essentially a bar loaded only at its ends. These forces must be equal in magnitude, opposite in direction. Further, they must have the same line of action. This means that truss elements carry only compression or tension.

We can compute the values of these axial loads either by considering the static equilibrium of each joint of the truss (method of joints) or by using the method of sections, where one isolates a portion of the truss and considers its static equilibrium. We will illustrate these methods here. Let us consider, for example the static equilibrium of joint E. Joint E is virtually a point (negligibly small in size in comparison with the truss structure) rather than an object with moment of inertia, and thus the static equilibrium corresponds to the condition of force balance. A free-body diagram of this joint is presented in Fig. 5.14b. We have shown the directions of unknown forces as if they were in tension. In tension, the truss member pulls the joint in the direction of the member. If the member is in compression, it pushes against the joint. Members that are in tension and those in compression can be identified with the symbols [T], and [C], respectively. If it turns out that our initial assignment of tension to both members is not correct for a particular member, the value of the member force will appear in negative.

Static equilibrium of joint E yields the following equations:

$$F_{EH}\,(0.8/1.0) + 16{,}667\ \mathrm{N} = 0 \Rightarrow F_{EH} = -20{,}833\ \mathrm{N}$$
$$\Rightarrow F_{EH} = 20{,}833\ \mathrm{N}\ [\mathrm{C}] \qquad (5.17a)$$

$$-F_{ED} - F_{EH}\cdot(0.6/1.0) = 0 \Rightarrow F_{ED} = 12{,}500\ \mathrm{N}$$
$$\Rightarrow F_{ED} = 12{,}500\ \mathrm{N}\ [\mathrm{T}] \qquad (5.17b)$$

Let us next consider the equilibrium of joint D:

$$F_{DH} = 0$$

$$-F_{DC} + F_{DE} = 0 \Rightarrow F_{DC} = 12{,}500\ \mathrm{N}\ [\mathrm{T}]$$

Note that if there are three truss elements coming together in a joint of a truss and two of the three share the same line of action, then the force on the third must be equal to zero.

Let us next compute the internal force in the bar member CH using the method of sections. We introduce an imaginary cut on the truss and consider the portion shown in Fig. 5.14d. Let us consider the equation of equilibrium in the vertical direction:

$$-20{,}000\ \mathrm{N} + 16{,}667\ \mathrm{N} - F_{CH}\cdot(0.8/1.0) = 0 \Rightarrow F_{CH} = -4{,}166\ \mathrm{N}$$

$$F_{CH} = 4{,}166\ \mathrm{N}\ [\mathrm{C}]$$

Thus, it is possible to evaluate the internal forces (forces carried by the truss elements) by considering the conditions of static equilibrium of truss joints and or of various sections of the truss. This is of course true if each joint (section) is statically determinate. For example, if we had connected points B and G or point D and G with a truss element, parts of the truss would be overbuilt or rigidified and would be statically indeterminate. The question that comes to mind is how do we determine the internal forces in a statically indeterminate system? This question has a rather complex answer. The following example, although admittedly simple, captures the essence of a statically indeterminate system.

Example 5.7. Axial Loading of a Rod That Is Fixed at Both Ends. Consider a rod whose ends are embedded into large concrete structures (Fig. 5.15). The length of the rod is L. A horizontal force P is exerted on the rod at a distance of $L/3$ from the fixed support B. The force P is much greater than the weight of the rod. Determine the reaction forces at the fixed ends B and D. Ignore the effect of the rod's weight on the support forces.

Solution: The condition of force balance in the horizontal direction requires that

$$H_B - H_D + P = 0 \tag{5.18}$$

This is the only equation of static equilibrium that relates these two forces. We have one equation and two unknowns. Clearly we need another equation to determine the two reaction forces uniquely. This additional equation concerns with the deformation of the beam. Because the beam is embedded in large and heavy structures at both ends, the length of the beam should not change even after the imposition of load P. That means part of the beam must be in compression and shorten a small amount and the other part be in tension and extend in length. Furthermore, the extent of compression must be equal to the extent of stretching.

It has been experimentally observed that, in ordinary construction materials, the length change in response to an axial load is proportional to the length of the structural element and the axial force exerted on it (Fig. 5.15b):

$$\Delta l = l \cdot F/(AE) \tag{5.19a}$$

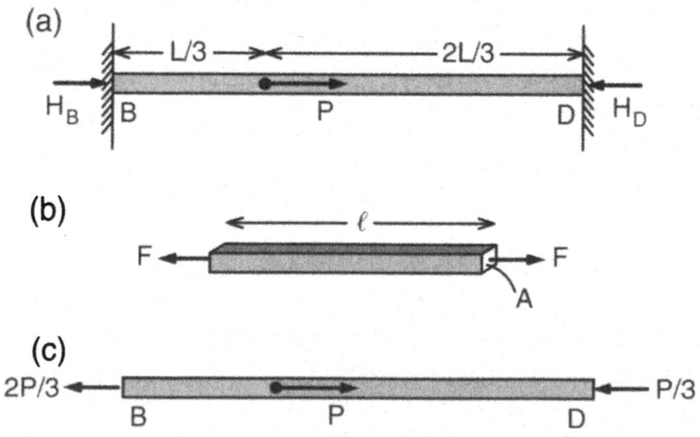

FIGURE 5.15a–c. A beam under the influence of external load P **(a)**. The segment BP is under tension **(b)** and PD under compression. The horizontal reaction forces can be computed by using the requirement that the length of the beam remains constant despite the load applied on it **(c)**.

in which Δl denotes the change in length, and l and A denote the original length and cross-sectional area of the axial member, respectively. The symbol F denotes the axial force exerted on the structural element; F is positive if it is tensile and F is negative if it is compressive. The parameter E is called the Young's modulus and has the units of force per unit area. It is a material property. A material is said to be stiff or compliant, depending on the value of E. When E is large, ΔL is small and therefore the material is stiff. Equation 5.19a is not a law of nature, but it is an experimental observation that appears to be valid for most materials undergoing small stretches or compressions. Typically, when a material is loaded gradually, the change in length is proportional to the load initially and the material will instantaneously go back to its unloaded configuration when the load is released. This is the elastic phase of loading when most materials act like a linear spring. Further increases in loading could complicate the relationship between the force exerted and the resulting change in length.

Equation 5.19 is called the basic law of elasticity or Hooke's law after the British scientist Robert Hooke (1635–1703). Hooke was an exceptional scientist and designer, with major contributions to the wave theory of light, theory of elasticity, thermodynamics, crystal structures, and gravity and the motion of planets. He was also involved in a bitter controversy with Newton, complaining that he was not given sufficient credit for his contributions to the formulation of nature's laws of motion. He wrote Eqn. 5.19b in a slightly different form:

$$P/A = E\ (\Delta L/L) \tag{5.19b}$$

The term on the left is called the force intensity, force per unit area, or axial stress. It is usually denoted by the symbol σ. The term $\Delta L/L$ is the dimensionless change in length or axial strain ϵ. Thus,

$$\sigma = E\ \epsilon \tag{5.19c}$$

Ligaments and tendons of the human body obey Hooke's law only at very low levels of strain. At higher levels of stress, the following nonlinear equation may be a better fit to describe the correlation between axial stress and strain:

$$\sigma = E\ (\exp\ (\epsilon) - 1) \tag{5.19d}$$

This equation shows that for a material to behave elastically (in a spring-like fashion) it is not necessary to have a linear relationship between stress and strain. All that is required is that for a given stress there is a unique strain. When stress goes to zero, strain also goes to zero. Tendons and ligaments are nearly elastic in the physiological loading range because the stress–strain curve obtained during the loading phase is approximately the same as the corresponding curve during the relaxation phase. Stress in a bone is typically three dimensional and direction dependent; that is, E would depend on the direction in which the force is applied. The skeletal muscle, when relaxed, can elastically be deformed. When activated, on the other hand, although it still stores elastic energy it will contract while in tension. The muscle property of being able to shorten under tension is a feature no nonliving material can replicate.

Returning to the problem at hand, we can now write the zero-elongation equation for the beam loaded with a point force P:

$$-(L/3) \cdot H_B/(AE) - (2L/3) \cdot H_D/(AE) = 0 \quad (5.20)$$

Solving Eqns. 5.18 and 5.20 for H_B and H_D, we find:

$$H_B = -2P/3; H_D = P/3$$

Because H_B turned out to be negative, it means that our initial supposition about its direction was wrong. The correct directions of the these support forces are shown in Fig. 5.15c. Note that the portion of the beam from B to P is in tension and the rest is in compression.

Example 5.8. A Cantilever Beam Carrying Its Own Weight. Let us consider a cantilever beam (Fig. 5.16a) that is embedded at one end in a body much larger in mass than the mass of the beam itself. The weight of the beam is equal to W and the span of the beam is L. Determine the support forces at point B.

Solution: The free-body diagram of the beam under consideration is shown in Fig. 5.16b. The equations of static equilibrium require that the following conditions be met:

$$H_B = 0$$

$$V_B - W = 0 \Rightarrow V_B = W$$

$$M_B - W\,(L/2) = 0 \Rightarrow M_B = W\,(L/2)$$

Thus, the surroundings at B exert not only a vertical force of W to the beam but also a counterclockwise external moment equal to W $(L/2)$. When the material of the beam is homogeneous, this moment is the result of the linear distribution of axial forces in the cross section (Fig. 5.16d). The top layers of the beam are under tension because they stretch as the beam deforms under its own weight. The bottom layers, on the other hand, contract and therefore are under compression. The axial force intensity (stress) σ at a vertical distance y from the central line of the beam is then given by the relation:

$$\sigma = (\sigma_o)\,[y/(h/2)] \quad (5.21)$$

in which σ_o is the maximum axial stress that occurs at $y = (h/2)$. This equation holds for small deformations of the beam where planar cross sections normal to the axis of the beam remain planar after deformation.

Similarly, one can show that the moment created by σ is given by the following relation:

$$dM = \sigma\,(b \cdot dy)\,y = (\sigma_o)\,[y^2/(h/2)]\,(b \cdot dy) \quad (5.22)$$

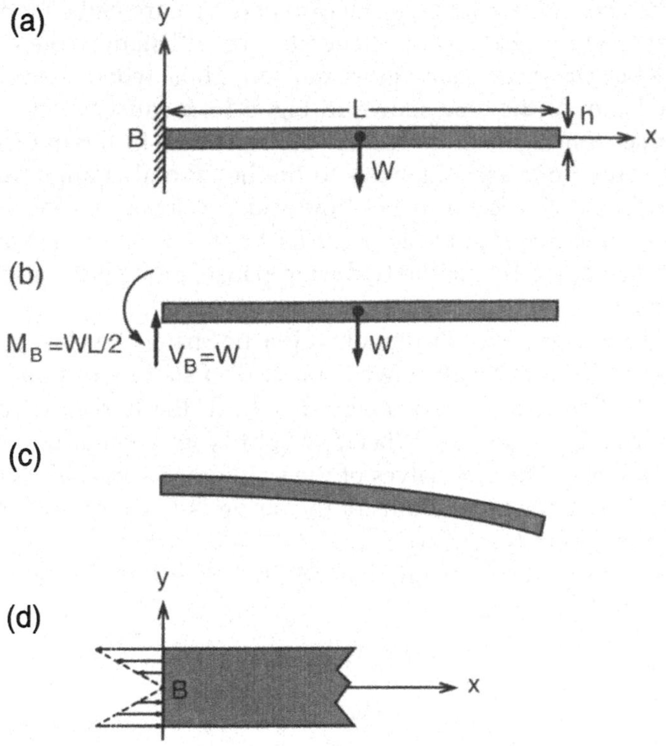

FIGURE 5.16a–d. A cantilever beam of length L carrying its own weight W **(a)**. The free-body diagram of the beam indicates all the forces acting on it **(b)**. The beam undergoes bending deformation under its own weight **(c)**. The stress distribution across the cross section is shown in **(d)**.

in which b denotes the width of the cross section, and dM denotes the moment created by axial forces applied at the cross-sectional area $(b \cdot dy)$. Integrating this expression over $y = -(h/2)$ to $(h/2)$ and equating it to M_B, we find:

$$\sigma_o = 6M_B/(bh^2) \tag{5.23a}$$

$$= 6W\,(L/2)/(bh^2) \tag{5.23b}$$

Note that the maximal axial stress is tremendous in comparison with the weight W divided by the cross-sectional area (bh). We could express the weight of the beam in terms of the mass density ρ times its volume and come up with yet another expression for σ_o:

$$\sigma_o = 3\rho\,(L^2/h) \tag{5.23c}$$

Thus, maximum axial stress increases with the square of the length of the beam. It is also inversely proportional to the thickness of the beam.

In reinforced concrete structures, the iron bars that are embedded in concrete resist much of the tension. Concrete can withstand large compressive forces but fractures easily under tension. Thus, when one constructs a concrete beam of the type shown in Fig. 5.16, it must be reinforced at the upper section of the cross section by iron rods. In the human body, when we abduct our arm sideways to the horizontal configuration and keep it there, the tension in the deltoid muscle times its lever arm balances the moment created at the shoulder by the weight of the arm. The internal forces of the human body during pause, movement, and motion are the subject of the next chapter.

Example 5.9. Static Equilibrium of an Elastic String. An elastic string of negligible weight and length L was attached to stationary points A and B (Fig. 5.17). The string was just about tight in the horizontal position, carrying virtually no tension. When a weight W was attached to the midpoint of the string, the two halves of the string made an angle θ with the horizontal axis. Young's modulus of the string is E. Determine the angle θ and the tension in the string.

Solution: The condition of force balance in the vertical direction requires that

$$-W + 2T \sin\theta = 0$$

$$T = W/(2 \sin\theta) \tag{5.24a}$$

Both the tension T and the angle θ are unknowns. Equations of static equilibrium are not sufficient to determine these parameters, and thus this simple system is statically indeterminate. We consider the material properties of the string to come up with an additional equation. Hooke's law dictates that

$$T = E\,(\Delta/L)\,A$$

in which E is the Young's modulus of the string, A is its cross-sectional area, and Δ is the extension of the string AB. The term (Δ/L) is the axial

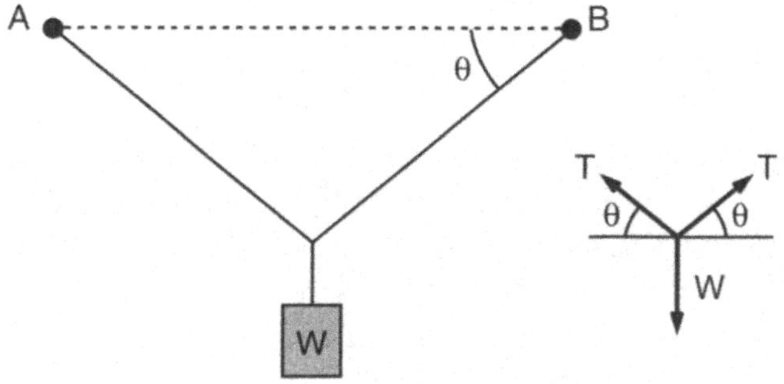

FIGURE 5.17. A cable of negligible weight supporting a weight W.

strain in the string caused by the placing of the weight W. Using trigonometry we can show that

$$(\Delta/L) = (L/\cos\theta - L)/L = (1 - \cos\theta)/\cos\theta$$

$$T = E\,A\,(1 - \cos\theta)/\cos\theta \tag{5.24b}$$

Eliminating T in Eqns. 5.24a and 5.24b, we derive the following equation for θ:

$$W/(2EA) = \tan\theta - \sin\theta \tag{5.25}$$

When the stiffness of the string is large, the angle θ must be small, and we can express $\tan\theta$ and $\sin\theta$ in powers of θ:

$$\tan\theta \cong \theta + \theta^3/3;\ \sin\theta \cong \theta - \theta^3/6$$

Hence

$$W = E\,A\,\theta^3 \text{ or } \theta = [W/(EA)]^{1/3}$$

The tension T can then be found by using Eqn. 5.24a.

Example 5.10. Rods, Weights, and Cords. The rod AB of length 4a is tied to stationary points through two inextensible strings (Fig. 5.18). A weight W is attached to the rod using five elastic strings. These strings all have the same cross-sectional area A and the same Young's modulus E. The weight of the rod and of the strings are negligible. Determine the tension in each of these strings.

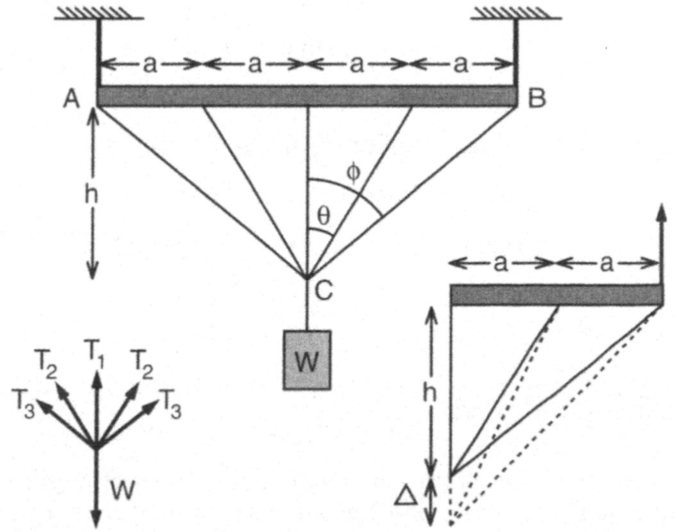

FIGURE 5.18. Rod AB of length 4a is attached to stationary points through two inextensible strings. The rod carries a weight W.

Solution: The free-body diagram of point C shown in the figure requires that

$$T_1 + 2T_2 \cos\theta + 2T_3 \cos\phi = W \tag{5.26}$$

in which the tensions T_1, T_2, and T_3 are the tensions in the elastic strings. The angles θ and ϕ are determined by the length parameters a and h. In this equation we have three unknowns. Their values cannot be determined uniquely using equations of statics alone. Thus, we focus on the deformation of elastic strings. Let Δ be the vertical displacement of point C in the direction of the gravitational force W. We assume that the strings are stiff so that Δ is small in comparison with the lengths of the strings. The extension of each of the elastic strings can then be defined as a function of Δ:

$$\Delta_1 = \Delta$$

$$\Delta_2 = [a^2 + (h+\Delta)^2]^{0.5} - [a^2 + h^2]^{0.5} \simeq \Delta \cos\theta$$

$$\Delta_3 = [4a^2 + (h+\Delta)^2]^{0.5} - [4a^2 + h^2]^{0.5} \simeq \Delta \cos\phi$$

Thus, the strains in the strings are

$$\epsilon_1 = \Delta_1/h = \Delta/h$$

$$\epsilon_2 = (\Delta/h)\cos^2\theta$$

$$\epsilon_3 = (\Delta/h)\cos^2\phi$$

We can express the tension in each string in terms of the strain:

$$T_1 = \mathrm{E}\,A\,(\Delta/h)$$

$$T_2 = \mathrm{E}\,A\,(\Delta/h)\cos^2\theta$$

$$T_3 = \mathrm{E}\,A\,(\Delta/h)\cos^2\phi$$

Substituting these values into Eqn. 5.26, we obtain:

$$T_1 = W/(1 + 2\cos^3\theta + 2\cos^3\phi)$$

$$T_2 = \cos^2\theta\, W/(1 + 2\cos^3\theta + 2\cos^3\phi)$$

$$T_3 = \cos^2\phi\, W/(1 + 2\cos^3\theta + 2\cos^3\phi)$$

5.6 Distributed Forces

A point force has a magnitude, direction, and a point of application. In actual life, most loads acting on objects are distributed over a surface or a volume. For example, snow load (kg/m^2) is a distributed surface load whereas gravity acts throughout the body of an object. The internal stress in a body may depend strongly on the specific distribution of external loads. For example, the soles of shoes are designed to more uniformly distribute ground forces on the base of the foot so as to avoid stress fracture or soft tissue damage.

Point forces lead to different internal stresses than distributed forces of the same magnitude and direction, at least in a small region surrounding the point of application of the force. In the study of motion and equilibrium of a stiff solid body, however, it is sufficient to replace distributed loads with a single force and or moment. Two-force systems are said to be statically equivalent if (a) the resultant force vectors are equal, and (b) the resultant moment of one force system with respect to a point in space is equal to that of the other. This equivalency relationship reflects the fact that when a force is moved up or down along its line of action, its effect on the motion of the body is not altered.

Example 5.11. Distributed Load Acting on a Beam. The force intensity acting on a beam shown in Fig. 5.19 is given by the following equation:

$$\mathbf{w} = \gamma b x^2 \mathbf{e}_2 \tag{5.27}$$

where $\mathbf{w}$ is the load per unit length and has the dimensions of N/m, γ is a constant, b is the width of the beam, and x is the distance along the $\mathbf{e}_1$ direction measured from point O. The loading is over the entire length L of the beam. Determine the equivalent force system.

Solution: The resultant force of the force system given by Eqn. 5.27 is

$$\mathbf{F} = \gamma b \int x^2 dx \; \mathbf{e}_2 = \gamma b \; (L^3/3) \; \mathbf{e}_2$$

The moment created by the two-force system must be equal:

$$X \gamma b \; (L^3/3) = \gamma b \int x^3 dx = \gamma b \; (L^4/4) \Rightarrow X = (3/4) \; L$$

in which X denotes the horizontal distance between point O and the point of application of the equivalent horizontal force. Note that X also represents the horizontal distance from point O to the centroid of the area occupied by the distributed load.

The equivalent force system to a distributed load need not be a point force, but it could be a force couple. A couple is composed of two forces that are equal in magnitude but opposite in direction. Because these forces

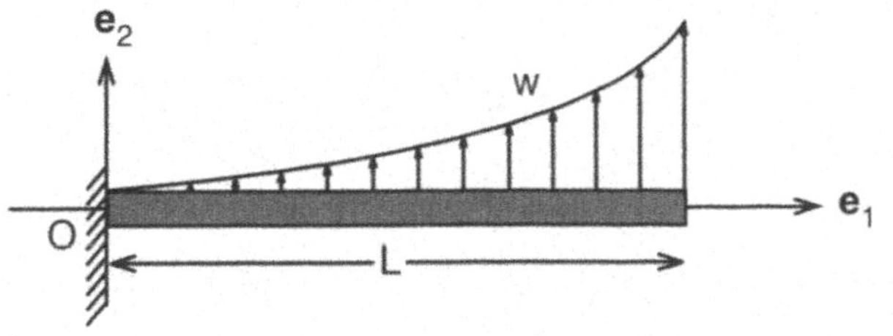

FIGURE 5.19. A distributed load w acting on a cantilever beam of length L.

do not share the same line of action, they exert a moment on the object but no resultant force. The axial stress distribution on the cross section of the cantilever beam of Example 5.8 could be represented by a force couple.

5.7 Summary

An object is in static equilibrium if (a) the resultant force acting on the object is equal to zero, and (b) the resultant moment acting on the object with respect to a point fixed in space is equal to zero. Thus, the conditions of equilibrium are

$$\Sigma \mathbf{F} = 0 \text{ and } \Sigma \mathbf{M} = 0$$

The moment **M** of a force **F** about point O is defined as

$$\mathbf{M} = \mathbf{r} \times \mathbf{F}$$

Although static analysis is strictly valid for objects in equilibrium, it could provide reasonable approximations for the forces involved in motions with small accelerations.

5.8 Problems

Problem 5.1. A woman with a tear in the anterior cruciate ligament of her left knee stands putting her weight on crutches and on her right foot as shown in Fig. P.5.1. She weighs 52 kg and has a height of 1.71 m. In this position her body and her crutches make angles of 63° and 80° with

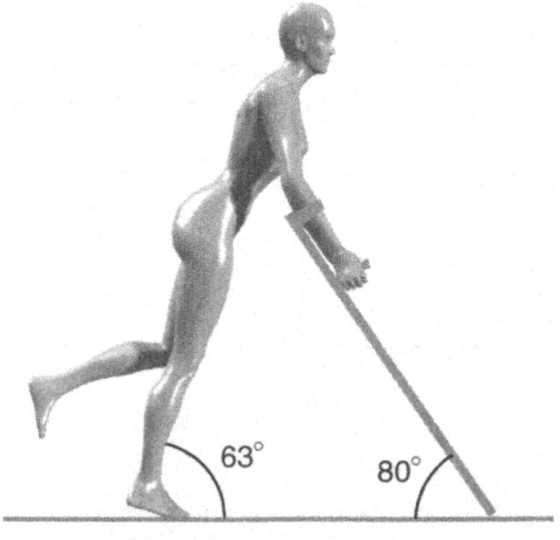

FIGURE P.5.1. A woman on crutches.

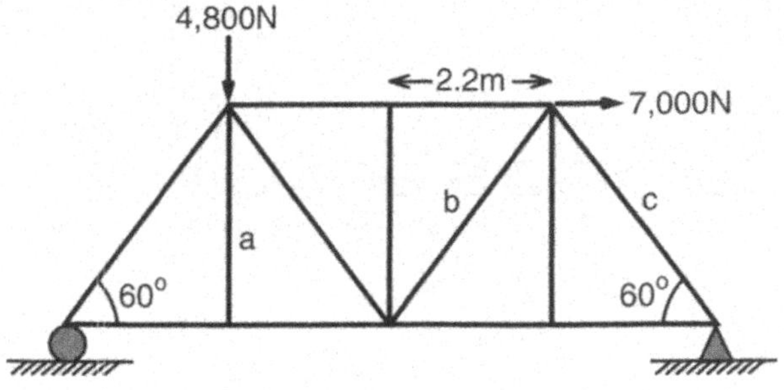

FIGURE P.5.2. A truss loaded with vertical and horizontal forces.

the horizontal plane, respectively. The distance between the point of application of the ground force on her right foot to her center of mass along the axis of her body is 0.89 m. The horizontal distance between the ground force acting on her right foot and the ground force acting on the crutches is 1.1 m. Determine the vertical ground force acting on the crutches. Assume that her body (and her right foot) is positioned symmetrically between the crutches.
Answer: 187 N.

Problem 5.2. Determine the forces carried by the truss elements a, b, and c of the truss shown in Fig. P.5.2. All horizontal elements of the truss have equal lengths.
Answer: a = 0, b = 4885 N (T), c = 4885 N (C).

Problem 5.3. Determine the equivalent point force and point couple systems to the distributed loads shown in Fig. P.5.3.
Answer: Let $\mathbf{e}_2$ be the unit vector in the direction opposite to the gravitational acceleration and let $\mathbf{e}_3$ be the unit vector outward normal to the plane of the paper. Then

$$\text{(a) } \mathbf{F} = -3{,}000 \text{ N } \mathbf{e}_2,\ \mathbf{M} = -2{,}400 \text{ N-m } \mathbf{e}_3$$

$$\text{(b) } \mathbf{F} = -14{,}400 \text{ N } \mathbf{e}_2,\ \mathbf{M} = -8{,}640 \text{ N-m } \mathbf{e}_3$$

$$\text{(c) } \mathbf{F} = -4{,}800 \text{ N } \mathbf{e}_2,\ \mathbf{M} = -2{,}880 \text{ N-m } \mathbf{e}_3$$

Problem 5.4. The man shown in Fig. P.5.4 is performing seated machine rows against a resistance of 60 kg. Determine the force his hands exert on the machine at $\theta = -20°$. Note that in this exercise both arms move back and forth in unison, and therefore, the two positions shown in the figure correspond to different instants of time. The lever is closest to the man's chest at $\theta = -20°$.
Answer: F = 490.5 N.

Problem 5.5. A circular rod of variable radius is held fixed at points B and D (Fig. P.5.5). An axial force P of 7,000 N is exerted on the rod at

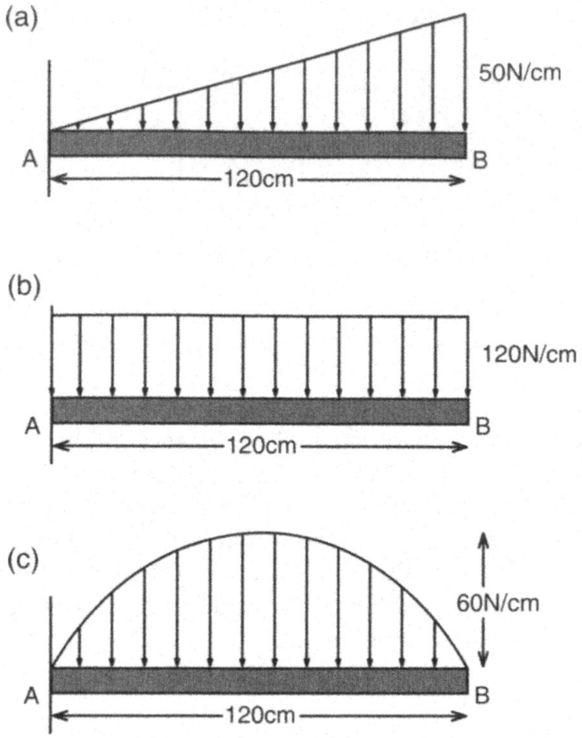

FIGURE P.5.3a–c. Distributed force systems acting on a beam.

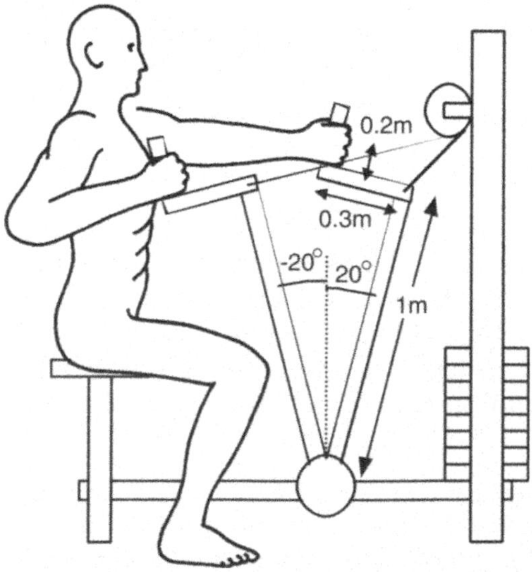

FIGURE P.5.4. A man performing seated machine rows.

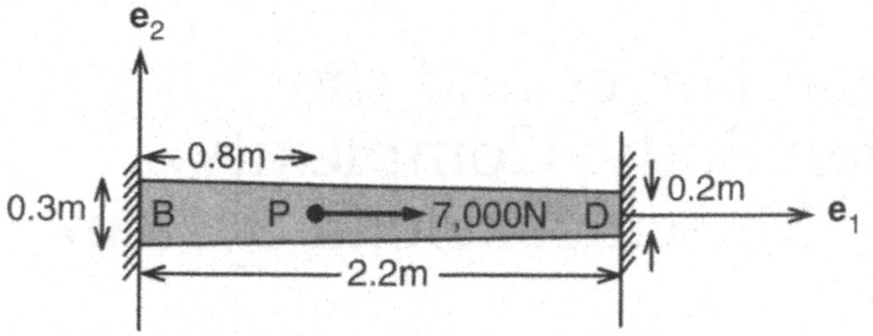

FIGURE P.5.5. A rod under axial load. The cross-sectional area of the circular rod varies linearly from one end to the other.

a distance of 0.8 m from point B. The length of the rod is 2.2 m. Its radius at B is 0.3 m and at D is 0.2 m. The rod is made of homogeneous material so that its Young's modulus E does not vary with position. Determine the horizontal support forces at B and D. Hint: Let $\mathbf{e}_1$ be the unit vector in the direction of P. Equation of motion in the $\mathbf{e}_1$ direction dictates that

$$7{,}000 \text{ N} - \text{B} - \text{D} = 0$$

where B and D are the horizontal support forces at B and D, respectively. They are positive when directed in $-\mathbf{e}_1$ direction. The section of the rod that lies between B and force P is in tension and the remaining section is in compression. As the distance between B and D does not change, elongation δ_1 of BP must be equal to the amount of compression δ_2 of PD. The length change δ of a rod under a uniaxial force is given by

$$\delta = \int \text{F}\, dx/(A\text{E})$$

where the integration is over the length of the rod, F is the axial force acting on the rod, dx is a small length element along the axis of the rod, $A = A\ (x)$ is the cross-sectional area, and E is Young's modulus. The parameter δ is positive when F is tension and negative when F is compression. Show that in this case $A\ (x) = \pi\ (0.3 - 0.045\ x)^2$. Furthermore, the constant length condition reduces to the following equation:

$$\text{B}\ \left[\int_0^{0.8}(0.3 - 0.045x)^{-2}\, dx\right] = (7{,}000 - \text{B})\ \left[\int_{0.8}^{2.2}(0.3 - 0.045x)^{-2}\, dx\right]$$

Answer: B = 5,075 N, D = 1,925 N.

Problem 5.6. Discuss the potential benefit of heel cushion cups in alleviating the heel pain that afflicts many runners.

6

Internal Forces and the Human Body: Complexity of the Musculoskeletal System

6.1 Introduction

External forces that act on a body induce physical stresses within the body. These stresses have wide-ranging consequences. The gravity causes compressive stresses in our musculoskeletal system, slowly shortening us as we get older. An accidental fall could induce such high-level stresses in a bone that it could fracture. Even without the application of an external force, an object can be under stress. Consider, for example, a thick rubber tube that is free of loading, except the action of gravity. Let us invert the rubber tube such that the inner cylindrical surface becomes outer and vice versa. At this configuration, the tube is stressed; the outer layer is under tension and the inner layer is under compression. Similar to the inverted cylindrical tube, ringlike specimens of animal aortas spring open after being cut in the radial direction, indicating that natural configuration of some biological tissues is not stress free.

Regardless of the state of motion, external forces acting on a body will cause internal stresses within the body. It is impossible to directly determine the system of forces carried by the muscles, the ligaments, and the joint articular surfaces in the human and animals. There is much redundancy in the human animal structure. The number of load-transmitting elements at a joint almost always exceeds the number of equations governing the motion of that joint. Consider for example the human knee shown in Fig. 6.1. This joint has a multitude of ligaments that hold it together and a large number of muscles acting on it. Even in the simplest weight-lifting exercises, multiple muscles will contribute to the lifting of the weight. How do we estimate the forces carried by various ligaments, muscles, and bones, given the external forces acting on the body?

The first step is to draw a free-body diagram of a body part in which all forces acting on it are clearly laid out. Then, one considers the equations of motion (or static equilibrium) to compute the unknown forces shown in the diagram. Typically, the unknown forces carried by a passive structures (capsules, ligaments, and the bones) intersecting at a joint

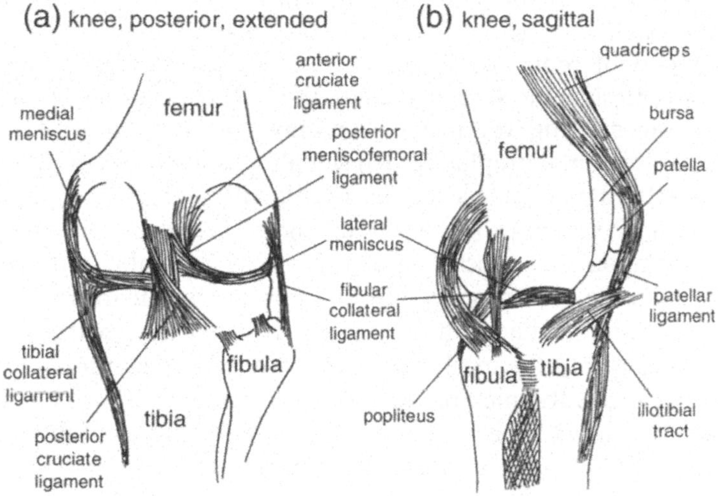

FIGURE 6.1a,b. The lateral view of the knee joint (**a**, posterior; **b**, sagittal) and some of the ligaments associated with it.

are lumped together and denoted as the total joint force. In the presence of large muscle moments acting on a joint, the moment created by passive structures on the rotational center of the joint may be negligible. This is however not true during impact loading where ligaments are maximally stressed to keep the joint intact.

In allocating forces to various muscle groups involved in posture, movement, and motion, two general approaches have been used—the reduction method and the optimization method. In the optimization technique, one assumes that muscles exert forces on bones near joints according to the minimization of some performance criterion. A large number of optimization criteria, including ad hoc principles of minimal muscle force, minimal muscle stress, and minimal energy expenditure or consumption, have been utilized. The method is attractive because sequences of movement in locomotion appear to suggest an intrinsic optimization process achieved through learning. This point can be illustrated with the consideration of vertical jumping where the position of the upper body is determined by the angles at the hip, knees, ankles, and metatarsal heads. The system has four degrees of freedom and thus our body allows the jumping task to be executed in a variety of ways. Yet, after practice, most movement tasks that lead to jumping seem to be performed in stereotyped manners. The sequence of muscular activation appears to be always in the order of upper body, upper legs, lower legs, feet, suggesting an optimization process. The biomechanics literature contains many examples of the application of optimization techniques in the study of the shoulder, hip, and knee. See review articles on the subject for further exploration of the merits and weaknesses of this approach.

The method of reduction is much simpler mathematically than the method of optimization. It is essentially the reduction of the number of unknown forces acting at a joint to a number of available equations of mechanics. Electromyographic (EMG) signals are often used as guides to choose muscles that actuate a certain movement. A common assumption in calculating the force produced by the agonist is to set the force produced by antagonistic muscles equal to zero. This results in a low estimate of the muscle force produced by the agonist. Despite the rather drastic simplifications one has to make in using the reduction method, it does provide insights into the mechanical function of various muscle groups. In many instances, the method actually captures the essence of the human body mechanics. The next section presents examples of the method of reduction in the study of muscle force during movement.

6.2 Muscle Force in Motion

The first step in the analysis of forces acting on a body segment is to draw a free-body diagram of the segment. To this end, the part of the body is considered as distinct from the entire body, and all forces acting on the part of the body are identified. Then the equations of motion are used to gather information about the unknown muscle forces acting on the body part. The following example on the kicking of a soccer ball illustrates this technique.

Example 6.1. Quadriceps Force Before Kicking. Digital analysis of movies capturing the kicking of a soccer ball during a penalty kick showed that the angular acceleration of the lower leg was maximal at the instant the foot struck the ball (Fig. 6.2). The lower leg was vertical at that instant. The angular velocity and angular acceleration just before the foot touched the ball were $\omega = 8$ rad/s and $\alpha = 400$ rad/s^2, respectively. The lower leg (including the foot) weighed 7 kg and its mass moment of inertia I^o about the center of rotation of the knee was determined to be 0.35 kg-m^2. The perpendicular distance d from the patellar tendon to the center of rotation of the knee joint was found to be 4 cm. The distance r from the center of mass of the lower leg to the center of rotation of the knee was 22 cm. Determine the force exerted by the quadriceps muscle before kicking.

Solution: The muscle that actuates the kicking motion is the quadriceps muscle. The hamstring muscle group acts as the antagonist and would prevent the rotation of the lower leg beyond that of extension. We neglect the force exerted by the antagonist. We assume that the center of gravity of the lower leg lies in the vertical line that passes through the center of rotation of the knee. Under these conditions, the only force that creates a moment with respect to the center of rotation is the force F_q that is ex-

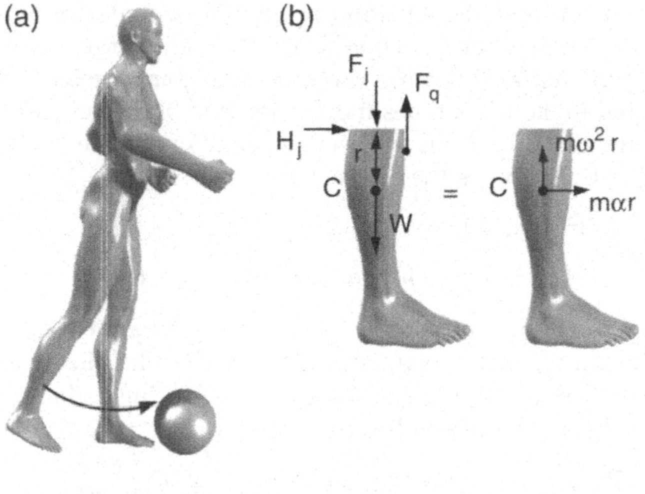

FIGURE 6.2a,b. A soccer player getting ready to kick a soccer ball **(a)**. The free-body diagram of the lower leg is shown in **(b)**.

erted by the quadriceps (see Fig. 6.2). Assuming that the knee stays stationary just before the takeoff, the conservation of angular momentum in the direction perpendicular to the plane of motion is

$$M = I^o \alpha \tag{6.1}$$

$$F_q \cdot 0.04\text{ m} = 0.35\text{ kg-m}^2 \cdot 400\text{ rad/s}^2$$

$$F_q = 3{,}500\text{ N}$$

Thus, using the principle of conservation of angular momentum we are able to make a low estimate of the quadriceps force.

Next let us consider the equation of motion in the vertical direction. The vertical forces acting on the lower leg are the quadriceps force, the weight of the lower leg, and the unknown vertical component of the knee joint force. The resultant of these forces must be equal to the mass of the lower leg times the acceleration of its center of mass in the vertical direction. Using the polar coordinates, presented in Chapter 2 we can show that this acceleration is equal to the square of the angular speed ω times the distance r from the center of mass to the rotational axis of the knee.

$$\Sigma F = m\,\omega^2\, r \tag{6.2}$$

$$F_q - 7\text{ kg} \cdot 9.81\text{ m/s}^2 - F_j = 7\text{ kg} \cdot (8\text{ rad/s})^2 \cdot (0.22\text{ m})$$

$$F_j = 3332.7\text{ N}$$

The free-body diagram in Fig. 6.2 shows that F_j pushes against the tibia and therefore is compressive in nature. Note that this force is nearly 50 fold greater than the weight of the lower leg.

Let us next consider the equation of motion of center of mass of the lower leg in the horizontal direction. The only external force acting on the lower leg in the horizontal direction is the knee joint force H_j. This force must be equal to the mass times the acceleration of the center of mass in the horizontal direction. This acceleration component is equal to the angular acceleration α times the distance r:

$$H_j = m\,\alpha\,r \tag{6.3}$$

$$H_j = (7\text{ kg}) \cdot (400\text{ rad/s}^2) \cdot 0.22\text{ m} = 616\text{ N}$$

Equations of motion do not tell us how the total joint force is shared between the various knee ligaments, tibia, and fibula. However, the example shows that the knee sustains high forces during kicking. The knee joint, situated between the body's two longest lever arms, is particularly susceptible to injury.

Example 6.2. Forces Produced by Biceps During High-Speed Forearm Curls. A 25-year-old man knelt facing a table on which his arm rested horizontally in front of his body. He was instructed to press his elbow into the table at all times. Motion of the elbow as the man flexed his forearm as fast as he could were recorded using a high-speed cinema camera (Fig. 6.3). At the instant when the forearm began to rise from the table, angular velocity ω was zero and angular acceleration α was 200 rad/s^2. Then, when the angle θ between the table and the forearm increased to 45°, the angular velocity and angular acceleration took on the following values: $\omega = 13$ rad/s and $\alpha = 100$ rad/s^2. The man performing the biceps curl weighed 62 kg. The mass of the forearm (including the hand) was 1.36 kg. The radius of gyration ρ of the forearm about the mass center was 11 cm. The distance L from the center of mass of the forearm to the center of rotation of the elbow was 18 cm. The length D of the forearm was 27 cm. Determine the moment produced by the biceps at the elbow. Determine the force produced by biceps at $\theta = 0°$ and 45°.

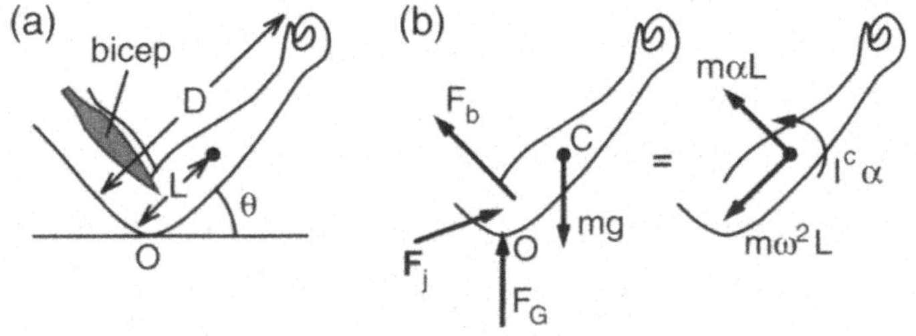

FIGIRE 6.3a,b. The quick flexing of the forearm against its own weight **(a)**. The free-body diagram of the forearm is shown in **(b)**.

Solution: The free-body diagram of the forearm is shown in Fig. 6.3b. The principle of conservation of angular momentum about the fixed point O of the elbow requires that

$$-mg\,L\cos\theta + \mathrm{M_b} = \mathrm{I^o}\,\alpha \tag{6.4}$$

in which m is the mass of the forearm about the elbow, L is the distance from its center of mass to the elbow, $\mathrm{M_b}$ is the moment created at the elbow by the muscles that flex the forearm, $\mathrm{I^o}$ is the mass moment of inertia of the forearm about the elbow, and α is its angular acceleration. In accord with the parallel axis theorem discussed in Chapter 4, the moment of inertia of the forearm about the elbow $\mathrm{I^o}$ is given by the relation:

$$\mathrm{I^o} = \mathrm{I^c} + m\,L^2 \tag{6.5a}$$

in which $\mathrm{I^c}$ denotes the moment of inertia about the mass center of the forearm. The radius of gyration ρ is related to $\mathrm{I^c}$ by the following equation:

$$\mathrm{I^c} = m\,\rho^2 \tag{6.5b}$$

Substituting the parameter values given as data into Eqns. 6.4 and 6.5, we find

$$\mathrm{M_b} = 2.4\cos\theta + 0.06\,\alpha$$

Thus, at $\theta = 0$, $\mathrm{M_b} = 14.4$ N-m, and at $\theta = 45°$, $\mathrm{M_b} = 7.7$ N-m. (6.6a)

Let us next determine the weight the subject must lift statically to generate the same value of muscle moment. Equations of static equilibrium require that

$$\mathrm{M_b} = mg\,L\cos\theta + m_1 g\,D\cos\theta \tag{6.6b}$$

in which m_1 is the mass of the weight lifted by the man and D is the length of his forearm. Using Eqns. 6.6a and 6.6b, we find that

$$m_1 = 4.5 \text{ kg at } \theta = 0°$$

$$m_1 = 3.2 \text{ kg at } \theta = 45°$$

A typical 25-year-old man in good physical condition can flex his forearm against a resistance of 10 kg; this shows that the biceps force that results from high levels of angular acceleration of the forearm is less than half the maximal force that can be generated by this muscle group. As we shall see later in this chapter, the smaller the force carried by an activated muscle, the faster is its rate of shortening.

A low estimate for biceps force $\mathrm{F_b}$ can be obtained by noting that its lever arm d is equal to 4 cm at $\theta = 0°$ and 3 cm at $\theta = 45°$. These estimates are $\mathrm{F_b} = 360$ N at $\theta = 0°$ and $\mathrm{F_b} = 257$ N at $\theta = 45°$.

For more information on the analysis of elbow forces from high-speed forearm movements, see Amis et al. (1980).

Example 6.3. Inverse Dynamics for Vertical Jumping. Determine the tension in the Achilles tendon during an instant of vertical jump performed by a volleyball player of 75 kg (Fig. 6.4). At the particular instant of the push-off under consideration, his center of mass had a vertical upward acceleration of 12 m/s^2. The angle the sole of his feet made with the horizontal axis was 30°. The angular velocity ω and angular acceleration α of his feet were measured to be -15 rad/s and -150 rad/s^2, respectively. The minus sign indicates that both the angular velocity and angular acceleration were clockwise. The length of the volleyball player's feet was 27 cm, and the length of the toe region of the feet was 7 cm. The moment arm of the Achilles tendon with respect to the center of rotation of the ankle was 4 cm. The foot weighed 1.5 kg. Compute the tensile force F_A carried by the Achilles tendon during the instant of push-off described above.

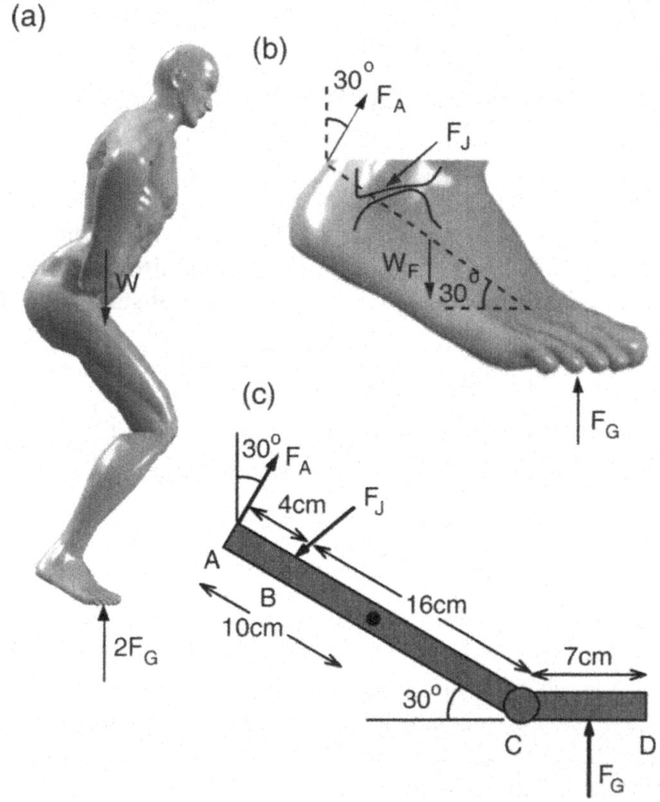

FIGURE 6.4a–c. A volleyball player performing a vertical jump. The external forces acting on the player are shown in **(a)**. The free-body diagram of one of his feet is illustrated in **(b)**. The simplified geometry of the foot and the forces acting on it are shown in **(c)**.

Solution: Considering the free-body diagram of the volleyball player shown in Fig. 6.4a, we find that

$$-W + 2\,F_G = m\,a_{cy} \tag{6.7a}$$

in which W is the weight of the volleyball player, F_G is the ground force exerted on each foot, and a_{cy} is the acceleration of the center of mass of the player in the vertical direction. Substituting the values of the input parameters into Eqn. 6.7a, we find

$$2\,F_G = 75\ \text{kg} \cdot (12 + 9.81)\ (\text{m/s}^2) \Rightarrow F_G = 818\ \text{N} \tag{6.7b}$$

If we had neglected the acceleration of the center of mass of the jumper, the ground force would be equal to half the weight (368 N).

Next, let us consider the free-body diagram of the foot. Unlike the whole body, the foot has a small mass, and the forces acting on it, like the ground force, are significantly greater than its mass times acceleration. Thus, although the feet are in motion, the inertial effects are neglected and equations of statics hold. The validity of this assumption can be confirmed by using the equations of dynamics as opposed to statics in determining the force carried by the Achilles tendon. The results not shown here indicate that neglecting the inertia of the foot results in errors of 1% or less in the tendon force.

According to the equations of statics, the net moment acting on the foot at the center of rotation of the ankle must be equal to zero. Thus, when we take the moment of forces acting on the foot with respect to point B we find

$$-F_A \cdot 0.04 - 1.5 \cdot 9.81 \cdot 0.06 \cdot \cos 30° + 818 \cdot (0.035 + 0.16 \cdot \cos 30°) = 0$$

$$F_A = 3531\ \text{N}$$

In this equation, we assumed that the distance between the center of mass of the foot and the ankle was 10 cm (see the free-body diagram in Fig. 6.4c). The force carried by the Achilles tendon is almost five times the body weight. Note that the weight of the foot contributed less than 0.3% to the tensile force in the Achilles tendon.

6.3 Examples from Weight Lifting

Weight lifting involves the rotation of a body segment against resistance. Because the exercise is done slowly (about one repetition per second), the inertial effects are neglected. We next present examples from the static analysis of weight lifting.

Example 6.4. Deltoids. Deltoids are a shoulder muscle group that is located on the upper side of the arms. Deltoids originate at the bones of the shoulder (clavicle and scapula) and end at the outer midsection of

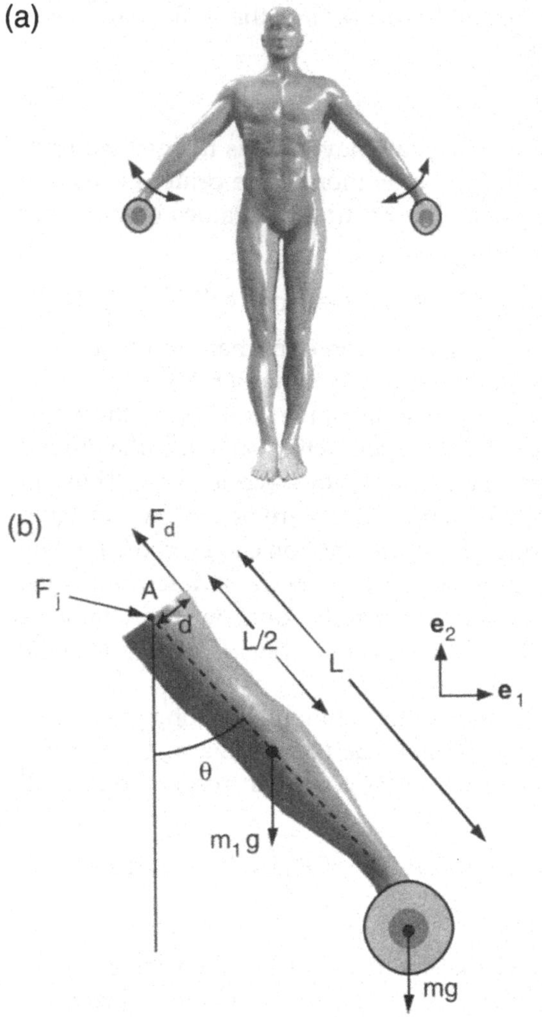

FIGURE 6.5a,b. A man performing standing lateral raise **(a)**. The free-body diagram of his arm is shown in **(b)**.

humerus. The primary function of this muscle group is to abduct the arms. Standing lateral raises (Fig. 6.5) is a major exercise for this muscle group. In performing this exercise, select a weight that allows you to warm up and learn the proper movement. Stand with chest out, back straight, and chin level. Starting with hands at your side raise the dumbbells upward to shoulder height, with elbows slightly bent. Lower the weight slowly to the starting position. Repeat movement.

Static Analysis of the Exercise: The free-body diagram of the arm is shown in Fig. 6.5b. The condition of zero moment at the center of rotation of the shoulder joint (point A) requires that

$$F_d \cdot d - m_1 g (L/2) \sin\theta - m g L \sin\theta = 0 \tag{6.8a}$$

in which F_d denotes the force produced by the deltoids, d is the lever arm of this force with respect to point A, m_1 is the weight of the arm (including the forearm and hand), L is the length of the arm, m is the mass of the weight carried, and θ is the angle of inclination of the arm with the vertical axis. Rearranging this equation we obtain:

$$F_d \cdot d = m_1 g\,(L/2) \sin\theta + mg\,L \sin\theta \tag{6.8b}$$

According to the data presented in Appendix 2, one set of upper arm, forearm, and hand constitute on the average 6.5% of the body weight of a young adult male. For a man who weighs 79 kg, this would amount to 5 kg; thus, $m_1 = 5$ kg. If the man is lifting 20-lb dumbbells, then $m \cong 9$ kg. If his upper limb is 73 cm long, the moment created by the deltoid muscle group is

$$F_d \cdot d = 150 \sin\theta \text{ (N-m)}$$

In general, the moment arm of the deltoids will change with the angle θ, but for simplicity let us assume that d = 7.5 cm for all angles under consideration. Then,

$$F_d = 2{,}005 \sin\theta \text{ (N)}$$

According to this equation, the maximum force carried by the deltoids occur when the arms are aligned horizontally ($\theta = \pi/2$). It is interesting to note that a muscle that weighs 1 to 2 kg can produce a force in excess of 200 kg in a young male. Also note that the moment created by the weights at the shoulder increases with increasing arm length. Thus, at first glance, it may appear as if the longer the arm, the higher must be the force produced by the deltoids. However, the moment arm d is probably larger in people with longer arms, and thus may counterbalance any increase in the muscle force caused by longer arm length.

Let us next consider the total joint force acting at the shoulder joint. Assuming that the deltoid muscle force acts in the direction of the arm, we have the following relations:

$$R_1 - F_d \sin\theta = 0 \tag{6.9a}$$

$$R_2 - (m_1 + m)\,g + F_d \cos\theta = 0 \tag{6.9b}$$

in which R_1 and R_2 represent, respectively, the horizontal and vertical components of the total joint force $\mathbf{F}_j$. Rearranging Eqn. 6.9 and substituting the input parameter values, one obtains

$$R_1 = 2{,}005 \text{ N and } R_2 = 137 \text{ N at } \theta = \pi/2$$

Thus, the total shoulder joint force is largely a compressive force that is comparable in magnitude to the force produced by the deltoids. It is likely that much of this force is resisted by the bone of the upper arm humerus. Passive soft tissue structures such as ligaments have low resistance to compression.

Example 6.5. Erector Spinae. Determine the tensile force produced by the erector spinae for the body positions shown in Fig. 6.6. Assume that the exercise is performed slowly so that equations of static equilibrium hold. The length L between the tip of the head of the athlete and his hip joint is 82 cm. The length L_c between the center of mass of the upper body and the hip joint is 44 cm. The upper body weighs 37 kg. The lever arm d of the erector spinae is 4 cm and is largely insensitive to the angle between the upper and lower body.

Solution: Consider the free-body diagram of the upper body shown in Fig. 6.6b. The resultant moment acting on the upper body must be equal to zero. If we take moments with respect to the point of application of the total joint force in the hip joint, following equation is obtained:

$$\mathrm{F_e} \cdot d - m\, g\, L_c \sin\theta = 0 \Rightarrow \mathrm{F_e} = m\, g\, (L_c/d) \sin\theta$$

in which $\mathrm{F_e}$ is the force produced by the erector spinae. Substituting the input parameter values into this equation, we find

$$\mathrm{F_e} = 2{,}823\ \mathrm{N} \quad \text{at } \theta = \pi/4$$

$$\mathrm{F_e} = 3{,}993\ \mathrm{N} \quad \text{at } \theta = \pi/2$$

$$\mathrm{F_e} = 2{,}823\ \mathrm{N} \quad \text{at } \theta = 3\pi/4$$

These results suggest that erector spinae is capable of generating forces that are four to five times the body weight.

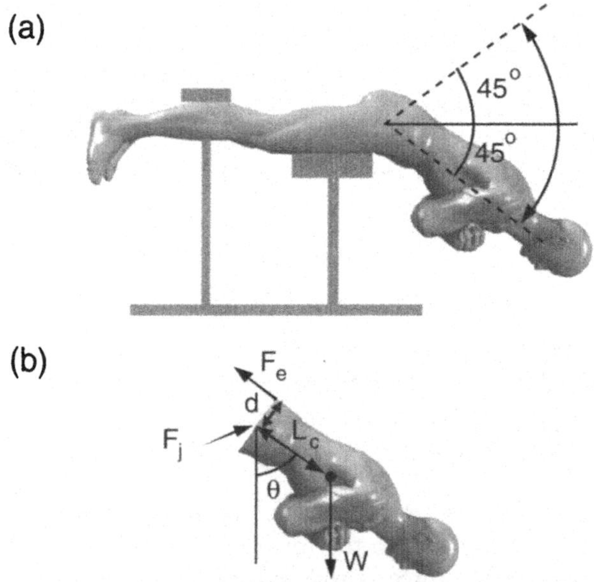

FIGURE 6.6a,b. A man performing back curls to strengthen his back muscle erector spinae **(a)**. The free-body diagram of his upper body is shown in **(b)**.

The static analysis presented here can be used to evaluate the forces involved in a large number of weight-lifting exercises. Some of the body dimensions required in the analysis can be obtained by direct measurement. Other input parameters such as the weight of a forearm or upper body can be evaluated by knowing the body weight and using it in the tables given in Appendix 2. Admittedly, the input data thus obtained will be approximations to direct measurements. However, the procedure gives a starting point for "back-of-the-envelope" computations. The lever arms of various muscle groups that move a body segment may be the hardest to measure. The data obtained by cadaver studies and by X-ray provide estimates for some of the important lever arms involved in movement and motion (Appendix 2).

6.4 Moment Arm and Joint Angle

Moment (lever) arm of a muscle acting on a joint changes with the angle between the two bones articulating in that joint. Consider, for example, the moment arms of the forearm flexors, biceps, and the brachioradialis muscles shown in Fig. 6.7a. The moment arm of biceps is nearly equal to zero when the angle between the upper arm and the forearm is 180°. The moment arm increases as the joint angle θ decreases from 180° toward 90°. What is the relation between the moment arm and the joint angle? This question can be addressed using geometric relationships. We illustrate the analysis by considering the dependence of the lever arm of biceps brachia on the joint angle. Figure 6.7b illustrates this muscle as a cord spanning through the humerus and inserting at the radius. Actually, the geometry chosen is a simplification of the topology of the biceps. This is a biarticular muscle in which both heads of the biceps arise at the scapula rather than at the proximal head of humerus. Additionally, a strong membranous band arising from the tendon of biceps attaches to the ulna. Nevertheless, Fig. 6.7b is useful in investigating the action of this muscle group on the elbow joint.

Let a denote the length of the humerus, b denote the distance between the center of rotation of the elbow joint and the point of insertion of biceps on radius, and θ be the angle between the humerus and radius (Fig. 6.7b). In this figure the length of the biceps muscle–tendon cord is represented by the letter c and the normal distance from the center of line of application of the biceps force to the center of rotation of the elbow joint by the letter d. Note that d is the moment arm of the biceps force with respect to the elbow joint. Using the cosine law, one can express the muscle–tendon length c as a function of a, b, and θ:

$$c^2 = a^2 + b^2 - 2ab \cos \theta \tag{6.10}$$

Because the tension in the muscle–tendon complex depends on the length of the complex, it is important to be able to compute this length as a function of the joint angle θ.

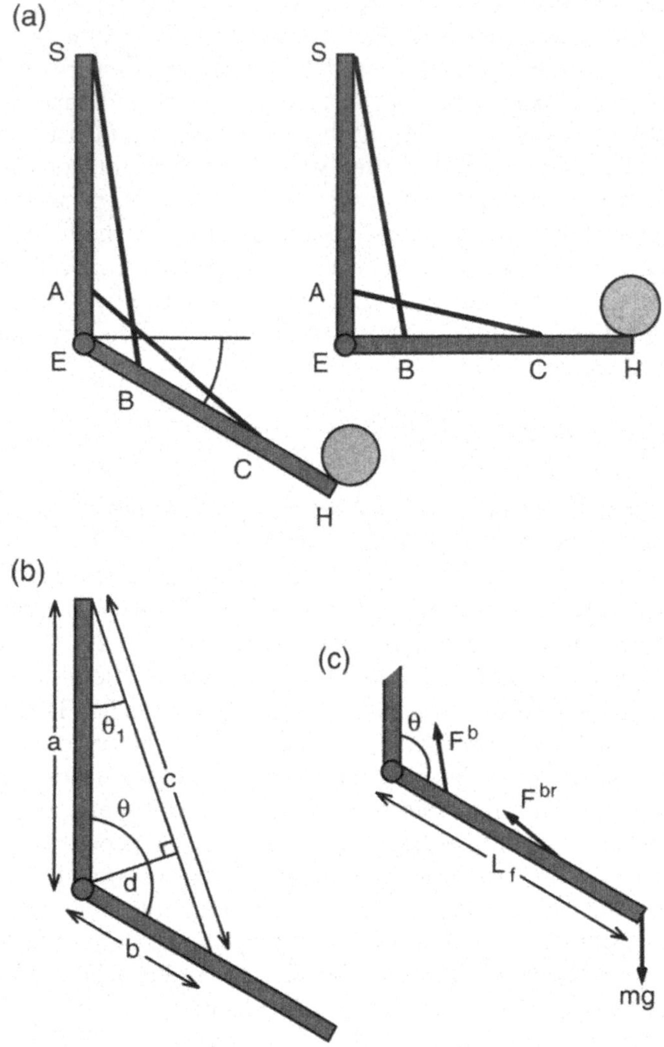

FIGURE 6.7a–c. Flexion of the forearm as a result of contraction of biceps brachii and brachioradialis **(a)**. The symbols S and E identify the centers of rotation of the shoulder and elbow muscle, respectively. The biceps muscle group is represented by a cord joining points S to B, and brachioradialis by a cord joining points A and C. The length parameters a, b, c, d, and angles θ and θ_1 used in the analysis are identified in **(b)**. Note that a and b refer to lengths along the adjoining bones whereas c is the length of the muscle–tendon complex. The moment arm is shown by the letter d in the figure. Forces acting on the forearm during flexion against a resistance of $m = 10$ kg are shown in **(c)**. Weight of the forearm is neglected in the free-body diagram.

Computation of d from the anthropometric data is also straightforward. Figure 6.7b indicates that

$$d = a \sin \theta_1 \tag{6.11}$$

in which θ_1 denotes the angle between the long axis of humerus and line of application of the biceps force. Using the cosine law, this angle can be expressed as a function of a, b, and c:

$$b^2 = a^2 + c^2 - 2ac \cos \theta_1$$

$$\cos \theta_1 = (-b^2 + a^2 + c^2)/2ac$$

$$\sin \theta_1 = (1 - \cos^2 \theta_1)^{1/2}$$

$$d = [4a^2c^2 - (-b^2 + a^2 + c^2)^2]^{1/2}/2c \tag{6.12}$$

Equations 6.10 and 6.12 can be used to determine the muscle–tendon length and the moment arm as a function of the joint angle.

Example 6.6. Moment Arm of Biceps. Consider a person flexing the forearm against a 10-kg weight as shown in Fig. 6.7c. Assume that the arm is weightless and that the biceps muscle is the one muscle involved in the flexion. The length a of the humerus is 32 cm and distance b between the point of insertion of biceps into radius and the elbow joint is 9 cm. The length of the forearm (including half the length of the hand) is 34 cm. Determine the moment arm of the biceps, and the biceps force at joint angle $\theta = 170°$, 135°, 90°, and 45°.

Solution: Using Eqn. 6.10 and the parameter values given for a and b, we compute the length c of the muscle–tendon complex as a function of joint angle θ. Then, we use Eqn. 6.12 to determine the moment arm d:

$$c = 40.89 \text{ cm}, \; d = 1.24 \text{ cm at } \theta = 170°$$

$$c = 38.89 \text{ cm}, \; d = 3.92 \text{ cm at } \theta = 135°$$

$$c = 33.24 \text{ cm}, \; d = 8.67 \text{ cm at } \theta = 90°$$

$$c = 26.41 \text{ cm}, \; d = 7.70 \text{ cm at } \theta = 45°$$

Next, we compute the moment exerted by the biceps force about the elbow joint. Because the motion occurs slowly, we can neglect the inertial effects and set the sum of moments acting on the forearm at the elbow equal to zero.

$$M_b - mgL_f \sin \theta = 0 \Rightarrow M_b = mgL_f \sin \theta \tag{6.13}$$

where M_b is the counterclockwise moment created by the biceps force at the elbow, $m = 10$ kg is the weight to be lifted, and $L_f = 34$ cm is the length of the forearm (Fig. 6.7c). The moment M_b can be expressed as follows:

$$M_b = d\, F_b \tag{6.14}$$

in which d is the moment arm and F_b is the magnitude of the biceps force. Using these two equations along with Eqns. 6.10 and 6.12, we compute the moment and the force produced by the biceps muscle as a function of the joint angle:

$$M_b = 579.2 \text{ N-cm}, F_b = 467.1 \text{ N at } \theta = 170°$$

$$M_b = 2358.5 \text{ N-cm}, F_b = 601.7 \text{ N at } \theta = 135°$$

$$M_b = 3335.4 \text{ N-cm}, F_b = 384.7 \text{ N at } \theta = 90°$$

$$M_b = 2358.5 \text{ N-cm}, F_b = 306.7 \text{ N at } \theta = 45°$$

If the moment arm were assumed constant throughout the range of θ, the biceps would have been predicted to produce the highest force at $\theta = 90°$. Obviously, this is not the case. A word of caution here: not all muscles of the upper and lower limbs experience as great a variation of the moment arm with the joint angle as does the biceps. For example, the moment arm of the triceps is much less dependent on the joint angle in comparison with biceps (not shown).

6.5 Multiple Muscle Involvement in Flexion of the Elbow

We have already seen that multiple muscles contribute to a certain task of movement or posture. Despite the presence of a number of muscle groups contributing to the same movement, it is common in biomechanics to consider one muscle group for the specified action, and compute the force (moment) that must be produced by this muscle to carry out the movement (against resistance). What is the magnitude of the errors involved in such back-of the-envelope type computations? We can address this question by assuming the following: (a) multiple muscles act on the joint and (b) the force generated by each muscle is proportional to the cross-sectional area at the midpoint at full activation. We illustrate this procedure by considering the flexion of the forearm against a resistance.

Example 6.7. Consider the flexion of the forearm considered in Example 6.6. Assume (as shown in Fig. 6.7c) that two muscle groups participate in the flexion, the biceps brachia and the brachioradialis. Assume further that the latter muscle originates at the humerus 4 cm away from the center of rotation of the elbow and inserts at the radius at 20 cm away from the center of rotation of the elbow. The cross-sectional areas of biceps and brachioradialis are, respectively, 12 cm^2 (A_b) and 4 cm^2 (A_{br}). Both muscles are maximally activated during flexion. Determine the contribution of each muscle group to the total joint moment.

Solution: The total joint moment is the moment created by the muscles to resist the clockwise moment created by the 10-kg weight. The joint moment was computed in Example 6.6 as a function of the joint angle θ. To determine the relative contribution of brachioradialis to the joint moment, we compute its moment arm with respect to the center of rotation of the elbow:

$$c_{br} = 23.9 \text{ cm}, d_{br} = 0.58 \text{ cm at } \theta = 170^\circ$$

$$c_{br} = 23.0 \text{ cm}, d_{br} = 2.46 \text{ cm at } \theta = 135^\circ$$

$$c_{br} = 20.39 \text{ cm}, d_{br} = 3.92 \text{ cm at } \theta = 90^\circ$$

$$c_{br} = 17.40 \text{ cm}, d_{br} = 3.25 \text{ cm at } \theta = 45^\circ$$

in which c_{br} and d_{br} represent, respectively, the length and the moment arm of the brachioradialis.

How do these two muscles share the load? We assume that the force produced by each muscle is proportional to their maximal cross-sectional area:

$$F_b/F_{br} = A_b/A_{br} = 3 \tag{6.15}$$

This assumption is equivalent to the statement that muscle fiber tension depends only weakly on the length of the fiber in physiological ranges of fiber length. Joint moment can be written as a summation of the moments contributed by the biceps and the brachioradialis:

$$M = F_b\, d_b + F_{br}\, d_{br} \tag{6.16}$$

As usual, the subscripts refer to the muscle groups involved in flexion. Equations 6.15 and 6.16 can now be used to compute muscle force as a function of the joint angle:

$$F_b = 404.1 \text{ N}, F_{br} = 134.7 \text{ N at } \theta = 170^\circ$$

$$F_b = 497.5 \text{ N}, F_{br} = 165.9 \text{ N at } \theta = 135^\circ$$

$$F_b = 334.3 \text{ N}, F_{br} = 111.4 \text{ N at } \theta = 170^\circ$$

$$F_b = 268.5 \text{ N}, F_{br} = 105.4 \text{ N at } \theta = 170^\circ$$

Comparing the biceps force computed in this example with that of Example 6.6, we find that biceps force was overestimated by 13% when the contribution of brachioradialis to the lifting of the weight was neglected. Inclusion of the triceps (which resists flexion) as opposed to brachioradialis in the analysis would show the evaluation of biceps force in Example 6.6 to be an underestimate.

6.6 Biarticular Muscles

Biarticular muscles act on two joints. These muscles include some of the major muscles of the upper and lower limbs. Hamstrings, a group of three muscles, constitute an important example for biarticular muscles. These muscles originate in the ischial tuberosity of the hipbone and insert into the bones of the lower leg. When the thigh and the hip are fixed, hamstrings flex the knee. They also extend the hip through the movement of the thigh. The third function of the hamstrings is to raise the trunk from a flexed position and keep it erect. Quadriceps is also a biarticular muscle group. They function mainly as the primary leg extender. The following example illustrates the effective functions of hamstrings, calfs, and quads during squats.

Example 6.8. An athlete is performing squats to strengthen knee muscles. The movement is slow enough to assume static equilibrium. Consider the four-link system shown in Fig. 6.8 to represent the athlete during squatting. The beam representing the hip is connected to the rod representing the upper leg at the hip joint. A tension-carrying cord representing the calf muscle connects the foot to the thigh. The quad muscle connects the thigh and the leg through a frictionless pulley mechanism representing the patella joint. The joints at H, K, and A are hinge joints. Determine the tension in hamstring, calf, and quads as a function of the angle the leg makes with the horizontal plane (θ).

Solution: For simplicity we assume symmetry with respect to the horizontal plane passing through the knee joint. Let P denote the force transmitted at the hip joint to each leg. The force P is then equal to half the weight of the upper body plus the weight used for the squat. For symmetry, we assume that the legs and the feet are weightless and that the entire weight is lumped into a single weight P acting at a distance c from the hip joint H. Consider the free-body diagram of the foot shown in Fig. 6.8b. The moment created by force P about A should be equal to the moment produced by the calf muscle at A:

$$-b \sin \phi \, F^c + c\,P = 0 \Rightarrow F^c = c\,P/(b \sin \phi) \qquad (6.17)$$

in which F^c denotes the force produced by the calf muscle group. Because of symmetry in the idealized structure, the calf muscle will produce the same tension as the hamstrings:

$$F^c = F^h \qquad (6.18)$$

Next, let us consider the free-body diagram of the feet and the lower legs (Fig. 6.8c). The moment of all external forces with respect to the knee joint must be equal to zero:

$$-(L \cos \theta - c)\,P - 2\,d^k\,F^h \sin \phi + d^q\,F^q = 0 \qquad (6.19)$$

where L represents the length of upper and lower leg. The first term on the left-hand side of Eqn. 6.19 is the moment of the P with respect to K. The second term is the moment created by the hamstring and calf muscles at the knee. Because of symmetry, the resultant of these two muscle

FIGURE 6.8a–e. The schematic diagram of a person performing squats **(a)**. The figure also shows the free-body diagrams **(b–e)** of body segments involved in this movement.

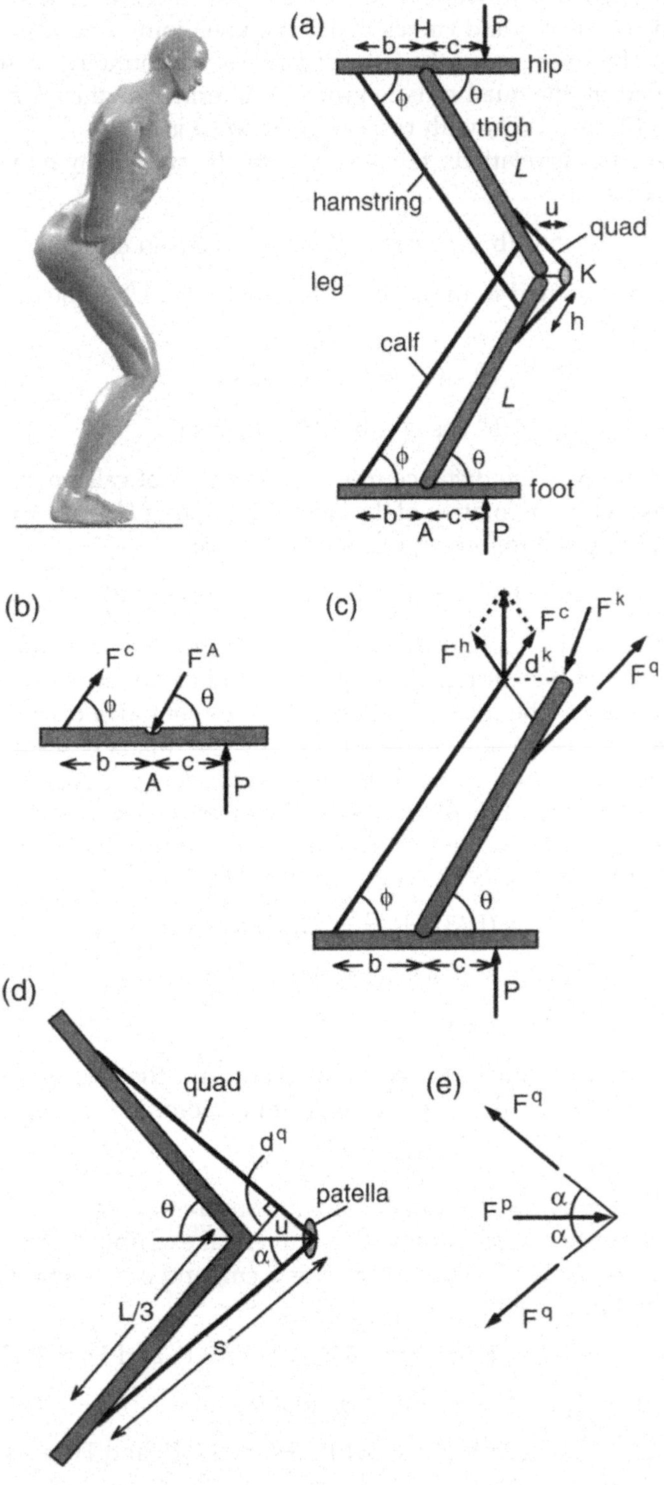
(a)
H
P
b
c
hip
φ
θ
thigh
L
hamstring
u
quad
leg
K
h
calf
L
φ
θ
foot
b
A
c
P
(b)
F^c
F^A
φ
θ
b
A
c
P
(c)
F^c
F^k
F^h
d^k
F^q
φ
θ
b
c
P
(d)
quad
d^q
θ
patella
u
α
L/3
s
(e)
F^q
F^p
α
α
F^q

forces acts vertically upward. The term $2\ F^h \sin \phi$ is the resultant of the tension in hamstring and calves about the knee joint, and d^k is its moment arm. The last term on the left-hand side is the counterclockwise moment created by the quad muscle group. The term d^q is the moment arm of the quad tension (F^q) with respect to the knee joint.

Using geometric relations, the moment arm d^k can be shown to be given by the relation:

$$d^k = (b + L \cos \theta - L \sin \theta \cos \phi / \sin \phi) \tag{6.20}$$

The angle ϕ is related to the angle θ and the length L^h of the hamstrings as follows:

$$L^h \sin \phi = (L + h) \sin \theta \tag{6.21a}$$

$$L^h \cos \phi = b + (L - h) \cos \theta \tag{6.21b}$$

in which h denotes the distance from the insertion of calf muscle on the femur to the center of rotation of the knee. Eliminating L^h from these equations, we obtain the following relationship for ϕ:

$$\tan \phi = (L + h) \sin \theta / [b + (L - h) \cos \theta] \tag{6.21c}$$

Now, that we have expressed ϕ as a function of θ, we can compute the moment arms of hamstrings about point H and calves about point A.

The next step is the determination of the moment arm d^q of the quad muscle group. The geometry associated with the quads is shown in Fig. 6.8d. We assume that the quads originate and insert at a distance of $L/3$ from the center of rotation of the knee and that the patella keeps the quads a distance u away from the femorotibial joint. We use the law of cosines successively to determine the moment arm d^q:

$$s^2 = (L/3)^2 + u^2 - (2L/3)\ u \cos (\pi - \theta) \tag{6.22a}$$

$$(L/3)^2 = s^2 + u^2 - 2\ s\ u \cos \alpha \tag{6.22b}$$

$$d^q = u \sin \alpha \tag{6.22c}$$

The compressive force acting on the quads and the knee joint can be found by considering the static equilibrium of point K (Fig. 6.8e):

$$F^p = 2\ F^q \cos \alpha \tag{6.23}$$

in which F^p represents the patellofemoral compressive force.

Let us illustrate this solution with a numerical example. Let us assume that $L = 40$ cm, $b = 10$ cm, $h = 8$ cm, $u = 5$ cm, and $c = 14$ cm. For body configuration at $\theta = 60°$:

$$\phi = 58°,\ \alpha = 45°,\ F^c = F^h = 1.5\ P\ ,\ F^q = 7.9\ P, \text{ and } F^p = 11.2\ P$$

When the body goes further down so that $\theta = 30°$:

$$\phi = 32°,\ \alpha = 22°,\ F^c = F^h = 2.64\ P\ ,\ F^q = 22.0\ P, \text{ and } F^p = 40.8\ P$$

These numerical results indicate that quads need to produce much more tension than the other two muscles involved in squatting: hamstrings and calves. Furthermore, the compressive force between the patella and the femur increases tremendously with increasing quads tension and increasing flexion of the knee. Repeated compressive force disrupts and destroys the cartilage coating of the articulating bone surfaces, leading to frictional resistance to the sliding motion of the kneecap during knee flexion and extension. This overuse injury is commonly observed among runners. The extensor mechanism of the knee is the most common site of chronic running injuries. Running-associated pain may be caused by excessive compressive stress acting on the articular cartilage of the patellofemoral joint. Recently, research has focused on how to restore damaged cartilage to the joints of the upper and lower limbs. The recent developments in biotechnology allow the growing of cartilage cells in culture dishes in a laboratory setting. Whether injection of fibrous elements or cartilage cells grown in vitro into the damaged cartilage of a joint will result in the restoration of this important tissue remains to be seen.

6.7 Physical Stress

To build a muscle, one must bring it to near exhaustion. What are the levels of force intensity (stress) in the muscle, when the muscle force peaks? To answer this question, one must evaluate (a) the force generated by a muscle and (b) the cross-sectional area of the muscle normal to the direction of the muscle force. Average axial force intensity or average normal stress at a cross section is defined as

$$\sigma_{av} = F/A \tag{6.24}$$

where σ_{av} denotes the average axial stress and F is the force acting on the cross-sectional area A. In parallel muscles, the cross-sectional area is not constant along the axis of the fiber. However, a single fiber can be represented in the form of a uniform circular cylinder (Fig. 6.9). Let A denote the cross-sectional area that is perpendicular to the axis of the cylinder. Then the axial stress acting on this cross-sectional area is uniform and is given by Eqn. 6.24. Let us now consider the stress that acts on a cross section whose normal vector makes an angle θ with the vertical axis (Fig. 6.9b). In this case, the cross-sectional area is given by the equation:

$$A_\theta = A/\cos\theta \tag{6.25}$$

Next, let us resolve the force F that acts on cross-sectional area A_θ into two components (Fig. 6.9c), one normal (F_n) and the other tangential (F_t)

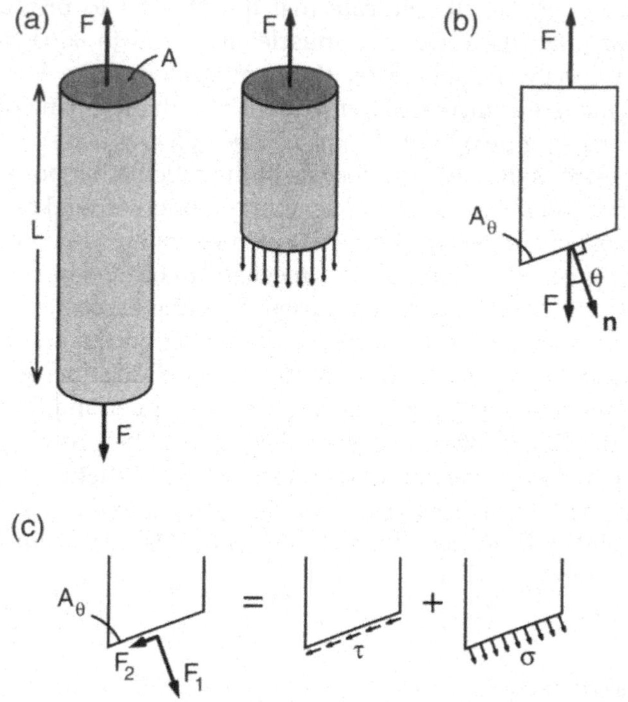

FIGURE 6.9a–c. A circular cylindrical specimen under the application of a tensile force of magnitude F **(a)**. The free-body diagram of the part of the specimen is shown in **(b)**. The stress distribution acting on an inclined cross section is shown in **(c)**.

to the area A_θ. The equations for these two force components are as follows:

$$F_n = F \cos \theta \tag{6.26a}$$

$$F_t = F \sin \theta \tag{6.26b}$$

Let us now define normal (σ) and shear (τ) stress components as follows:

$$\sigma = F_n/A_\theta = (F/A) \cos^2 \theta \tag{6.27a}$$

$$\tau = F_t/A_\theta = (F/A) \sin \theta \cos \theta \tag{6.27b}$$

It is clear that the concept of stress is a bit more complex than the entities that can be represented as vectors. Two distinct directions (not just one) are associated with stress, the direction of the force and the direction of the surface area on which the force acts. Stress is a measure of force intensity on a small area element. It is a tensor of a second order. Scalars such as temperature and mass are said to be tensors of the zero order; vectors are tensors of first order.

Equation 6.27 shows that both the normal and the shear stress vary considerably with the orientation of the cross-sectional area. If we take the derivatives of Eqns. 6.27a and 6.27b with respect to θ and set them equal to zero, we can determine those directions in which either σ or τ is maximum (minimum). When $\theta = 0$, then the normal stress takes its largest value ($\sigma = F/A$) whereas the shear stress becomes equal to zero. On the other hand, when $\theta = \pi/4$, then the shear stress assumes its maximum value ($\tau = 0.5\ F/A$). The evaluation of the maximal values of shear and normal stress is important in the consideration of failure of a material. Some materials fail easily under tension (compression) and others fail readily under shear stress. Concrete is not resilient to tension whereas steel can withstand both tension and compression. Ceramics are known to have low resistance to shear stress. If the circular cylinder under consideration had low resistance to shear but high resistance to tension, it would fail in the form of a tear that occurs at 45° relative to the axis of the cylinder.

As illustrated by the previous examples, in many cases Newton's second law allows us to compute contact forces and the resultant internal forces at a cross section. Once we know the resultant force carried by a structural member, we can compute the average stress by dividing the value of the force with the cross-sectional area of the planar cross section. The question that comes to mind is how close is the average stress to the actual stress. The large body of literature in solid mechanics allow us to provide some insights. First, in the presence of cracks or holes, the average stress is not an accurate indicator of the actual stress. Stress will intensify in the vicinity of a fracture or rupture. Also, if the cross section cuts across a number of different materials with different stiffness, stress in the material with higher stiffness may be greater than the one with lower stiffness. We can illustrate this by considering a cylindrical specimen consisting of two materials under the action of a tensile force (Fig. 6.10). Imagine this to represent a long limb of the human body, a long bone surrounded by soft tissue. The average stress normal to the axis of the cylinder is given by Eqn. 6.24. What are the average stresses carried by the two concentric cylinders? The total force acting on the cross section must be equal to F, and this leads to the following equation:

$$F = \sigma_1\ \pi\ r_1^2 + \sigma_2\ \pi\ (r_2^2 - r_1^2) \tag{6.28}$$

in which r denotes the radial distance and the index refers to concentric cylinders made of materials 1 and 2. We need additional information to determine the values of σ_1 and σ_2. We assume that planar cross sections that are normal to the axis of the specimen remain plane and normal to the axis. Thus, every line element parallel to the axis of the specimen undergoes the same extension Δ. Axial strain ϵ is defined as follows

$$\epsilon = \Delta/L \tag{6.29}$$

in which L denotes the length of the cylinder.

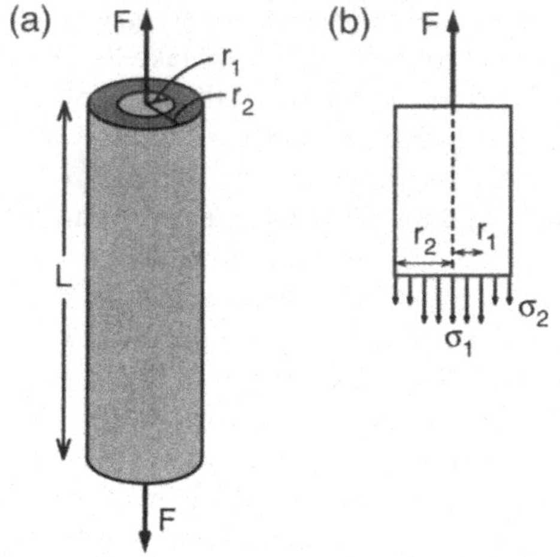

FIGURE 6.10a,b. A circular cylindrical specimen made of two different materials is under tensile force **(a)**. Stress distribution on a cross section that is normal to the long axis of the specimen is shown in **(b)**. The material occupying the core of the cylinder is stiffer than that of the outer shell.

If both materials are linearly elastic, then according to Hooke's law we have

$$\sigma_1 = E_1 \epsilon \qquad (6.30a)$$

$$\sigma_2 = E_2 \epsilon \qquad (6.30b)$$

in which E_1 and E_2 denote, respectively, Young's modulus of materials 1 and 2. Combining Eqns. 6.28, 6.29, and 6.30, we obtain:

$$\sigma_1 = E_1\ F/[E_1\ \pi\ r_1^2 + E_2\ \pi\ (r_2^2 - r_1^2)] \qquad (6.31a)$$

$$\sigma_2 = E_2\ F/[E_1\ \pi\ r_1^2 + E_2\ \pi\ (r_2^2 - r_1^2)] \qquad (6.31b)$$

Notice that $(\sigma_1/\sigma_2) = E_1/E_2$. If the cross-sectional areas of the two materials are comparable and if E_1 is much greater than E_2, then material 1 carries much of the force applied on the specimen. This would be the case of a relaxed limb that is under tension; bone would carry much of the applied load. Consider a human thigh under traction. Approximately 30% of the cross-sectional area is bone and the rest is composed of muscle and fat tissue. Because the fat tissue is much more compliant than muscle, it would carry practically no force. The cross-sectional area that effectively carries the traction force must be that of the cross-sectional area of the muscle and the bone.

6.8 Musculoskeletal Tissues

Skeletal muscle is the actuator of human and animal movement. In the following, we briefly review the physical properties of a muscle fiber, measured in vitro, and then move on to the properties of whole muscles. We also discuss the mechanical properties of tendons, ligaments, and bones.

Skeletal Muscle

Consider a living relaxed skeletal muscle fiber that is aligned vertically, with one end fixed the other free, as shown in Fig. 6.11. The length L_0 of the fiber is measured in this force-free configuration. Then, a weight of W is attached to the free end. The fiber will elongate rapidly in response to the applied load and then will appear to reach a steady-state configuration. The length L of the fiber is measured at that instant. The experiment is continued in this fashion with the addition of extra load and al-

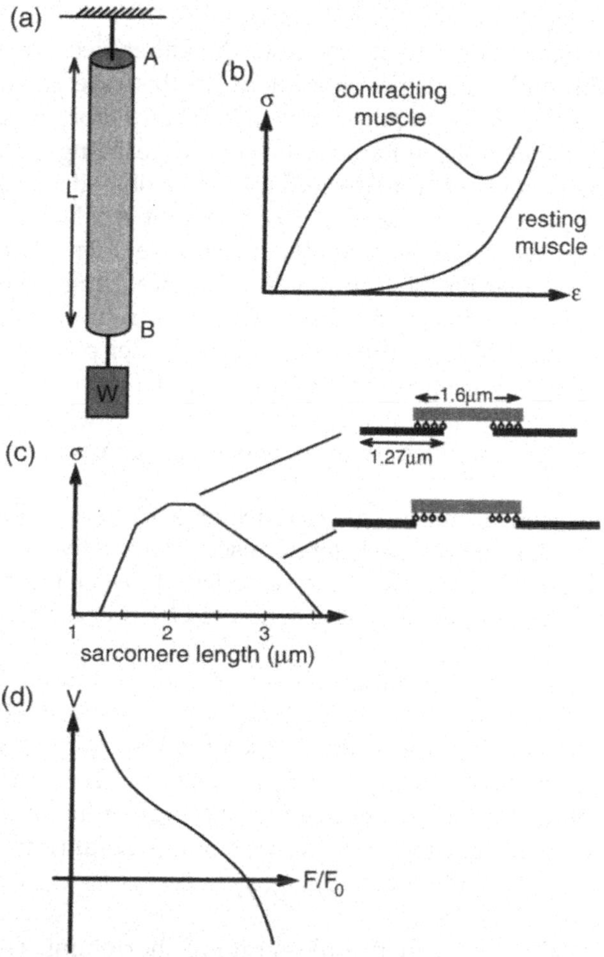

FIGURE 6.11a–d. A muscle fiber under tension **(a)**. The isometric stress–strain relation of the fiber during passive relaxed state and fully contracted isometric state is shown in **(b)**. The fiber stress versus sarcomere length in the midregion of the fiber is illustrated in **(c)**. The force–velocity relation of the muscle fiber during isotonic contraction is represented in **(d)**.

lowing the fiber to reach a steady state corresponding to the new load. Two parameters are defined, the average stress and average strain:

$$\sigma = F/A \text{ and } \epsilon = \Delta/L_o$$

where F denotes the imposed weight W. Because the material is homogeneous and the fiber is a circular cylinder, the stress would be expected to be uniform on the cross section of the fiber. When the stress σ is plotted against the strain ϵ, the resulting curve takes the shape shown in Fig. 6.11b. Experiments indicate that a relaxed muscle fiber can be stretched with relative ease in the physiological range but that large increases in length require considerable tensile force exerted on the fiber and thus might lead to rupture of the fiber structure.

Next, let us maximally activate the fiber (by either electrical or chemical stimulation) and repeat the experiment. In this case, in the physiological range of muscle length, the fiber will typically shorten (rather than elongate) under the application of weight before reaching a steady-state length. The total fiber stress σ that keeps the active fiber at a certain length is plotted against the strain ϵ (Fig. 6.11b). This plot is called an isometric length tension curve. Three important features are evident from this plot. One is that the longer the fiber in the physiological range, the higher is the steady-state force it can produce. This is a reflection of the microstructure of the skeletal muscle (Fig. 6.11c). Under conditions of constant length, muscle force generated is proportional to the extent of overlap between the actin and myosin filaments. The degree of overlap decreases after sarcomere length approaches a certain value characteristic of the muscle.

In the active state, a muscle can produce tensile stress of the order of 10 to 40 N/cm^2. The stress in a skeletal muscle in the active state is much higher than in the passive state at the same length. Also, when an active muscle is stretched further, the force generated by it begins to drop, signaling failure.

Another primary experiment on the contraction of muscle fibers concerns the rate of shortening of a contracting fiber against a resistance. Consider a fiber in isometric contraction. When the load applied on it is reduced, the fiber begins to shorten while still in tension. This is the so-called isotonic experiment often encountered in the muscle literature. A measure of the rate of shortening is the dimensionless parameter V:

$$V = (dL/dt)/L \tag{6.32}$$

in which V is called the dimensionless rate of shortening. The time derivative of fiber length L appearing in this equation is evaluated at 100 ms after the beginning of the shortening process. The parameter V is typically plotted against the ratio of the load carried by the fiber (F) to the load carried at isometric state at the same fiber length (F_o). Typical results on a skeletal muscle fiber are shown schematically in Fig. 6.11d.

The figure indicates that the smaller is the tensile force carried by a muscle, the higher is the rate at which it shortens; a muscle will contract fastest against zero load. A fast fiber contracts faster than a slow fiber. In the human, the maximal muscle fiber shortening rate ranges from about 1 to 10 fiber lengths per second. The rate at which a muscle can shorten during contraction determines how quickly one can flex arms and legs; it plays an important role in virtually all athletic events, from short-distance running to sports that require throwing and hitting skills.

Next, let us focus on parallel muscles where muscle fibers are parallel to the long axis of the fiber. One such muscle is schematically shown in Fig. 6.12. Such a muscle usually has a belly in the middle. Thus, the average tensile stress acting on a cross section that is perpendicular to the axis of the muscle decreases from one end of the fiber toward the midsection. The muscle is much stiffer overall at its ends because it is in this

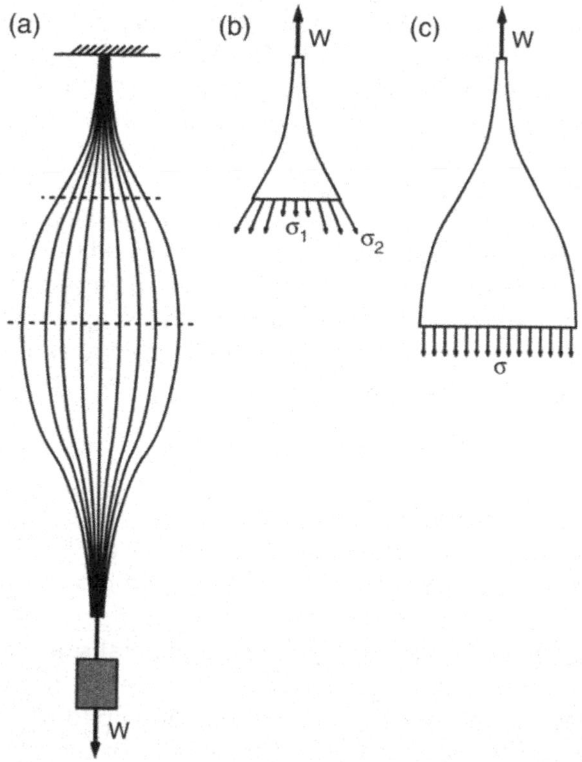

FIGURE 6.12a–c. Contraction of a parallel fiber muscle under the application of a weight **(a)**. The stress distribution on a cross section close to the ends of the muscle and at midsection are illustrated in **(b)** and **(c)**, respectively.

region that the fibers of the dense connective tissue of the muscle converge and become interwoven to each other to form tendon. Much of the force carried by the muscle is carried by this connective tissue in the end regions. In the belly region of the muscle, however, the cross-sectional area is large and the average tensile stress is small compared to the tensile stress carried by the adjacent tendons. Because the major muscles of the human body must produce forces comparable to the body weight, the cross-sectional area of a skeletal muscle in the belly region can be of the order of tens of centimeters squared. It is the belly part of the muscle that undergoes significant shortening during muscle contraction. The shape of the parallel muscle suggests that, in human body structure, the form (shape) may follow function.

Ligaments and Tendons

Muscles transmit forces to bones via tendons. The moment created by a muscle about a joint on which it acts depends on the moment arm, and that is determined by the geometry of insertion of its tendon to the bone. Unlike the muscle, the tendon has more or less a constant cross-sectional area. In vitro, tendons have been stretched under the application of tensile loads. The stress–strain relationship is highly nonlinear. The tendon is more compliant at low loads than at high loads. Tendons typically have linear properties to a strain of about 9%. Physiologically, they appear to operate at a stress of 5 to 10 N/mm^2.

Ligament properties are similar to those of tendons. Ligaments like tendons carry tension and again like tendons can store elastic energy. They will stretch small amounts under the application of tensile force, and under normal circumstances, they return to their resting length upon the lifting of load. Their ultimate tensile strength is comparable to that of tendons.

Bones

Bones are the stiffest of all musculoskeletal tissue. Long bones are irregular hollow cylinders filled with a loose cellular tissue (the marrow) containing blood and other matter. The compact bone is a composite of organic and inorganic material, the organic phase being nearly all collagen. The inorganic phase consists of water and a mineral salt called hydroxyapatitie. Electron microscopy has shown that the hydroxyapatite is in the form of very fine needles only 15 nm wide and up to 10 times as long. These needles are brittle but strong. The bone matrix in which they are embedded lowers the stiffness and protects the needles from breaking. The average value of Young's modulus (stiffness coefficient) is 20,000 N/mm^2, about one-tenth of that for steel. If the bones taken from a cadaver are dried and then tested, they fail at a tensile stress of about 100

N/mm^2 at a strain of 0.004. Bones, as is fitting for their irregular shapes, carry complex stress systems. Let us illustrate this with the following example.

Example 6.9. Bone Fracture. A man whose humerus had fractured earlier believed that he was now healed but that he had a stiff elbow. He lay on a bed and asked a friend to gently help him flex his elbow (Fig. 6.13a). His friend placed one hand on the forearm just below the elbow and the other just above the wrist and pushed in opposite directions with a force of 15 N. The distance between his hands was 19 cm. Apparently, he had pushed too strongly; the humerus failed at its weakest point, the original fracture site, shown as BB′ in the figure. Determine the maximum tensile stress that occurred at the fracture site during bending. Assume that the normal stress varied linearly along the cross section of the humerus. The diameter of the humerus was 7 cm and the thickness of the compact bone was 1.5 cm at the site of the fracture.

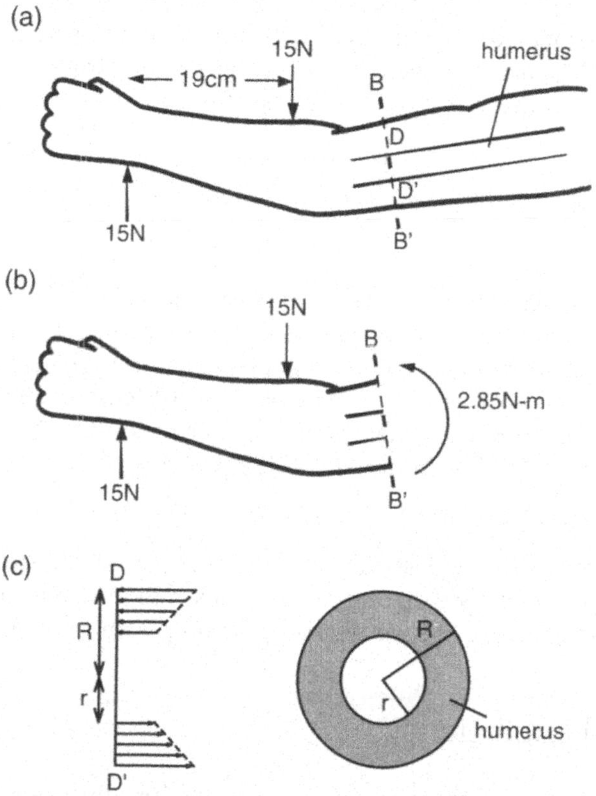

FIGURE 6.13a–c. Fracturing of the humerus bone of a person with stiff elbow, resulting from incorrect manipulation at the forearm **(a)**. The free-body diagram of part of the arm is shown in **(b)**. The stress distribution on a cross section of the humerus is illustrated in **(c)**. The original fraction was at BB′.

Solution: The free-body diagrams of the arm is shown in Fig. 6.13b. For simplicity in analysis we neglected the weight of the arm. Because the elbow was stiff, it did not bend during the manipulation of the forearm. The force couple exerted on the forearm was transmitted to the upper arm. According to the free-body diagram, the magnitude of the net moment acting on a cross section of the humerus is given as follows:

$$M = 15\text{ N} \cdot 0.19\text{ m} = 2.85\text{ N-m}$$

What kind of stresses does a bending moment cause on a cross section normal to the long axis of the bone? We had seen earlier (in Chapter 5) that bending moment caused axial stress in a cantilever beam. If the humerus could be considered as a linearly elastic solid, the stress distribution would be linear (Fig. 6.13c). The stress is minimum at the inner edge and maximum at the outer edge. This maximum stress (σ_o) is related to the moment M acting on the cross section by the formula:

$$\sigma_o = M\,(h/2)/J_x \tag{6.33}$$

in which h denotes the maximum dimension of the cross section in the y direction, and J_x, axial moment of inertia, is a geometric parameter that depends only on the shape of the cross section.

$$J_x = (bh^3/12) \tag{6.34a}$$

for a rectangle with width b and height h, and

$$J_x = (\pi/4)\,(R^4 - r^4) \tag{6.34b}$$

for an annulus with inner radius r and outer radius R.

The cross section of the humerus occupied by compact bone could be represented as an annulus with outer radius equal to 3.5 cm and thickness equal to 1.5 cm. Then, $h = 7.0$ cm and $J_x = 105\text{ cm}^4$. Under these conditions the maximum normal stress σ_o corresponding to the cross-sectional moment of 2.85 N-m would be equal to

$$\sigma_o = 285\text{ N-cm }(3.5\text{ cm})/(105\text{ cm}^4) = 9.5\text{ N/cm}^2$$

Although this value is less than the yield stress for failure of normal young adult bone, the yield stress in the crack region would be expected to be much less because the fracture had not completely healed at the time.

6.9 Limb-Lengthening

Bones of a living person may behave quite differently from bone taken from a corpse and dried. Living bone is self-repairing. Alterations in the distribution of stress in a bone could yield in significant growth or remodeling. In the low-gravity situation of space flight, the compressive stresses acting on the bones are much less than that on earth, and bones lose thickness and strength. On the other hand, on earth, the bones of the leg, which carry the weight of the body, thicken with age.

Orthopaedic surgeons have begun exploiting the relationship between bone stress and bone growth to correct skeletal abnormalities. In the 1940s, in an isolated hospital in Siberia, Professor Gavriil Ilizarov came up with an ingenious method to treat limb length inequality, congenital limb deficiency, and other types of bone or joint deformities. Limb correction (lengthening) is reshaping of a limb involving little invasive treatment. The procedure was described at length by Dr. Ilizarov in an article entitled "Clinical Application of the Tension-Stress Effect for Limb Lengthening" that appeared in 1990 in *Clinical Orthopaedics and Related Research.* Briefly, an external fixator (much like a bone scaffold) is applied on the affected bone (Fig. 6.14). The fixator is composed of a series of stainless steel cir-

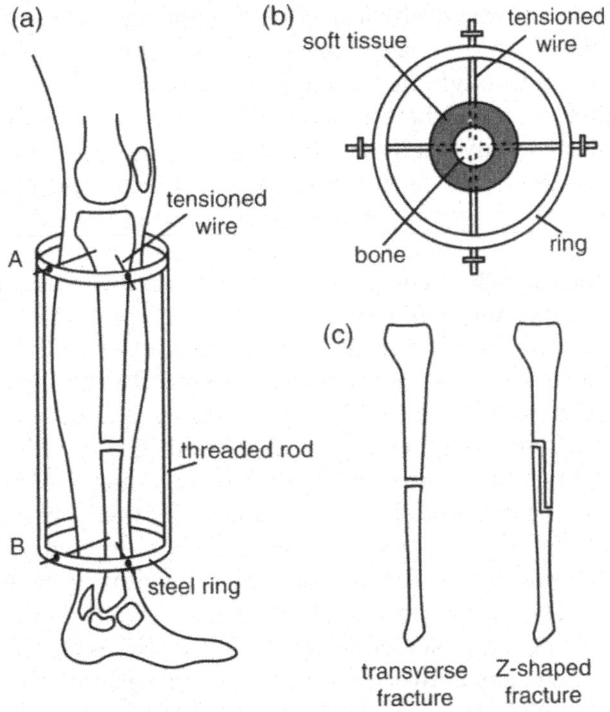

FIGURE 6.14a–c. Schematic drawing of limb-lengthening procedure as applied to the leg of a man **(a)**. Two steel pins are inserted at cross sections *A* and *B* on the two ends of the femur **(b)**. While *A* is kept stationary with respect to the thigh, *B* can be moved along the axis of the bone. The bone is separated into two parts in the midregion and the distance between *A* and *B* is slightly increased. Soon new bone will form in the cracked region. The two different types of fractures used in the operation are shown in **(c)**.

cular rings that are attached together by threaded rods. Each ring is attached to the limb through the bone by taut steel wires and thicker titanium pins. Two wires that are perpendicular to each other cross through the bone, keeping the bone stationary relative to the external fixator.

Then, the bone is cut in half through a small incision with minimal damage to the surrounding muscle tissue. The distance between the two rings is increased slightly. In response to the sudden relaxation of stress, the surface of separation acts much like a growth plate. In the fracture gap, fibroblasts secrete and arrange new collagen fibers along the long axis of the bone. Osteoblasts use these newly formed fibers as templates to form new bone. The gap is filled by this fresh tissue in 4 to 5 days. After that period, the distance between the pins is increased at small increments (1 mm/day). This procedure is called distraction. Elongation of the freshly formed bone tissue stimulates further growth and remodeling. With time, the new bone gains the ultimate strength of the old bone. When this procedure is applied to the thigh or lower leg of a patient, he or she continues to have the use of the leg while the bone growth and remodeling occurs. However, the load that would normally be carried by the bone is transmitted through the pins to the scaffold.

The procedure described here uncovered the fact that a surgeon can create a growth plate anywhere in bone with appropriate conditions of bone fracture, fixation of the separated bone segments, and distraction. The formation of new bone only in a fracture surface (or growth plate in children and adolescents) may be a reflection of the contact inhibition of growth of biological cells. Cells grown in a culture will continue to divide until they cover the entire surface area of the plate. Cell division ceases with the establishment of multiple physical contacts with the neighboring cells. Cancer cells, on the other hand do not observe contact inhibition, and they grow to form multiple cell layers in a culture dish.

Example 6.10. Limb-Lengthening. A 17-year-old girl with 5-cm tibial shortening underwent a single-fracture limb-lengthening. Determine the physical stress in the tibia and the surrounding soft tissue when the body is in upright configuration. The girl weighed 54 kg. The initial length of her lower leg (L) was 22 cm. The average outside radius of the lower leg (R_m) was 3.5 cm. The average radius of the tibia (R_b) was 1.3 cm. Young's modulus of the soft tissue surrounding the bone (E_m) was 100 N/cm^2. The initially imposed gap between the adjacent surface of fracture (Δ) was 2 mm.

Solution: The free-body diagram of the lower leg in the standing position is shown in Fig. 6.15. The lower leg is modeled as a concentric circular cylindrical column composed of bone and the surrounding soft tissue. A compressive force that is equal to half the body weight is assumed to act on the flat top surface of the column. With this assumption, we have lumped the mass of the lower leg with the rest of the body and po-

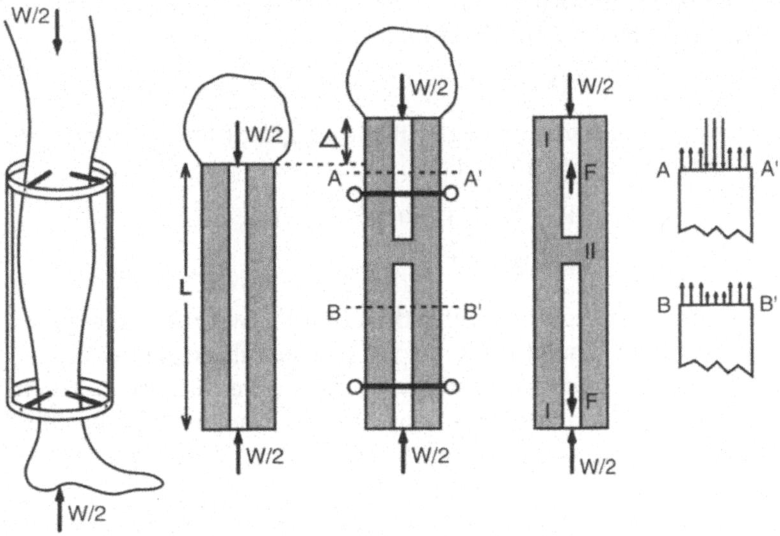

FIGURE 6.15. The free-body diagram of a leg during the process of bone and soft tissue lengthening.

sitioned it on top of the column. As shown in the figure, the net effect of the external fixator on the bone can be represented by a pair of equal and opposite forces acting in the vertical direction. The magnitude of this force is denoted by F. The part of the cylinder identified with the symbol I in the figure is under a uniform compressive force of W/2. In this region, at section AA′, the stress distribution must obey the following relation:

$$-\pi R_b^2 \sigma_b + \pi (R_m^2 - R_b^2) \sigma_m = -W/2 \tag{6.35}$$

where σ_b and σ_m represent the axial stress in the bone and the soft tissue, respectively.

How do we determine the relative values of σ_b and σ_m? The stress in the soft tissue can be computed using Hooke's law:

$$\sigma_m = E_m \epsilon = E_m \Delta/L = 100 \text{ N/cm}^2 (0.2/22) = 0.91 \text{ N/cm}^2$$

Substituting this value into Eqn. 6.35, we obtain

$$-3.14 (1.3 \text{ cm})^2 \sigma_b + 3.14 [(3.5 \text{ cm})^2 - (1.3 \text{ cm})^2] (0.91 \text{ N/cm}^2) = -(27 \text{ kg}) (9.81 \text{ m/s}^2)$$

$$\sigma_b = 44.2 \text{ N/cm}^2$$

Thus, the bone carries 50-fold-greater stress than the surrounding soft tissue. Next, we consider the stress distribution at cross section BB′. Because the bone is separated into two parts in that region, it carries no stress and the resultant force on this cross section from the stress in the soft tissue must be equal to (F - W/2):

$$\pi(R_m^2 - R_b^2)\,\sigma_m = -W/2 + F$$

Because σ_m is uniform throughout the length of the cylinder, we can use this equation to determine F:

$$F = W/2 + \pi(R_m^2 - R_b^2)\,\sigma_m$$

$$F = 294.97\ \text{N}$$

Thus, the force F is slightly greater than half the weight of the girl.

As the new bone tissue fills the gap in between the separated bone segments, the bone will begin to carry low levels of tensile stress. When the bone is distracted stepwise, this new bone will be stretched by the step of distraction.

6.9 Summary

The forces and the moments acting on the joints of the upper and lower limbs were considered. Main forces considered in analysis are those produced by body weight, muscles, and externally applied contact forces. In the case of impact problems, the forces carried by ligaments must be considered. Static analysis is valid strictly for bodies at equilibrium or moving at constant speed. For a body to be in static equilibrium, the sum of the forces acting on the body must be equal to zero. In addition, to ensure rotatory equilibrium, the sum of the moments exerted by external forces must be equal to zero. When a limb undergoes a rotation, the forces and moments acting on it do not add up to zero. Nonetheless, for sufficiently slow limb movements, static analysis provides reasonable accurate results. The equations of static equilibrium are used to develop estimates for the unknown muscle forces acting on a joint.

The concept of physical stress was also introduced. Stress is a measure of force intensity on a surface area. It is associated with two directions: the direction of the unit vector that is normal to the surface area and the direction of the force applied to it. The component of stress that is perpendicular to the area is called the normal stress. The component tangential to the area is denoted as shear stress. A material will crack at a point at which the stress intensity reaches a certain point. Some materials have large resistance to compression (concrete) and others to tension (steel). The long bones of our skeletal system can withstand compression better than tension. The bones also fracture under sufficiently high levels of shear stress.

The growth and remodeling of bones, ligaments, and tendons are modulated by physical stress. Orthopaedic surgeons have devised methods to lengthen limbs and do other corrections by appropriately altering the state of stress in the bones and muscles involved.

6.10 Problems

Problem 6.1. A man is performing dumbbell kickbacks to strengthen his triceps (Fig. P.6.1). He weighs 55 kg. His forearm and his hand together constitute 2.5% of his body weight. The length of his forearm is 26 cm. The lever arm of the triceps with respect to the center of rotation of elbow is 2.4 cm. The weight he carries is equal to 5 kg. Assuming that the man performs the dumbbell kickbacks slowly, determine the triceps force as a function of angle θ his forearm makes with the vertical axis. *Answer:* $F_T = 604 \sin\theta$ N.

Problem 6.2. The man performing the dumbbell kickbacks in Problem 6.1 wants to know how the rate of rotation of his forearm affects the ease of the exercise. So, he begins to raise his forearm from vertical to horizontal position as quickly as he can, still using the same weight. When his forearm makes 45° with the vertical axis, the angular velocity and angular acceleration of the forearm becomes -8 rad/s and -28 rad/s^2, respectively. A minus sign associated with angular velocity refers to clockwise rotation of the right hand. Determine the force produced by the triceps at the elbow at the instant considered. $I^c = 0.021$ kg-m^2. *Answer:* $F_T = 478$ (N).

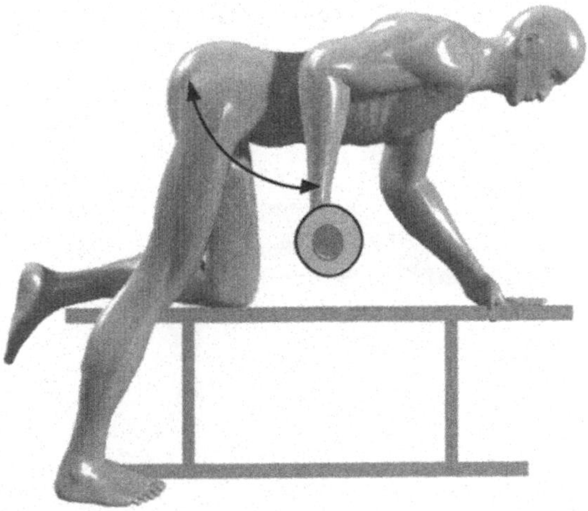

FIGURE P.6.1. A man performing dumbbell kickbacks.

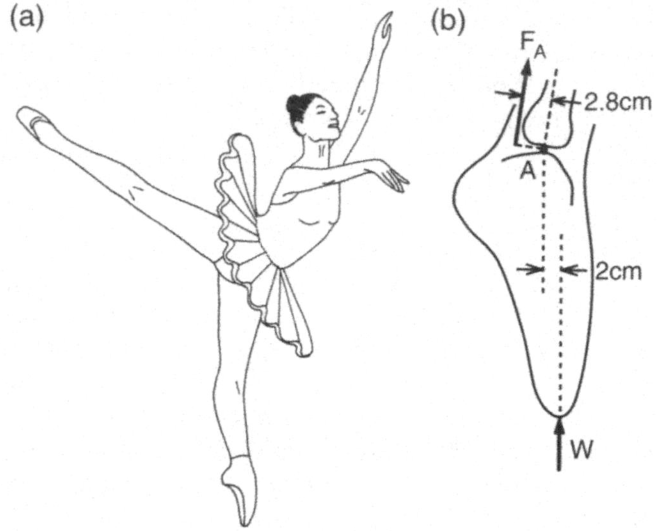

FIGURE P.6.3a,b. A dancer in arabesque position **(a)**. The free-body diagram of the foot that is touching the floor is shown in **(b)**.

Problem 6.3. A dancer weighing 46 kg takes the arabesque position shown in Fig. P.6.3. Determine the force in her Achilles tendon. Various dimensions concerning the structure of her foot are marked in the figure. *Answer:* $F_A = 322$ N.

Problem 6.4. The dancer shown in Fig. P.6.4 is performing a leg-swinging spin. His movements are slow and lyrical and therefore equations of sta-

FIGURE P.6.4. A dancer performing a leg-swinging spin.

tic equilibrium hold. He weighs 66 kg and his height is 1.71 m. Using the data provided in Appendix 2 (section on body segment properties), determine the moment created by the weight of the raised leg at the hip joint of the dancer. Which muscle group is principally responsible for keeping the lower limb horizontal as shown in the figure? Can you provide an estimate of the force generated by the contracting muscle?

Problem 6.5. A number of arm and leg movements are shown in Fig. P.6.5. What are the principal muscle groups used in each exercise? Determine the moment created by these muscles at the positions marked in the figure. For the computations, use your body dimensions and the tables given in Appendix 2. As to the weight (resistance) used in the movements, per-

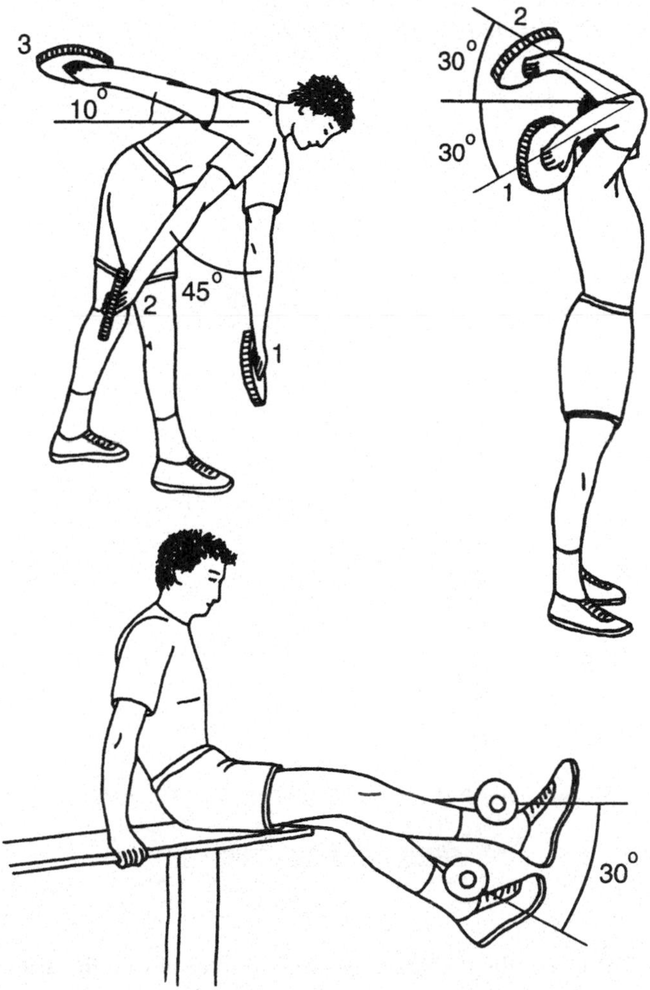

FIGURE P.6.5. Some arm and leg movements used for physical conditioning.

form the exercises yourself and determine the resistance for which you can comfortably repeat the movement at least six times. If you need additional data, consult an anatomy book and make educated guesses.

Problem 6.6. A weight lifter lies on a bench with arms extended horizontally on each side holding dumbbells of 10 kg on each hand (Fig. P.6.6). The lifter then gently pulls the arms toward the chest all the way to the vertical position. The pectoralis muscle is indicated in the figure as a cord connecting points A and B. The line connecting the points A and S makes 20° with the horizontal plane. Assuming that pectoralis muscle is responsible for this flexion action, compute the force generated by this muscle at $\theta = 0°$, 45°, and 90°. Evaluate the muscle length for each configuration.

Answer: $\theta = 90°$, F = 0, l = 18.9 cm
$\theta = 45°$, F = 1181 N, l = 22.6 cm
$\theta = 0°$, F = 4841 N, l = 24.8 cm.

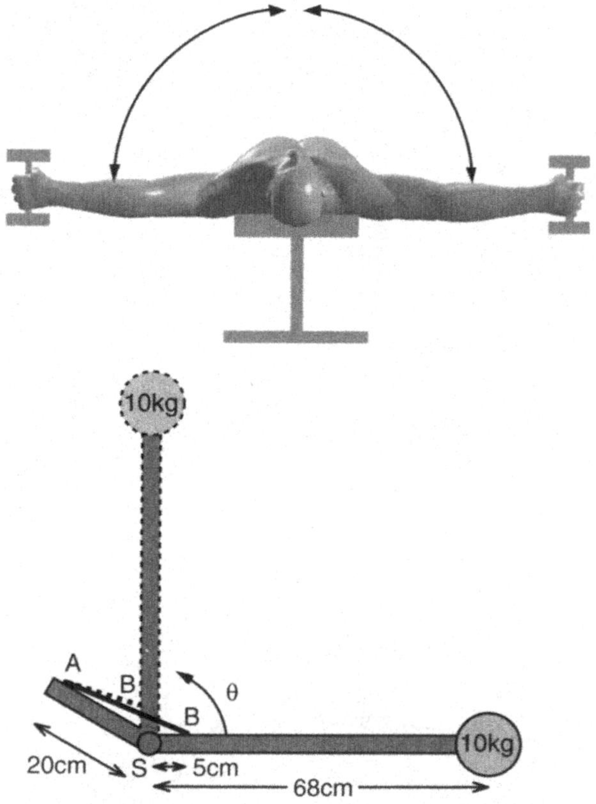

FIGURE P.6.6. The use of the pectoralis muscle group in flexing the arms while in supine position. The geometric parameters such as the points of origin and insertion of pectoralis (points *A* and *B*) are identified.

Remark: The force produced by the pectoralis muscle to lift the 10-kg dumbbell is enormous. In the human body, practically all muscular moment arms are short in proportion to the levers they move. As a result, muscles generate forces that are significantly greater than the load to be moved. This seeming disadvantage of the human body is in fact an advantage in certain circumstances. This advantage can be clearly illustrated in this problem. Having a short moment arm for the pectoralis means that a very small movement of the short end of the lever is magnified in direct proportion to the length of the arm. Thus, a small shortening of the muscle would move the dumbbell 90°. As the dumbbell moves a much larger distance than the attachment point of pectoralis to humerus, it is clear that contraction of the chest muscle can produce large velocities at the wrist, which is quite useful in tennis and many other sports involving throwing motion.

Problem 6.7. When an individual stands on one foot, the loaded knee supports the body weight (Fig. P.6.7). The line of application of this force is at a distance of 14 cm away from the center of rotation of the knee. The person weighs 68 kg. Determine the tension in the muscle–tendon complex composed of the gluteus maximus, the tensor fasciae latae, and the illiotibial band. The moment arm of this complex with respect to the center of rotation of the knee was measured to be 5 cm.
Answer: T = 1868 N.

Problem 6.8. Consider a man stretching his body by holding onto a bar with both hands. His body assumes the vertical position and his feet are not touching the ground (Fig. P.6.8a). The same man does the reverse exercise next; with his feet attached to a horizontal bar, he assumes the

FIGURE P.6.7. An individual standing on one foot. The weight acts eccentrically on the knee.

FIGURE P.6.8a,b. Stretching of the whole body by holding onto a horizontal bar **(a)** or by suspending with feet up **(b)**.

vertical position, again stretching his body (Fig. P.6.8b). In the third configuration the man stands still on his feet. Determine the axial force at his neck, waist, thighs, and ankles. Determine the average normal stress at these locations. The man weighs 68 kg and he is 1.78 m tall. The cross-sectional areas of his neck, waist, thighs, and ankles are 66, 280, 88, and 41 cm^2, respectively. Use the data given in Appendix 2 to determine other input parameters you may need in your computations.

Problem 6.9. A woman wearing high-heeled boots is climbing stairs. The heel of one boot gets stuck on a small hole on the staircase (Fig. P.6.9). Her body weight (52 kg) pushes the woman forward and the sharp edge of the staircase hits her lower leg. The lever arm of her body weight with respect to the edge of the staircase d is 22 cm. The angle her lower leg makes with the vertical axis (θ) is 40°. Determine the maximum tensile stress acting on the cross section of the tibia just above where the edge of the staircase hit her leg. The inner and outer diameters of compact bone segment of the tibia at that cross section are 2 and 3.2 cm, respectively. Hint: Consider the free-body diagram shown in Fig. P.6.9b. The equations of the static equilibrium require that on cross section BB′

$$F_1 = W \cos \theta$$

$$F_2 = W \sin \theta$$

$$M = W d$$

Assume that this internal force system is carried by the tibia. Both the axial force F_1 and the moment M lead to normal stresses in the bone. The tangential force F_2 leads to shear stress and therefore is not considered further. The normal stress distribution from F_1 is uniform in the cross section. The normal stress distribution from M, on the other hand, varies linearly as shown in Fig. P.6.9c and can be computed by using Eqns. 6.33 and 6.34b. To calculate the normal stress at a certain point in the cross section, one must add both contributions. In the addition, tensile stress is considered positive, compressive stress negative. *Answer:* $\sigma = (4{,}111 - 80)\ \mathrm{N/cm^2} = 4{,}031\ \mathrm{N/cm^2}$.

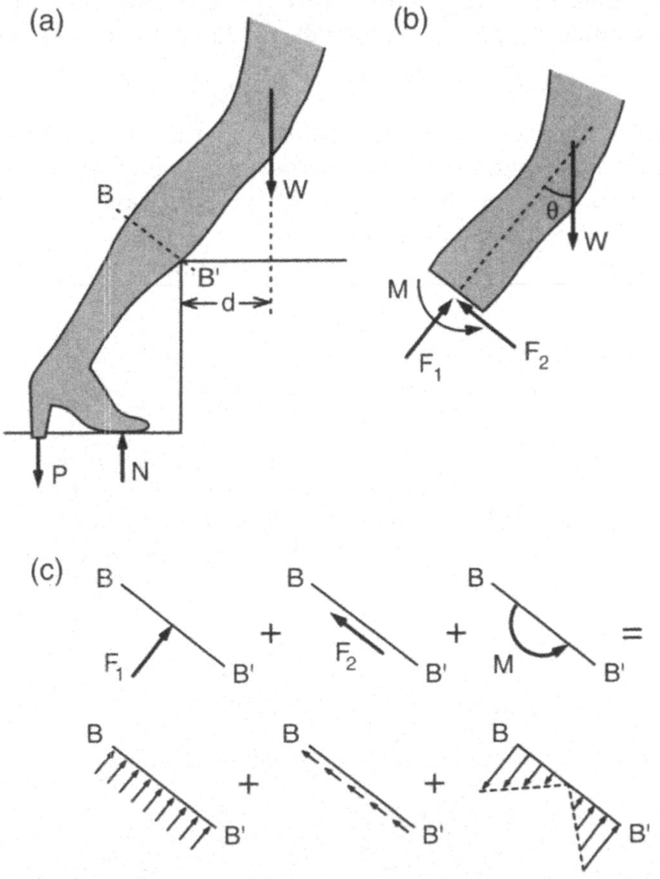

FIGURE P.6.9a–c. While she is climbing stairs, the high heel of a woman's boot gets stuck in a small hole **(a)**. The free-body diagram of the woman's leg minus the lower part is illustrated in **(b)**. The resultant stress distribution on the cross section *BB'* is schematically shown in **(c)**.

Problem 6.10. Determine the axial stress versus axial strain curves of three specimens obtained from materials such as various fabrics, strings, springs, or leaves and branches. Fix the specimen at one end and apply weights at the other. Videotape the elongation of the specimen under the application of the load. Quick-capture and digitize the data using a computer and plot stress versus strain. How would you determine the cross-sectional areas of your specimens? If a specimen has a constant depth, like that of a fabric, one could use the parameter (force divided by the width of the specimen) as a substitute for stress.

Problem 6.11. An elastic spring of stiffness k and force-free length L_o is attached to a plate at one end (Fig. P.6.11). The spring reaches steady-state length of L_1 under the application of a large weight W_1 at its free end. At that stage, the weight W_1 is replaced with a smaller weight W while keeping the extended length of the spring constant. After the exchange of the weights, the spring is allowed to go free. Show that the velocity v of the contracting spring is given by the following expression:

$$v = [g/(k\ m)^{1/2}]\ (m_1 - m)\ \sin\ ((k/m)^{1/2}\ t)$$

where $m = W/g$ and $m_1 = W_1/g$. The velocity v is a function of time and the ratio m/m_1. When the spring mass system is first let go, its velocity is equal to 0. To come up with a single velocity value for each different (m/m_1) ratio, try two different definitions: (1) velocity v^* is the maximum velocity of mass m during the contraction of the spring, and (2)

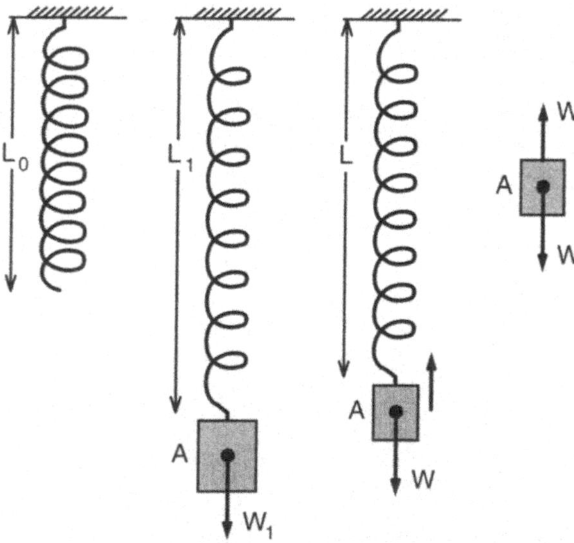

FIGURE P.6.11. Contraction of a spring from a prestretched position. The weights W and W_1 correspond to masses m and m_1, respectively.

velocity v^* is the velocity of m at 0.1 s after the spring–mass system is released. Use the following parameter values in plotting v^* as a function of the relative load m/m_1: $m_1 = 0.5$ kg, $L_o = 20$ cm, $k = 100$ N/m. Compare your results with the force–velocity relationship of a contracting skeletal muscle fiber. Note that the contraction velocity V that was defined in the text is dimensionless. It would be equivalent to the term v^*/L_1 in this problem.

Problem 6.12. A 17-year-old girl with 5-cm tibial shortening underwent a single fracture limb lengthening (Fig. P.6.12). The girl weighed 54 kg. The initial length of her lower leg (L) was 22 cm. The average outside radius of the lower leg (R_m) was 3.5 cm. The average radius of the tibia (R_b) was 1.3 cm. The Young's modulus of the soft tissue surrounding the bone (E_m) was 100 N/cm^2. The initial gap between the adjacent surface of fracture (Δ) was 2 mm. What is the physical stress at section

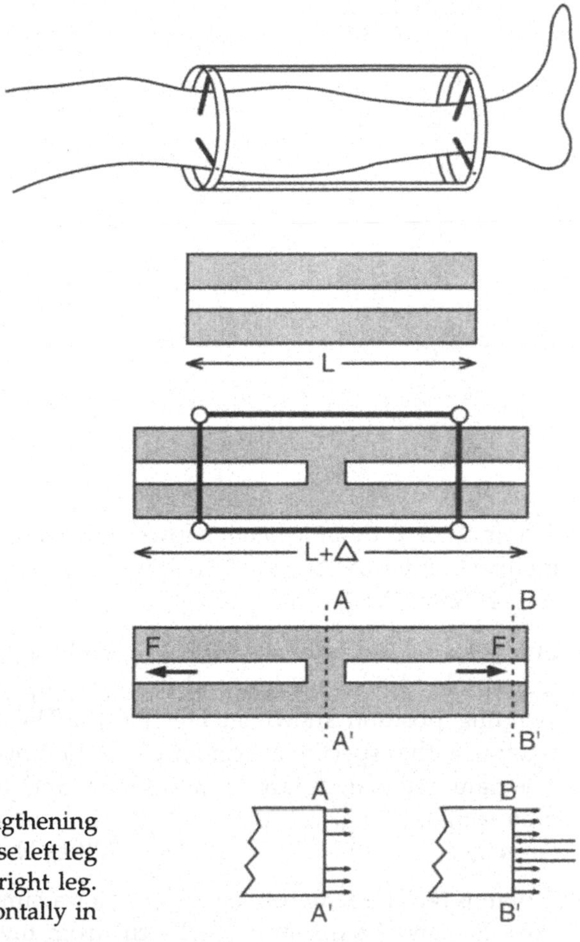

FIGURE P.6.12. The limb-lengthening procedure on a patient whose left leg was 5 cm shorter than the right leg. The leg is positioned horizontally in the figure.

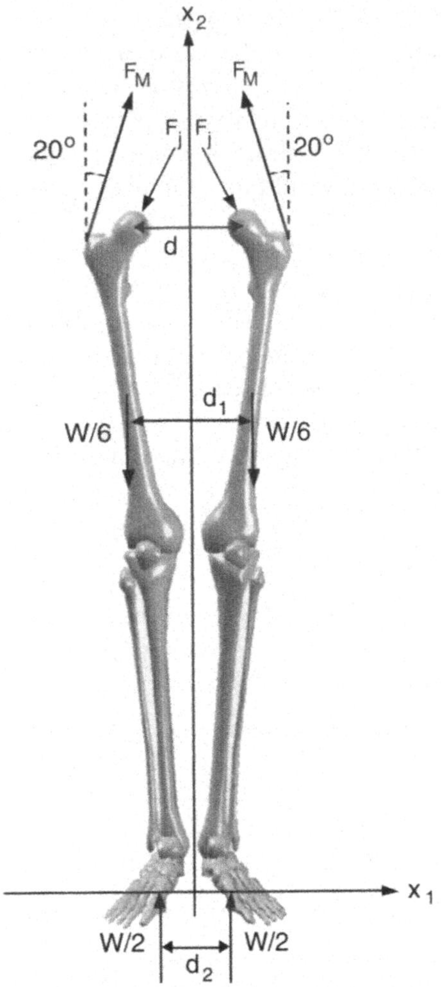

FIGURE P.6.13. The force acting on the thigh bone femur during the standing position.

AA′ in the soft tissue when the lower leg is positioned horizontally immediately after the bones of the lower leg were cut into two?
Answer: $\sigma = 0.91$ N/cm^2.

Problem 6.13. Determine the force F_M produced by the principal abductor muscle gluteus medius and the total hip joint force F_j during the standing position shown in Fig. P.6.13. The lever arm c of gluteus medius with respect to the center of rotation of the hip is equal to 7 cm. The parameters d, d_1, and d_2 are 36, 37, and 16 cm, respectively. The man weighs 73 kg.
Answer: $F_M = 520$ N, $F_J = -178\,\mathbf{e}_1 - 728\,\mathbf{e}_2$ N.

Problem 6.14. The femorotibial joint is not a simple hinge, but the bone force F_R acts at a distance $d = 2.4$ cm from the center of rotation (Fig.

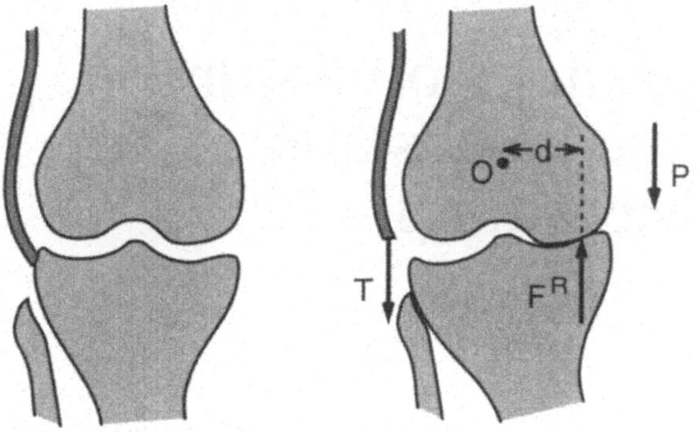

FIGURE P.6.14. Anterior view of the knee joint. The stress-bearing area is dual-cup shaped. The joint force transmitted from femur to tibia lies at a distance $d = 2.4$ cm away from the center of rotation of the knee.

P.6.14). Compute the joint force F_R and the tension T in the gluteus maximus for an individual standing on one foot. The lever arm of the weight P = 540 N is about $L = 14$ cm. The lever arm of the tendon of gluteus maximus $h = 8.1$ cm.

Answer: T = 597 N, F_R = 1137 N.

7

Impulse and Momentum: Impulsive Forces and Crash Mechanics

7.1 Introduction

A large force acting on a body during a very brief period of time may instantaneously alter the velocity of the body. Such forces often occur when two bodies collide. A runner crushes down upon his or her heel with a briefly sustained but intense force that often reaches many times the body weight. Each heel strike sends shock waves through the body, causing accelerations as high as 15 *g*. Running is not the only mode of motion during which impulsive forces act on humans. Accidental falls are the leading cause of death from injury among persons aged 65 years and older. Nearly 300,000 hip fractures occur in the United States in a year. Finally, it must be noted that car-crash injury is still the number one killer for adults under the age of 35 years. To that end, the mechanics of objects impacting on padded surfaces are of great interest in biomechanics. The risk of injury from striking an automobile dashboard is of obvious importance in everyday life. In this chapter, we present an introduction to impact mechanics. We illustrate how the impact forces affect the movement and motion of the human body. The analysis is based on the mathematical relationship between impulse and momentum.

7.2 Principle of Impulse and Momentum

The impulse ζ of a force **F** over a time interval $\Delta t = t_f - t_i$ is the integral of the force **F** over the time interval:

$$\zeta = \int_{ti}^{tf} \mathbf{F}dt \qquad (7.1)$$

The entity ζ is a vector and has the units of N-s. If the force does not change direction during the time period Δt, the magnitude of the impulse is equal to the area under the $\|\mathbf{F}\|$–time curve (Fig. 7.1).

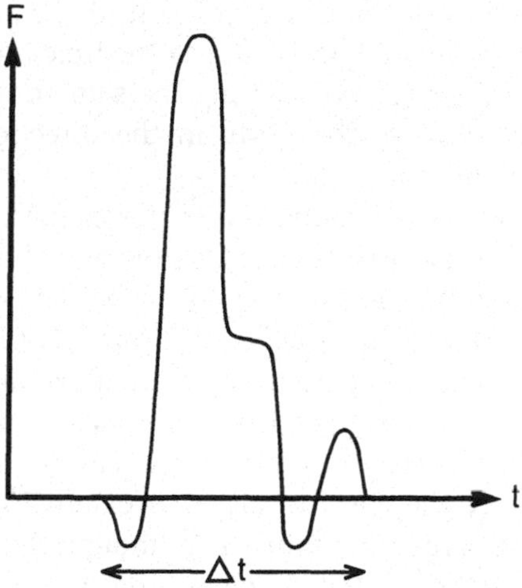

FIGURE 7.1. The plot of a magnitude of impulsive force F against time. The magnitude of the impulse ζ generated by F is equal to the area under the force–time curve.

The impulse acting on an object is directly related to the change in linear momentum. The linear momentum **L** of an object at time t was previously defined as

$$\mathbf{L} = m\,\mathbf{v}^c \tag{7.2}$$

in which m is the mass of the object and $\mathbf{v}^c$ is the velocity of its center of mass at time t (Fig. 7.2). According to Newton's second law, the equation of motion of the center of mass of an object is

$$d\mathbf{L}/dt = \Sigma\mathbf{F}$$

where $\Sigma\mathbf{F}$ denotes the resulting force acting on the object. Integrating the equation of motion for the center of mass between time $t = t_i$ and $t = t_f$, we obtain the following relationship:

$$m\,\mathbf{v}^c{}_f - m\,\mathbf{v}^c{}_i = \Sigma\zeta \tag{7.3}$$

in which $m\mathbf{v}^c{}_f$ and $m\mathbf{v}^c{}_i$ denote, respectively, the linear momentum of the body immediately after and before the time interval $\Delta t = t_f - t_i$, and $\Sigma\zeta$ is the linear impulse imparted to the system by external forces during that

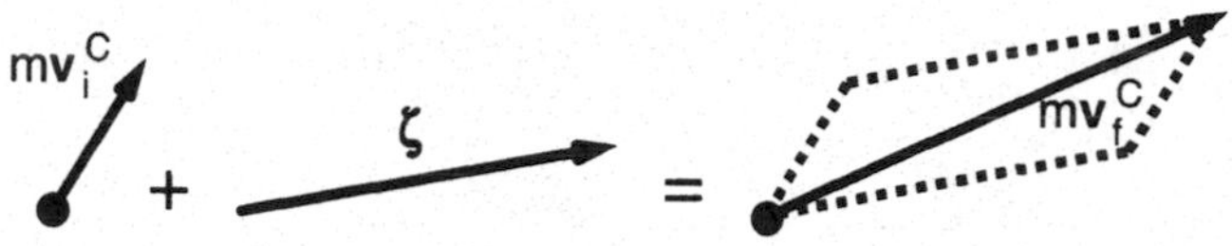

FIGURE 7.2. The effect of impulse on the linear momentum of an object. The vectorial sum of the impulse and the linear momentum before the impulse must be equal to the linear momentum after the impulse.

period of time. According to this equation, the change in the linear momentum of a body during the time interval Δt is equal to the impulse acting on the body during the same time interval (Fig. 7.2). The linear momentum of the body in the direction normal to the impulse remains unaffected.

In some situations an external force acting on a body is large compared to other forces exerted on the body but the time interval during which the force acts is small. A force that becomes very large during a very small time interval is called an impulsive force. When an impulsive force acts on a body, there may be an appreciable alteration in velocity during the period of application of the force. If some of the external forces acting on the object during a time interval $(t_f - t_i)$ are impulsive, we may neglect entirely the effect of all other external forces on the motion of the object in the same time interval. Although the velocities may be altered as a result of impulse, the change in the spatial position of an object is negligible.

Example 7.1. Use of Seat Belts in a Car Crash. A car traveling at a speed of 120 km/h hits a thick concrete wall. Because of the deformation of the front of the car during collision, it takes 0.5 s for the car to come to a complete stop. Determine the average impact force on a front-seat passenger who is (i) buckled and (ii) not buckled. Experments with dummies indicate that if a passenger were not to wear a seat belt, he or she would hit the windshield and that collision would take place in 1 ms. In this particular case, we assume the weight of the front-seat passenger to be 60 kg.

Solution: The case is schematically shown in Fig. 7.3. Because the impulsive force from the collision is much greater than other forces acting on the front passenger (the weight of the passenger and the contact forces between the seat and the passenger), impulse ζ in both cases is equal to

$$\zeta = 60 \text{ kg } [0 - (120 \text{ km/h})] \, \mathbf{e}_1 = -2000 \text{ kg} \cdot \text{m/s } \mathbf{e}_1$$

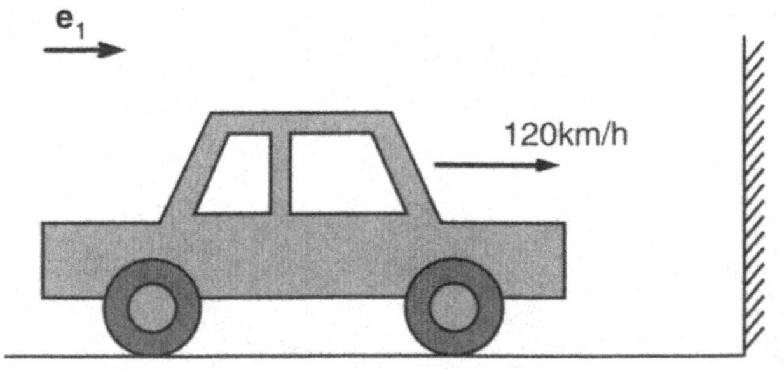

FIGURE 7.3. Collision of a car with a concrete wall. A front-seat passenger wearing a seat belt would decelerate as the front of the car was being deformed by the forces of impact. However, a passenger not wearing a seat belt would hit the front of the colliding car with the oncoming velocity.

in which $\mathbf{e}_1$ is the unit vector along the direction of motion before the crash occurs. This impulse on the passenger wearing the seatbelt occurs in 0.5 s and therefore the average crash force acting on the passenger wearing the seat belt is equal to

$$\mathbf{F}_{av} = -2{,}000\ (\text{kg} \cdot \text{m/s})/(0.5\ \text{s})]\ \mathbf{e}_1 = -4{,}000\ \mathbf{e}_1\ (\text{N})$$

Let us now consider the average acceleration of the passenger wearing the seat belt during the crash. The only significant horizontal force acting on the passenger is the force of impulse by the seat belt. This force divided by the mass of the passenger must be equal to its acceleration (deceleration in this case) and thus:

$$\mathbf{a}_{av} = -66.7\ \mathbf{e}_1\ (\text{m/s}^2)$$

This acceleration is approximately sevenfold greater than the gravitational acceleration, and studies on cadavers have shown that at this level of acceleration, the front-seat passenger might escape the crash without a significant head injury. Note that the longer it takes for the car to deform and absorb the shock, the better it is for the safety of the passenger. It is clear that cars loaded in the front with energy-absorbing materials such as steel will have lower values of average acceleration than a light and small car.

Next let us consider the front-seat passenger who is not wearing a seat belt. Even after the front of the car hits the wall and begins decelerating, because the passenger is not tied to the car, he or she will still be moving forward at 120 km/h. The passenger soon will hit the front windshield with this velocity, and the duration of the crash is 1 ms in this case. Therefore, the average crash force and the average acceleration during the crash become

$$\mathbf{F}_{av} = -2{,}000{,}000\ \mathbf{e}_1\ (\text{N})$$

$$\mathbf{a}_{av} = -33{,}350\ \mathbf{e}_1\ (\text{m/s}^2)$$

This value is about 3,400 times the gravitational acceleration, and surveys of car crashes indicate significant injury to the head at acceleration values above 200 g. It would have been a miracle if the passenger had survived.

When two objects collide, the impulse forces they exert on each other are equal in magnitude but opposite in direction. Writing Eqn. 7.3 for objects A and B and then summing them, one can show that

$$m^A\ \mathbf{v}^A{}_f + m^B\ \mathbf{v}^B{}_f = m^A\ \mathbf{v}^A{}_i + m^B\ \mathbf{v}^B{}_i \qquad (7.4)$$

in which m^A and m^B are the masses of objects A and B, and $\mathbf{v}^A$ and $\mathbf{v}^B$ denote their velocity of center of mass (Fig. 7.4). The subscripts i and f refer to times t_i and t_f, respectively. According to this equation, the sum of linear momentum of two objects during impact must be conserved. Impulsive force acts as an internal force for a system of colliding objects.

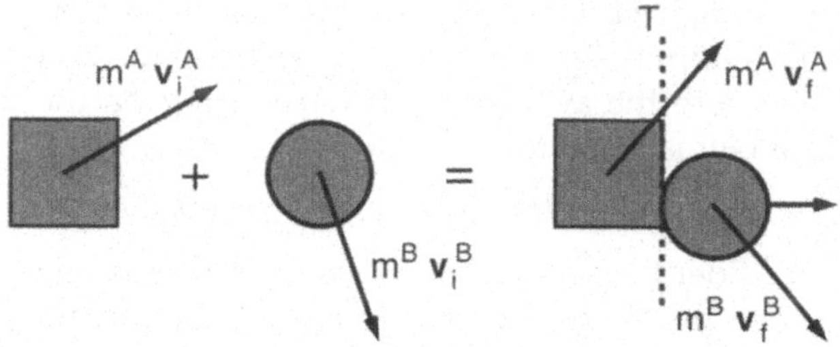

FIGURE 7.4. The conservation of linear momentum during collision of two objects A and B. The symbols m^A and m^B denote, respectively, the masses of objects A and B. The vector **v** denotes the velocity of the center of mass of an object. The subscripts i and f refer to times immediately before and after the collision, respectively. The vector **n** is the unit vector that is normal to the tangent plane T at the point of contact. Note that the sense of direction of **n** is chosen arbitrarily.

Let us illustrate this equation with a simple example from classical mechanics. Suppose that a rigid body is dropped from a height h upon a mass-spring system (Fig. 7.5). Almost at the instant of contact the mass attached to the spring acquires momentum, the falling body loses momentum, and the two bodies begin to move as one. The conservation of momentum requires that

$$m^A \mathbf{v}_o = (m^A + m^B)\, \mathbf{v}_f$$

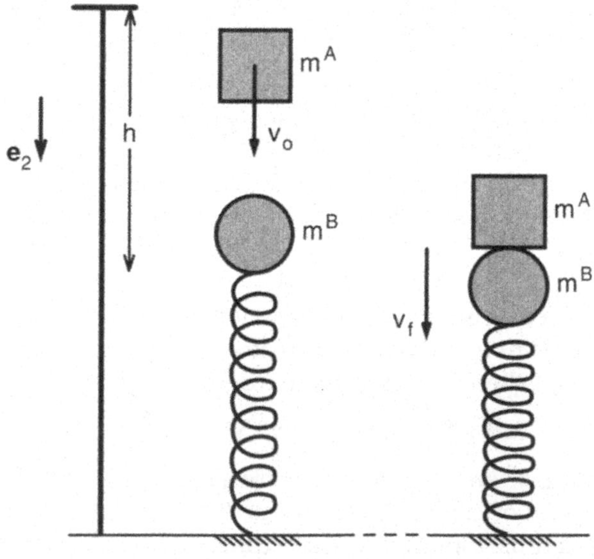

FIGURE 7.5. The free fall of a mass (m^A) onto a mass–spring system with mass m^B.

where m^A is the falling mass, m^B is the mass attached to the spring, $\mathbf{v}_o$ is the velocity of the center of mass of object A just before impact, and $\mathbf{v}_f$ is the velocity of the combination after the impact. Because the mass A falls from a distance h its velocity before the collision is given by the equation:

$$\mathbf{v}_o = (2hg)^{1/2}\,\mathbf{e}_2$$

in which g denotes as usual the magnitude of the gravitational acceleration and $\mathbf{e}_2$ is the unit vector in the direction of gravitational force. The velocity of the two masses A and B right after the impact is then given by the following relation:

$$\mathbf{v}_f = [m^A/(m^A + m^B)]\,(2hg)^{1/2}\,\mathbf{e}_2$$

Example 7.2. Quarterback Hits a Concrete Wall. Gus Ferrotte, the former quarterback of the Washington Redskins, scored a touchdown against the New York Giants in 1997 and then in the excitement of the moment roamed head-on to the nearby concrete wall (Fig. 7.6). Gus had a helmet on but nevertheless strained his neck. He also became a subject of jokes. To explain his behavior, he said to a television reporter, "I would hit heads with other football players after a touchdown and it never caused strain in my neck." Is there a difference between hitting a football player on the head and hitting a concrete wall?

Solution: Let us assume that Gus is represented by an object of mass m_1, and the football player he would hit head on by m_2. The masses m_1 and m_2 are comparable in magnitude. Assuming that Gus had a velocity $v_o\,\mathbf{e}_1$ before the collision and the other player was at rest and further assuming that after Gus hits him, they move in unison with velocity $v\,\mathbf{e}_1$, we find that:

$$m_1 v_o = (m_1 + m_2)\,v \tag{7.5}$$

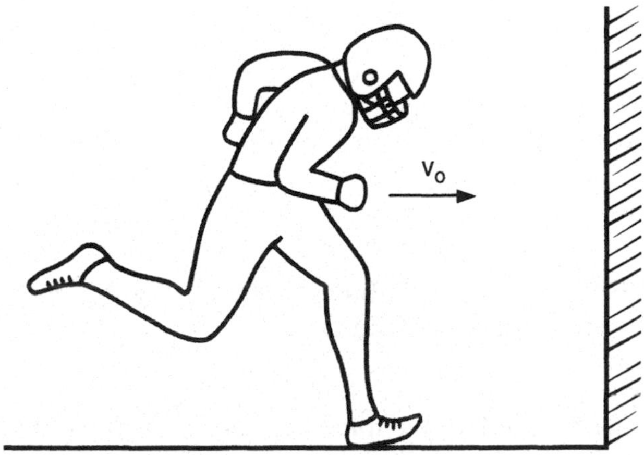

FIGURE 7.6. A quarterback running toward a concrete wall with the intention of hitting it head-on in celebration of his touchdown.

If the two football players were about the same weight, Gus's speed would decrease by one-half as a result of the collision. Thus, according to Eqn. 7.3, the impulse acting on him is given by the following relation:

$$\zeta = -(m_1 v_o/2)\,\mathbf{e}_1$$

Next, let us represent the concrete wall as an object with mass M. Because M is much greater than m_1, when Gus hits the concrete wall, he loses all his linear momentum during collision; hence, in this case

$$\zeta = -m_1 v_o\,\mathbf{e}_1$$

Thus, the impulse of collision with a concrete wall is twice as great as that of colliding head-on with another player.

7.3 Angular Impulse and Angular Momentum

The moment of momentum of a body B with respect to a point O that is fixed on earth was defined by the following integral over the mass of the body:

$$\mathbf{H}^o = \int \mathbf{r} \times \mathbf{v}\, dm$$

in which $\mathbf{r}$ and $\mathbf{v}$ are, respectively, the position and velocity of mass element dm in the inertial reference frame E (Fig. 7.7). We have shown (Chapter 3) in Eqn. 3.39b that

$$\mathbf{H}^o = m\,(\mathbf{r}^c \times \mathbf{v}^c) + \mathbf{H}^c$$

in which the superscript c refers to the center of mass. According to the conservation of moment of momentum:

$$d\mathbf{H}^c/dt = \Sigma\mathbf{M}^c \tag{7.6a}$$

$$d\mathbf{H}^o/dt = \Sigma\mathbf{M}^o \tag{7.6b}$$

in which $\Sigma\mathbf{M}^c$ and $\Sigma\mathbf{M}^o$ denote the resultant external moment about the center of mass and the fixed point O, respectively.

Integrating Eqn. 7.6 between times t_i and t_f we obtain:

$$\mathbf{H}^c_f - \mathbf{H}^c_i = \int \Sigma\mathbf{M}^c\, dt \tag{7.7a}$$

$$\mathbf{H}^o_f - \mathbf{H}^o_i = \int \Sigma\mathbf{M}^o\, dt \tag{7.7b}$$

Integrals in Eqns. 7.7a and 7.7b represent the angular impulses of the external forces about points C and O:

$$\Lambda^c = \int \Sigma\mathbf{M}^c\, dt \text{ and } \Lambda^o = \int \Sigma\mathbf{M}^o\, dt \tag{7.7c}$$

In the presence of impulsive forces, we need only to consider impulsive forces and moments when evaluating Λ^c or Λ^c. Because the time duration of an impulsive force is very small, contributions of regular forces to the angular impulse can be neglected.

FIGURE 7.7. A rigid body in planar motion.

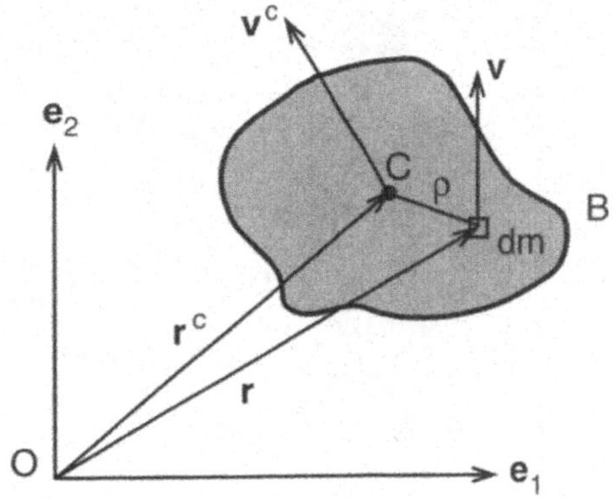

For rigid bodies that are symmetric with respect to the plane of motion, we have already seen in Chapter 4 that

$$\mathbf{H}^c = \mathrm{I}^c\, \omega\, \mathbf{e}_3 \tag{7.8a}$$

$$\mathbf{H}^o = \mathrm{I}^o\, \omega\, \mathbf{e}_3 \tag{7.8b}$$

in which ω is the angular velocity of the rigid body, $\mathbf{e}_3$ denotes the unit vector perpendicular to the plane of motion, and I^c and I^o are the mass moment of inertia with respect to the center of mass and point O, respectively. Combining Eqns. 7.8 with Eqns. 7.7, we obtain:

$$\mathrm{I}^c\,(\omega_f - \omega_i) = \Lambda^c \tag{7.9a}$$

$$\mathrm{I}^o\,(\omega_f - \omega_i) = \Lambda^o \tag{7.9b}$$

At this point, it is useful to recapture the principal assumptions used in the study of impact and collision.

1. Velocities and angular velocities may change greatly during the brief time interval of impact.
2. Positions of the bodies do not change appreciably during impact.
3. Forces (and moments) that are nonimpulsive (force and moments that do not assume very large values during the time of collision) are neglected.

As examples of negligible forces during impact we cite gravitational force and the force exerted by a stretched or compressed spring. Only the impulsive forces and moments produce sudden changes in velocities and angular velocities.

Example 7.3. A man standing still is struck by an impact force (Fig. 7.8). Assume that the man can be represented by a uniform rod of length L and mass m. The distance between the point of application of the impact force and the center of the rod is d. Determine the velocity of the center of mass and the angular velocity just after the blow.

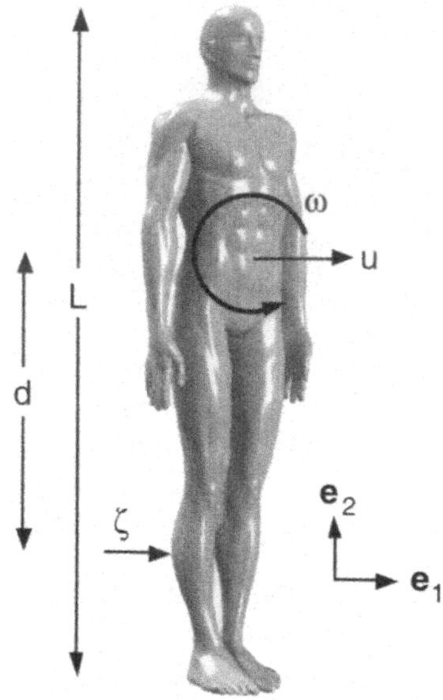

FIGURE 7.8. A man standing at rest is hit with an impulsive force.

Solution: If $u\ \mathbf{e}_1$ and $\omega\ \mathbf{e}_3$ are the velocity of the center of mass and the angular velocity just after the blow, the conservation of linear and angular momentum leads to the following equations:

$$m\ u = \zeta \Rightarrow u = \zeta/m \tag{7.10a}$$

$$m\ (L^2/12)\ \omega = d \cdot \zeta \Rightarrow \omega = d \cdot \zeta/[m\ (L^2/12)] \tag{7.10b}$$

Consistent with impact analysis, contributions of the body weight and the contact forces to the velocity after the impact were neglected in comparison to the contribution of the impulsive force.

Even when the magnitude of the impulse cannot be measured, it is possible to deduce a mathematical relationship between u and ω. Eliminating ζ from Eqns. 7.10a and 7.10b, we find

$$u = (L^2/12)\ \omega/d$$

Suppose the blow is struck at the end of the rod, so that $d = L/2$; then $u = (L/6)\ \omega$.

Example 7.4. A body of arbitrary planar shape is attached to a fixed axis through A as shown in Fig. 7.9. A blow of impulse ζ is exerted on the body at an angle ϕ with the vertical axis. Determine the angular velocity of the body immediately after the blow and the impulsive reaction at the hinge.

Solution: Let ω_o and ω denote the angular velocity before and after the impulse. The change in angular momentum is related to the angular impulse as follows:

$$m\ (h^2 + k^2)\ (\omega - \omega_o) = d \cdot \zeta \sin \phi \tag{7.11a}$$

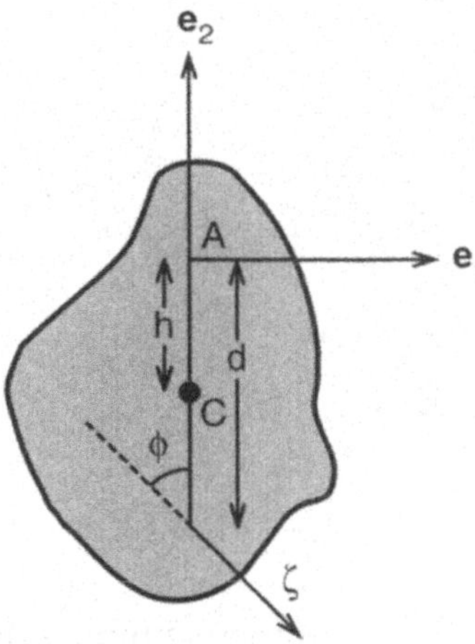

FIGURE 7.9. The effect of an impulse on the rotation of a rigid body around an axis that passes through point A.

in which h is the distance between the center of mass C and the hinge A, k is the radius of gyration of the object B with respect to C, and d is the distance between the point of application of the impulse and the hinge A. Rearranging Eqn. 7.11a, we obtain

$$\omega = \omega_o + d \cdot \zeta \sin \phi / [m\,(h^2 + k^2)] \tag{7.11b}$$

To compute the impulse acting on the hinge, we use the principle of conservation of linear momentum (see Eqn. 7.3):

$$m\,h\,(\omega - \omega_o) = \zeta \sin \phi + \int A_1 dt \tag{7.12a}$$

$$0 = \int A_2 dt \tag{7.12b}$$

in which A_1 and A_2 are the components of the impulsive reactive forces the hinge exerts on the object in consequence of the blow. The left-hand side of Eqn. 7.12b is zero because the velocity of the center of mass in the $\mathbf{e}_2$ direction is equal to zero.

Example 7.5. Thoracic Injury Potential of Basic Competition Tae Kwon Do Kicks. In a competition, a tae kwon do master hits the other on the chest with a tae kwon do kick. Consider a simple mechanical analysis in which the leg of the tae kwon do master is represented by a rod of mass m and length L (Fig. 7.10). The rod rotates about its center with angular velocity ω_o, and the end of the rod marked as A collides with the stationary mass M. The velocity of the center of mass of the rod is equal to zero before the collision. After the impact, the points of contact move with the velocity v. At this time the linear velocity of the center of mass of the kicking leg is u and its angular velocity is ω. Determine the velocities u, v, and ω. Determine the impulse exerted by the tae kwon do master.

Solution: Because points C and A belong to the same rigid body (rod), their velocities are related by the following equation:

$$v = u + \omega\, L/2 \tag{7.13a}$$

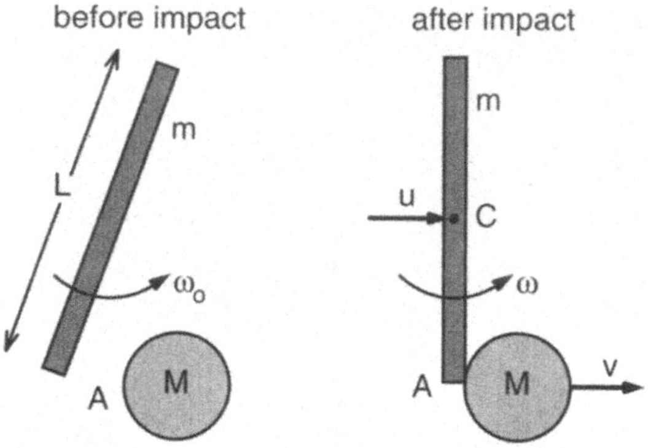

FIGURE 7.10. A tae kwon do master hits his partner on the chest with a tae kwon do kick. In the impact analysis the kicking leg is represented by a rod of mass m and length L. The partner is represented as a sphere with mass M.

The linear momentum of the system before the collision must be equal to the linear momentum after the collision:

$$0 = m\,u + M\,v$$

$$= m\,u + M\,u + M\,\omega\,L/2$$

$$\omega = -u\,(m + M)/(M\,L/2) \tag{7.13b}$$

Additionally, the moment of momentum about C of the system before the collision must be equal to the corresponding moment of momentum after the collision

$$(mL^2/12)\,\omega_o = (mL^2/12)\,\omega + M\,v\,(L/2) \tag{7.13c}$$

Combining Eqns. 7.13a and 7.13b, we find

$$v = -(m/M)\,u$$

Using this expression together with Eqn. 7.13b in Eqn. 7.13c, we obtain:

$$u = -\omega_o\,(L/2)\,(M)/(4M + m)$$

$$v = \omega_o\,(L/2)\,(m)/(4M + m)$$

$$\omega = \omega_o\,(M + m)/(4M + m)$$

Because the lower limb constitutes about 17% of the body weight, $m = 0.17\,M$, and thus:

$$u = -0.24\,\omega_o\,(L/2)$$

$$v = 0.04\,(L/2)\,\omega_o$$

$$\omega = 0.28\,\omega_o$$

These results indicate that right after the kick the center of mass of the kicking leg moves backward, whereas the man who is hit moves forward.

Let us next compute the impulse generated at the collision. The conservation of linear momentum for mass M leads to

$$\zeta = 0.04\,M\,\omega_o\,(L/2)$$

For $M = 75$ kg, $L = 1.02\ m$, and $\omega_o = 22$ rad/s, the impulse ζ turns out to be 33.6 N-s. If the impulsive force of the kick were to last 20 ms, then the average impulsive force would have been 1680 N.

Example 7.6. Impact Force Acting on a Runner's Heel. Runners hit the ground with one leg at a time, then use the heel of the colliding leg as a pivot to push forward (Fig. 7.11). The angular velocity of the leg before hitting the ground is $-\omega_o\ \mathbf{e}_3$. The velocity of runner's trunk before the impact is $v_o\ \mathbf{e}_1$. The length of the leg is equal to L. The mass of one leg constitutes 17% of the body mass m. For simplicity, we assume the weight of the leg to be negligible and that the weight of the runner is lumped on the hip. Determine the impulse exerted by the ground on the runner's heel. What is the angular velocity of the leg immediately after the impact?

Solution: When the heel A becomes fixed on the ground, the leg rotates about A. While the impulsive force is acting at A, the rate of change of

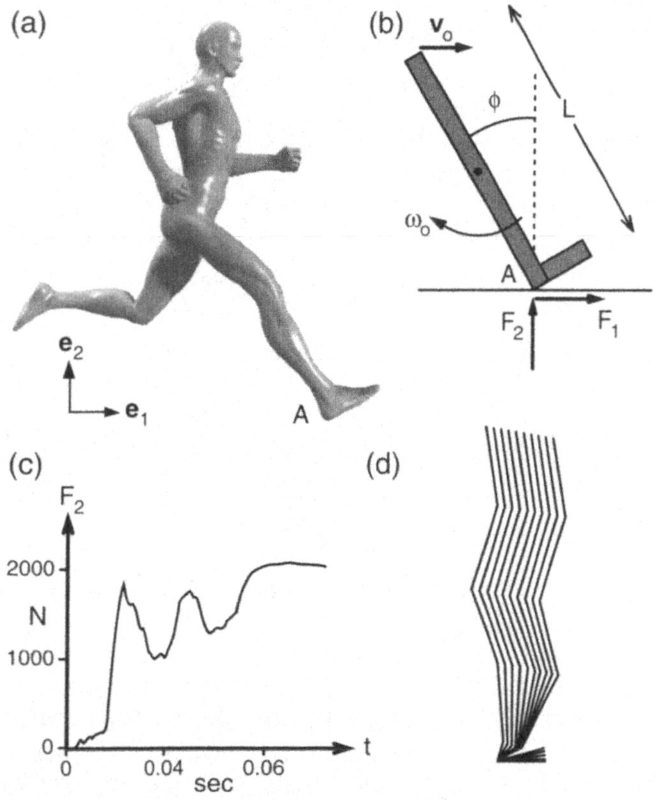

FIGURE 7.11a–d. The force of impulse during running **(a)**. The impulsive ground force acting on the heel of a runner is shown in **(b)**. The vertical component of the ground force is plotted as a function of time **(c)**. Stick figures for the first part of the contact phase are shown in **(d)**.

moment of momentum of the runner about A is zero. The moment of momentum A before the impulse (H_o $\mathbf{e}_3$) is given by the expression:

$$H_o = -m\, v_o\, L \cos\phi$$

where ϕ is the angle the leg makes with the vertical axis at the impact.
And after the impulse

$$H = -m\, L^2\, \omega$$

in which $-\omega$ is the angular velocity of the leg after the impact. Because there is no change of moment of momentum about A, we have $H = H_o$, which gives us

$$\omega = v_o \cos\phi\ /L$$

To determine the impulse during the impact, we need to know the velocity of the center of mass after the impact. When A becomes fixed and the leg rotates about A, the velocity of the center of mass becomes

$$\mathbf{v} = \omega\, L \cos\phi\, \mathbf{e}_1 + \omega\, L \sin\phi\, \mathbf{e}_2$$
$$= v_o \cos^2\phi\, \mathbf{e}_1 + v_o \cos\phi \sin\phi\, \mathbf{e}_2$$

This equation shows that the impact alters both the speed and the direction of velocity of the center of mass of the runner. As a result of the impact, the runner gains an upward velocity. Also, the speed of his center of mass is reduced to $v_o \cos\phi$ during the course of the impact.

To find the components of impulse, we consider the change in linear momentum:

$$\zeta_1 = m\,(\omega\, L \cos\phi - v_o) = m\, v_o\,[\cos^2\phi - 1]$$
$$\zeta_2 = m\,\omega\, L \sin\phi = m\, v_o \cos\phi \sin\phi$$

Let $L = 1.06$ m, $\phi = 35°$, $m = 71$ kg, $v_o = 6$ m/s, and $\omega_o = 25$ rad/s. Then we find:

$$\omega = 4.6 \text{ rad/s}$$
$$\zeta_1 = m\,(\omega\,(L/2)\cos\phi - v_o) = -140.2 \text{ N-s}$$
$$\zeta_2 = 200 \text{ N-s}$$

Assuming the duration of the impact to be 0.1 s, the mean impulsive force in the vertical direction is found to be 2,000 N. The vertical ground reaction force has received the greatest attention of biomechanics researchers because of its magnitude. Particular interest in the first early phase of the ground force (impact force) has been motivated by the concern about the transmission of shock waves upward through the musculoskeletal system. Runners who initially contact the ground with their heels tend to elicit a high force of short duration that has been termed as the impact peak. Magnitude of the impact force is considered as a potential indicator of overuse injuries in running, degenerative changes in joints, and low back pain. At running speed of 6 m/s this force is about three body weights, which is in agreement with our estimate. A typical time history of vertical ground force measured by using a force plate in the study of Bobbert et al. (1992) is shown in Fig. 7.11c. In this graph the impact force peaks at 1850 N. The data were collected on a runner whose steady-state speed was 4.5 m/s. The data indicate that the impact force increases with running speed.

We might ask the question whether the peak ground force measured during running qualifies to be called impulsive force? According to our definition, an impulsive force must be so much greater than any other force acting on a body so that the velocity of the body changes sharply during the brief period the impulsive force acts. Also, the duration of application of impact force must be so short that the position of the body does not change appreciably during impact. Bobbert et al. (1992) filmed the running events as they collected data on the ground reaction force. Stick figures representing a runner's body are shown in Fig. 7.11d for the first part of the contact force in heel–toe running. The leftmost figure shows the segment orientations on the last frame before touchdown. The time duration between two subsequent stick figures is 5 ms. As can be seen in the figure, there is a small change in the configuration of the runner during the first 100 ms of the contact. These observations suggest that the impact must occur in less than 0.1 s. Otherwise, impact force is not an impulsive force in the strictest sense—neither it is too large compared to the body weight nor does the configuration of the runner remains constant during the course of its application.

A note on the foot as a shock observer may be in order. The foot sustains impact forces and reduces potential injury to the body by deforming upon striking the ground. If the foot were a rigid object, the ground reaction force acting on it would be of great magnitude and short duration. The bones of the foot, however, are tied together by flexible ligaments and the movement of these bones relative to each other is also controlled by tendons. The presence of the deformable heel pad reduces the vertical force of impact. As the foot deforms in response to the force of impact, the ligaments and tendons stretch, absorbing much of the shock. As a result, the impulse is caused by a more sustained force of smaller amplitude.

7.4 Elasticity of Collision: Coefficient of Restitution

The examples of collision discussed in the previous sections of this chapter had two common features: (a) the impulse of collision occurred in the direction of common normal to the contact area, and (b) after the collision the velocity at the point of contact was the same in both solids. There are many cases in real life where these assumptions do not hold. For example, in running, the impulse of collision has both a tangential and normal component at the surface of contact. Condition b is not satisfied when a ball dropped onto a hard surface rebounds from the surface. In this section we relax these two assumptions and consider impulse in more general terms.

Let A and B denote two rigid bodies colliding in the time interval $t_f - t_i$. The times t_i and t_f represent the initial and final instants of a time interval during which the spatial positions of the two bodies remain essentially the same. Let P and P′ designate points that come into contact with each other during the collision of two bodies A and B. Let **n** be the unit vector in the direction of the common normal to the contact surface. How do we determine the unit normal **n**? Let T denote the common tangent plane to the surfaces of A and B at the point of contact. The vector **n** is a unit vector that is normal to T. When a runner hits the asphalt with one foot, the unit vector **n** could be chosen as the unit normal vector to the asphalt pointing outward from the ground. This is because both the foot and the running shoe are much more deformable than the asphalt, and therefore, at the point of contact, the shoe would assume the curvature of the asphalt ground. Note that although the direction of **n** is uniquely determined by the tangent plane T, the sense of direction of **n** is arbitrary.

Let us next resolve the velocities at the point of contact into two components, one in the direction of the normal **n** and the other in the tangent plane T:

$$\mathbf{v}^{\mathrm{P}} = u^{\mathrm{P}}\,\mathbf{n} + \mathbf{w}^{\mathrm{P}} \tag{7.14a}$$

$$\mathbf{v}^{\mathrm{P}'} = u^{\mathrm{P}'}\,\mathbf{n} + \mathbf{w}^{\mathrm{P}'} \tag{7.14b}$$

where u^{P} and $u^{\mathrm{P}'}$ denote the projections of the velocities of the contact points P and P′ along the **n** direction. The velocities $\mathbf{w}^{\mathrm{P}}$ and $\mathbf{w}^{\mathrm{P}'}$ are in the tangent plane T.

A parameter e called coefficient of restitution is introduced as a measure of the capacity of colliding solids to rebound from each other:

$$\mathrm{e} = (u^{\mathrm{P}}{}_f - u^{\mathrm{P}'}{}_f)/(u^{\mathrm{P}'}{}_i - u^{\mathrm{P}}{}_i) \tag{7.14c}$$

where the subscripts i and f refer to the times t_i and t_f. Thus, for example, $u^{\mathrm{P}}{}_f$ denotes the value of u^{P} at time t_f. The parameter e takes values from 0 to 1. When two solids stick at the point of impact, the collision is said to be plastic and e = 0. The coefficient of restitution is very nearly zero if one of the colliding solids is made of soft clay. In the opposite end, when e = 1, the velocity of approach has the same magnitude of velocity of separation in the **n** direction. The collision is said to be elastic when e = 1. In this case the impact of collision does not lead to dissipation of mechanical energy. A ball dropped from a height h onto a hard planar surface will rebound to the same height after the collision if e = 1. It must be emphasized that although parameter e is relatively easy to measure, its interpretation as a measure of deformability of the colliding bodies is difficult because it depends upon the materials, geometry, and initial velocities.

What happens during collision to the velocities of contact points in the tangential plane of contact? To address this question, let us resolve the impulse ζ into two components:

$$\boldsymbol{\zeta} = \boldsymbol{\nu} + \boldsymbol{\tau} \tag{7.14d}$$

where $\boldsymbol{\nu}$ is parallel to the common normal $\mathbf{n}$ and $\boldsymbol{\tau}$ is the projection of the impulse in the tangent plane at the point of contact. We assume that there is no slipping at time t_f if and only if

$$\|\boldsymbol{\tau}\| < \mu \|\boldsymbol{\nu}\| \tag{7.14e}$$

where μ, the coefficient of friction, is assumed constant for the colliding solids. Equation 7.14e is satisfied during running where the impacting foot does not slip on the ground. Most accidental falls occur, on the other hand, when this inequality cannot be satisfied. In such cases there is slip at time t_f and the impulse of collision in the tangential plane is related to the impulse in the normal direction by the following equation:

$$\|\boldsymbol{\tau}\| = \mu \|\boldsymbol{\nu}\| \tag{7.14f}$$

Note that the sense of direction of the tangential impulse $\boldsymbol{\tau}$ must be opposite to the tangential component of the velocity of separation.

Example 7.7. Kicking a Rock. An angry young man kicks a spherical rock (Fig. 7.12). The mass of his leg is M and the mass of the rock is m. At the instant the man hits the rock, his leg makes an angle of θ with the vertical axis. The angular velocity ω_o of his leg just before he hits the rock is in the counterclockwise direction. The length of his leg is L and the radius of the rock is R. Assuming e = 0.5, determine the subsequent motion and the impulse.

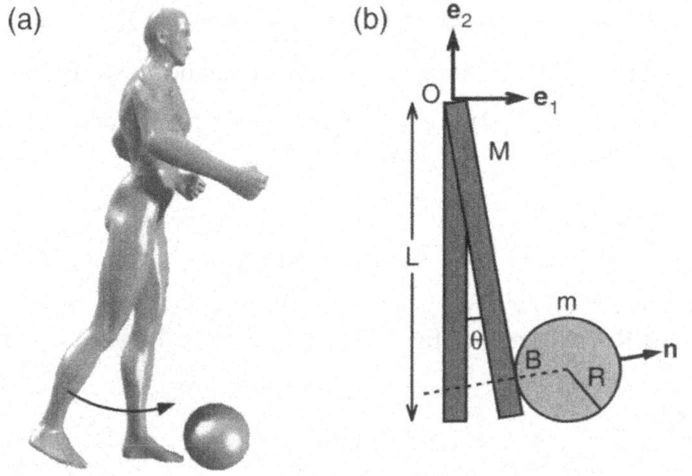

FIGURE 7.12a–b. A young man kicks a rock in anger **(a)**. The kicking leg is represented by a rod of mass M and length L and the rock by a sphere of mass m and radius R **(b)**.

Solution: Let us represent the lower limb involved in the kicking with a uniform rod of mass M and length L. We will assume that the rod is fixed at point O representing the hip joint. We will determine the location of the impact as well as its magnitude and direction. Let B and B′ denote the points of contact on the foot and on the rock during the course of impact. The collision occurs at B, a distance $R\,(1 - \sin\theta)$ from the ground. The geometry shown in the figure dictates that the point B is at a distance $L - R\,(1 - \sin\theta)$ from point O.

The direction of collision **n** is given by the equation

$$\mathbf{n} = \cos\theta\,\mathbf{e}_1 + \sin\theta\,\mathbf{e}_2$$

The velocity of point B before the collision is

$$v_o\,\mathbf{n} = [L - R\,(1 - \sin\theta)]\,\omega_o\,\mathbf{n}$$

The velocity of B′ before the impact is zero. Then, according to the equation of restitution (Eqn 7.14c), we have

$$\mathrm{e}\,\{[L - R\,(1 - \sin\theta)]\,\omega_o - 0\} = (u - v) \tag{7.15a}$$

in which u and v denote, respectively, the velocity of the rock and the rod at points B′ and B immediately after the collision. Because the rod is fixed at point O even after the impact, the velocity v is related to the angular velocity of the rod after the collision by the following relation:

$$v\,\mathbf{n} = [L - R\,(1 - \sin\theta)]\,\omega\,\mathbf{n} \tag{7.15b}$$

Because the impulsive force between the leg and the rock is an internal force, the moment of momentum for this system (rod plus sphere) should not be affected by the collision, and thus

$$(M\,L^2/3)\,\omega_o = (M\,L^2/3)\,\omega + m\,[L - R\,(1 - \sin\theta)]\,u \tag{7.15c}$$

Solving Eqns. 7.15a, 7.15b, and 7.15c for ω, v, and u, we find:

$$\omega = \omega_o\{(M\,L^2/3) - \mathrm{e}\,m\,[L - R\,(1 - \sin\theta)]^2\}/\{(ML^2/3) + m\,[L - R\,(1 - \sin\theta)]^2\}$$

$$v = [L - R\,(1 - \sin\theta)]\,\omega$$

$$u = [L - R\,(1 - \sin\theta)][\mathrm{e}\,\omega_o + \omega]$$

Assuming $\mathrm{e} = 0.5$, $\theta = \pi/6$, $\omega_o = 15$ rad/s, $M = 12$ kg, $m = 7$ kg, $L = 1.08$ m, and $R = 0.16$ m, we find

$$\omega = -1.5\ \mathrm{rad/s}$$

$$v = -1.5\ \mathrm{m/s}$$

$$u = 9\ \mathrm{m/s}$$

As the man hits the rock in the direction of common normal at the point of contact, neither the impulse nor the velocity of approach had a component in the tangential plane. The impulse ζ acting on the leg (rod) is given by the equation:

$$[L - R\,(1 - \sin\theta)]\,\zeta = (M\,L^2/3)\,(\omega_o - \omega)$$

$$\zeta = 4.67\,(15 - 1.5) = 63\ \text{N-s}$$

7.5 Initial Motion

If one of the supports of a body at rest suddenly gives way or is removed, the reactions at the other supports are instantaneously altered and the accelerations of the body are changed. To determine the altered reactions immediately after the removal of the support, we note that all the displacements and velocities resulting from the new accelerations are infinitely small and therefore can be neglected. The sudden removal of a support produces just the same effect as the sudden application of a force that is exactly the reverse of that exerted by the support while all other forces remain the same. A suddenly applied force is in no way comparable to an impulsive force. A finite force may alter the acceleration immediately but will require a finite time to generate a finite velocity. An impulsive force, on the other hand, produces finite changes of velocity instantaneously.

Example 7.8. A Cord Is Cut. A uniform horizontal bar of length L and mass m is supported by two vertical cords (Fig. 7.13). The distance between the two cords is d. If one of the cords is suddenly cut, determine the tension in the other cord immediately after the cutting.

Solution: Let T be the tension immediately after the cutting and T_o before. From equations of statics we know that T_o must be equal to half the weight of the bar ($T_o = mg/2$). Let α be the angular acceleration of the bar immediately after the breaking of the cord. The conservation of moment of momentum about the center of mass dictates that

$$T\,(d/2) = -(mL^2/12)\,\alpha \tag{7.16a}$$

The equation of motion of the center of mass in the vertical direction is

$$T - mg = m\,(d/2)\,\alpha \tag{7.16b}$$

Solving Eqns. 7.16a and 7.16b for α and T, we find

$$\alpha = -(6\,d\,g)/(3d^2 + L^2) \tag{7.17a}$$

$$T = (mg\,L^2)/(3d^2 + L^2) \tag{7.17b}$$

If the cords were attached at the ends of the bar before one was cut, then $L = d$, and $T = mg/4$. As the bar aligns itself in the vertical direction after the cord is cut, the tension T in the other cord eventually increases to become equal to mg.

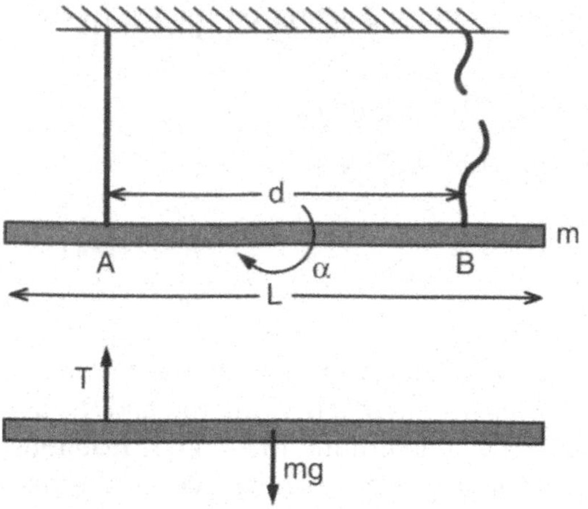

FIGURE 7.13. A horizontal bar is suspended from a ceiling using two cords. The sketch shows the bar at the instant immediately after one of the cords is cut.

Example 7.9. Resisting Gravity. A man of mass m makes a bet that he can hang by his hands from a parallel bar at least for a minute (Fig. 7.14). After 37 s, however, he can no longer stand the pain in his shoulders and lets one hand go. Determine the force exerted by the bar on the other hand of the man. Assume that the distance between the two hands of the man on the parallel bar is d and that the radius of gyration is k.

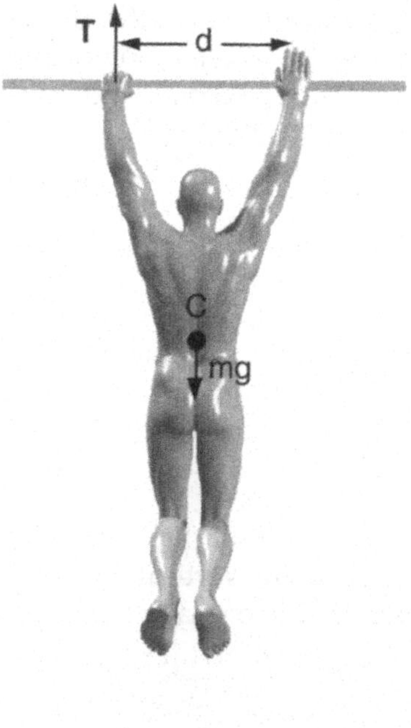

FIGURE 7.14. A man grabs a bar with both hands and remains suspended in air for about 35 s. To relieve the tension in his shoulders, he lets go of one hand. With one hand released, he acquires a clockwise angular acceleration α.

Solution: Let T be the vertical force exerted by the bar on the man immediately after he lets go his one hand, and $2T_o$ before. Because the resultant force acting on a body at rest is equal to zero, $T_o = mg/2$.

The conservation of angular momentum of the body about the center of mass requires that

$$-T\,(d/2) = m\,k^2\,\alpha \tag{7.18a}$$

where α denotes the angular acceleration of the man. The equation of motion of the center of mass of the whole body in the vertical direction is

$$T - mg = m\,(d/2)\,\alpha \tag{7.18b}$$

Note that in this equation the term $(d/2)\,\alpha$ represents the acceleration of the center of mass. Solving these two equations for α and T, we find:

$$\alpha = -2\,d\,g/(d^2 + 2k^2)$$

$$T = 2m\,g\,k^2/(d^2 + 2k^2)$$

Thus, immediately after the release of one hand, the man gains angular acceleration in the clockwise direction. The smaller the distance between the hands, the greater is the force exerted on the holding hand.

7.6 Summary

Impulse $\boldsymbol{\zeta}$ of a force $\mathbf{F}$ during a time interval $t_f - t_i$ is defined as the time integral of the force over $t_f - t_i$. The conservation of linear momentum before and after an impulse requires that

$$m\,\mathbf{v}^c{}_f - m\,\mathbf{v}^c{}_i = \Sigma\boldsymbol{\zeta}$$

in which m is the mass of the body, $\mathbf{v}^c{}_f$ and $\mathbf{v}^c{}_i$ are, respectively, the velocity of the center of mass at t_f and t_i, and $\Sigma\boldsymbol{\zeta}$ is the resultant impulse acting on the body.

Similarly, the conservation of angular momentum of a rigid body for which the plane of motion is a plane of symmetry yields the following equation:

$$\mathrm{I}^c\,(\boldsymbol{\omega}_f - \boldsymbol{\omega}_i) = \int \Sigma\mathbf{M}^c\,dt = \Sigma\,\boldsymbol{\Lambda}^c$$

in which I^c denote the moment of inertia with respect to the mass center, and $\boldsymbol{\omega}_f$ and $\boldsymbol{\omega}_i$ are the angular velocities of the body before and after impulse. The term $\Sigma\mathbf{M}^c$ is the resultant moment about the center of mass. The term $\boldsymbol{\Lambda}^c$, called angular impulse, is the time integral of the moment $\mathbf{M}^c$.

A force that becomes very large during a very small time is called an impulsive force. Such a force can alter the velocity of a solid during the brief period it acts. Thus in the presence of impulsive forces, the contribution of finite forces to linear and angular impulse are neglected. A parameter called coefficient of restitution is introduced as a measure of the capacity of colliding bodies to rebound from each other. (See Section 7.4 for the mathematical definition of the coefficient of restitution.)

If one of the supports of a resting body suddenly gives way or is removed, the reactions at the other supports are instantaneously altered. The reason for the alteration is the change in acceleration of the body. However, in this case, the body does not immediately gain velocity as a result of a support giving way or being removed.

7.7 Problems

Problem 7.1. The frequency of crack formation during impact of a cadaver head against a flat, rigid surface was measured in a number of studies. A series of free fall (drop) tests using embalmed cadaver heads showed that a free fall of greater than 50 cm frequently resulted in the fracture of the skull. Consider a similar experiment and drop grapefruits and watermelons from various heights and determine the frequency of fracture. Would a grapefruit be a good model for human head? Does the size of the fruit have an effect on the frequency of crack formation? Note that serious brain injury may occur even in the absence of rupture of the skull. Large accelerations of the head may result in abrupt changes in local pressure in the brain and can cause excessive shearing deformation. Football players have concussions without any apparent damage to the skull.

Problem 7.2. The specific gravity of human head averages about 1.097 g/cm^3. This value is slightly higher than that of water. Determine the specific gravity of a grapefruit and a watermelon by determining its weight and dividing it by the volume of water it replaces when tossed into a bucket full of water.

Problem 7.3. Brain injury caused by a blunt impact is often associated with changes in internal pressure and the development of shear strains in the brain. Positive pressure increases are found in the brain behind the site of impact on the skull. These increases are thought to contribute to the local contusion of the brain tissue. To correlate the acceleration of the head with the level of injury to the brain, the Gadd Severity Index (GSI) was introduced (see Bronzino, 1995). This parameter is a measure of the impulse generated during a head-on collision. It is defined as

$$\text{GSI} = {}_0\int^t a^{2.5}\, dt$$

where a is the instantaneous acceleration of the head and t is the duration of the pulse. If the integrated value exceeds 1,000, severe injury is predicted to result. Consider a fall where the head of a person hits

the ground with a vertical velocity of 8 m/s. The impact lasts 0.05 s and the impact force is uniform during the course of impact. Determine the value of GSI for this head impact.
Answer: GSI = 16191.

Problem 7.4. If a person were not wearing a seat belt in a car when the car hit a wall or a large tree, the overall effect is that of a person hitting a massive wall with the velocity of the car before collision. In that sense a collision may be considered equivalent to falling from a height h onto a concrete sidewalk. Suppose the speed of the car before the collision is 90 km/h. Determine the height of a free fall that would give the same velocity before collision.
Answer: $H = 31.8$ m.

Problem 7.5. In a study of hip fracture etiology, young healthy athletes weighing 70 kg performed voluntary sideways falls on a thick foam mattress. The mean value for the vertical impact velocity of the center of mass of a falling athlete was 2.75 m/s. Assuming that there was no rebound immediately after the impact, compute the vertical impulse due to the fall.
Answer: $\zeta = 192.5$ N-s.

Problem 7.6. A uniform rod of mass m and length L can turn freely around point A (Fig. P.7.6). It is held in its highest position and is then allowed to fall. On reaching its lowest position, it encounters a fixed object. The object remains stationary while the rod rebounds. The coefficient of restitution e between the rod and the object is equal to 0.4. Find the impulse exerted on the rod. What is the rebound velocity at the site of the collision?

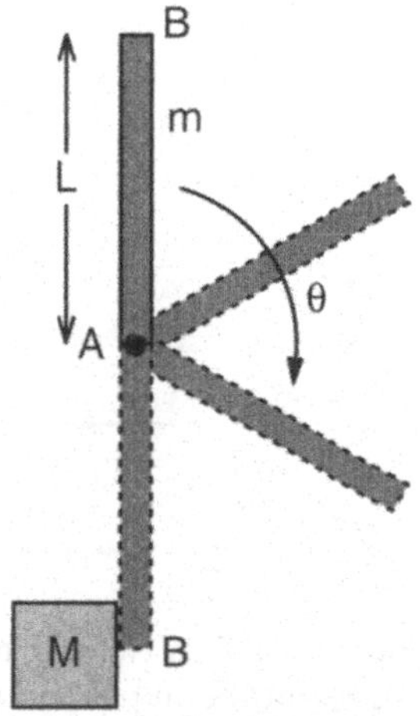

FIGURE P.7.6. The rotation of a uniform rod around point A and the resulting impact with a stationary object.

Hint: To compute the velocity of the rod before collision, derive a differential equation for angular speed using conservation of angular momentum.
Answer: $v = 0.4\,(6gL)^{1/2}$, $\zeta = 0.7\,m\,(6gL)^{1/2}$.

Problem 7.7. Consider sideways fall of an individual whose weight is 60 kg and height is 1.60 m. Represent the falling person as composed of two equal uniform rods AB (lower body) and BD (upper body), hinged together at B (hip joint) as shown in Fig. P.7.7. Both rods have mass m = 30 kg and length L = 0.8 m. Immediately before the impact, the point A (feet) was moving in the $-\mathbf{e}_1$ direction with speed equal to 1.2 m/s. The angular velocities of the lower body and the trunk immediately before the impact were measured as -7.5 rad/s $\mathbf{e}_3$ and -4.8 rad/s $\mathbf{e}_3$. The angles the lower body and the trunk made with the $\mathbf{e}_1$ direction were 0 and $\pi/6$, respectively. The lower body remained at rest on the ground immediately after the impact, whereas the upper body began rotating in the counterclockwise direction with angular velocity equal to 3.2 rad/s $\mathbf{e}_3$. Determine the impulse of the impact acting on the individual.
Answer: $\zeta = 24\,\mathbf{e}_1 + 353\,\mathbf{e}_2$ (N-s).

Problem 7.8. A man hits a ball of radius R and mass m with a cylindrical rod of length L and mass M (Fig. P.7.8). Before the impact the ball had an initial velocity $v_o\,\mathbf{e}_2$ and angular velocity $\omega_o\,\mathbf{e}_3$ as shown in the figure. The rod, on the other hand, was rotating around the stationary point A with angular velocity $-\Omega_o\,\mathbf{e}_3$. The distance between the point of impact and point A is denoted as d. Assume the force of impulse to be perpendicular to the contact area. Determine the angular velocity Ω of the rod and velocity of the center of mass v of the ball immediately after the impact. Use the following parameter values in your computations: M = 10 kg, L = 1 m, d = 0.8 m, Ω_o = 5 rad/s, m = 5 kg, R = 0.1 m, v_o = 10 m/s, and the coefficient of restitution e = 0.8.

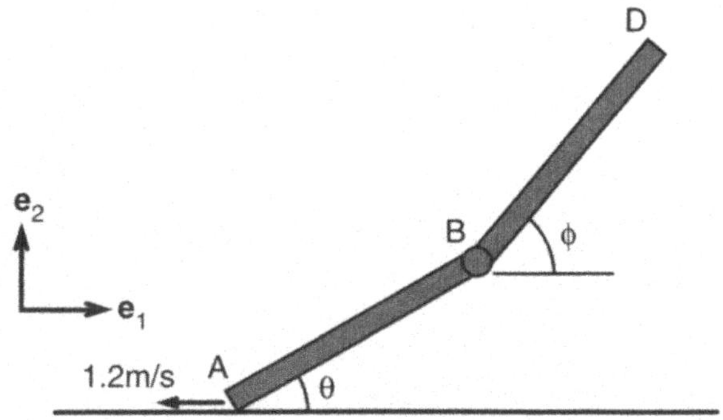

FIGURE P.7.7. A two-bar model of a sideways fall. The uniform rods AB and BD represent the upper and lower body, respectively.

FIGURE P.7.8. The collision of a ball with a cylindrical rod that is free to rotate around point A.

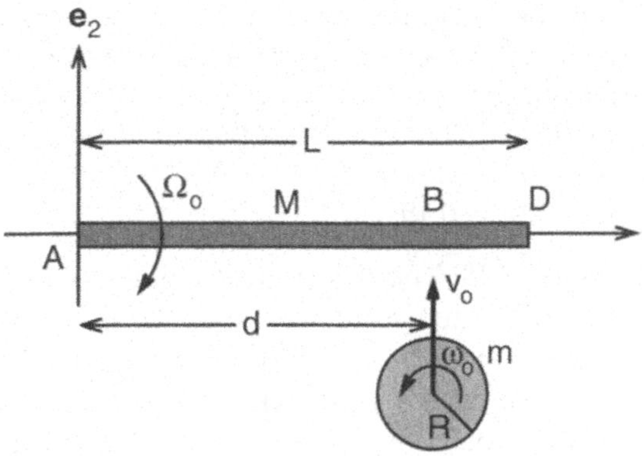

Hint: Because the impulse passes through the center of the spherical ball, it cannot change the ball's angular velocity ω_o. Also, the moment of momentum of the system (the rod and the ball) about point A (H^o) should not change as a result of the impact:

$$H^o = -(ML^2/3)\,\Omega_o + (2mR^2/5)\,\omega_o + m\,v_o\,d$$
$$= (ML^2/3)\,\Omega + (2mR^2/5)\,\omega_o + m\,v\,d$$

The equation for the coefficient of restitution yields

$$e = (d\,\Omega - v)/(v_o + d\,\Omega_o)$$

Answer: $\Omega = 10.4$ rad/s, $v = -2.9$ m/s.

Problem 7.9. A large mass m falls on a spring with downward velocity v_o. The spring constant is denoted as k. Determine the force exerted by the spring on the mass m as a function of time. What is the impulse of this force? Assume the downward direction to be positive.
Answer: $F = -(km)^{1/2}\,v_o \sin[(k/m)^{1/2}\,t] - mg\,(\cos[(k/m)^{1/2}\,t] - 1)$

Problem 7.10. The horizontal velocity of the center of mass of a 70-kg runner immediately before he placed a heel on a flat surface was found to be 2.5 m/s. The horizontal ground force acting on the heel of the runner followed the relation:

$$F_1\,\mathbf{e}_1 = -3{,}000 \sin(\pi t/0.2)\,\mathbf{e}_1$$

in which F_1 has the unit of N and time t has the unit of seconds (s). Determine the horizontal velocity component of the center of mass of the runner at 0.05, 0.1, and 0.15 s after the heel strike. Note that experimental values of the horizontal ground force during running is predominantly biphasic. During the initial phase (termed braking), the direction of the horizontal ground force opposes forward motion. During the latter phase (termed propulsion), its direction leads to forward acceleration. The relative magnitudes of the braking and propulsive impulses for a given trial can serve as an objective measure for verifying

whether a runner satisfies the so-called constant velocity criterion. In constant velocity running, the forward and backward impulses exerted by the ground must be equal in magnitude. The ground force equation presented above satisfies this condition.
Answer: $v = 1.7$ m/s at $t = 0.05$ s, $v = -0.2$ m/s at $t = 0$, and $v = -2.2$ m/s at $t = 0.15$ s.

Problem 7.11. The risk of head injury from striking an automobile dashboard is often correlated with the maximum linear acceleration of the head during the collision. To better understand the mechanics of collision, a team of researchers dropped rigid balls of different masses from a height of h onto an elastic surface with spring constant k. They found that the maximal displacement of the surface during the collision was given by the following relationship:

$$\Delta = (2m\, g\, h/\mathrm{k})^{0.5}$$

Determine a relationship between the mass of the fallen object and its acceleration at maximal displacement.

Hint: Write down the equation of motion of the object in the vertical direction and substitute k Δ for the spring force. Your result should predict that the smaller the mass of the object, the greater the peak acceleration during impact. Based on this observation, some researchers argued that children may be at greater risks than adults when striking a padding surface.
Answer: $\mathrm{a} = (2ghk/m)^{1/2} - g$.

Problem 7.12. An individual is running on a treadmill whose speed is set to v_o (Fig. P.7.12). The velocity of the center of mass at an instant im-

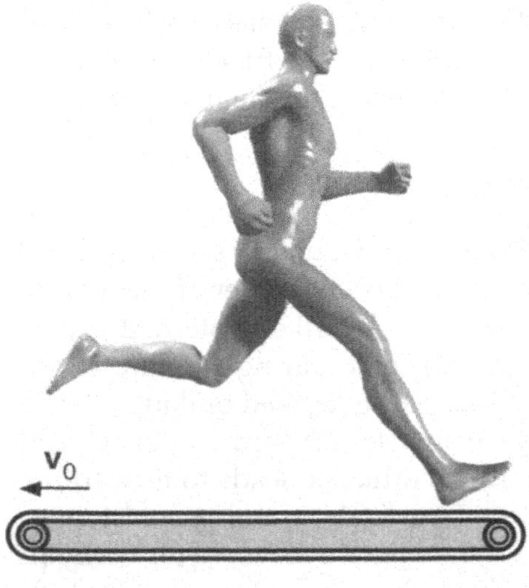

FIGURE P.7.12. A runner on a treadmill.

mediately before one of the legs hit the tank-treading rubber platform is measured to be zero. The angular velocity of the leg at that very instant is $-\omega_o$. Once the foot hits the platform, it begins to move with it. Assuming that the leg can be represented as a weightless uniform rod of length L with the lumped mass m of the body attached to it at the hip, determine the impulse exerted by the treadmill on the runner. Determine the velocity of the hip immediately after the impact. Does the impact force increase with increasing v_o?

8

Energy Transfers: In Pole Vaulting, Running, and Abdominal Workout

8.1 Introduction

Humans and animals expand energy at rest and in activity. About a quarter of the chemical energy used in muscle contractions goes into performing work against external forces. The rest is converted into heat. The primary method of assessing energy expenditure during an activity is through the evaluation of exchange of oxygen and carbon dioxide. Glucose and fat metabolism depend on the availability of oxygen. The amount of oxygen and carbon dioxide exchanged in the lungs normally should equal to that used and released by the body tissues in converting food energy into heat and mechanical work. The carbon and oxygen contents of carbohydrates, fats, and proteins differ dramatically, and therefore the amount of oxygen used during metabolism depends on the type of food fuel being oxidized.

The rate at which body uses energy is called the metabolic rate. At rest, the body usually burns a mixture of carbohydrate and fat. The average resting energy expenditure for a 70-kg man is about 2,000 kcal/day. This value reflects the minimum amount of energy required to carry out the body's essential physiological functions. The basal metabolic rate is directly related to the fat-free mass of the body because preserving fat requires almost no energy expenditure. The other factors that affect the basal metabolic rate are surface area of the body (the larger the surface area, the higher the rate of heat loss across the skin), age (metabolic rate decreases with age), body temperature, stress, and various hormones.

The body's ability to gauge muscle needs for oxygen during exercise is not perfect. At the beginning of exercise, the oxygen transport system (respiration and circulation) does not immediately supply the needed quantity of oxygen to the active muscles. The oxygen consumption requires several minutes to reach the required steady-state level while the body's oxygen requirements increase markedly the moment exercise begins. As a result, after the completion of the exercise, even though muscles are no longer actively working, oxygen demand does not immediately decrease.

This consumption exceeding what is usually required when at rest is referred to as the oxygen debt.

The amount of energy expended for different activities varies with the intensity and the type of the exercise. Some activities such as bowling or archery require only slightly more energy than when at rest. At the other extreme, sprinting requires so much energy expenditure that it can be maintained for only a few seconds. The energy expenditure per minute during high-speed running and crawl swimming is probably the highest among athletic activities, followed in order by handball, basketball, weight lifting, cycling, and so on. The metabolic cost of running increases with increasing speed of running.

The oxygen consumed during an athletic activity increases in proportion to the effort. For example, the oxygen uptake per minute is proportional to the speed of running. Eventually, as the speed of running further increases, the body reaches a limit for oxygen consumption. Even though the work intensity continues to increase, the oxygen consumption peaks and remains constant or drops. The peak value of oxygen consumption is called the maximum oxygen uptake. This parameter is regarded as a measure of cardiorespiratory endurance and aerobic fitness.

Part of the expenditure of energy during an athletic activity beyond that of the resting level results from the additional demands imposed on the heart and the rest of the circulatory system. The remaining part correlates with the state of activation of the skeletal muscles of the body, and may be intrinsically related to the work done by the muscles against the environment. In the following, we present an introduction to the study of energy transfer during athletic activity.

8.2 Kinetic Energy

Kinetic energy is a measure of the state of motion of an object. Kinetic energy is zero when the object is at rest. It is a scalar variable and is usually represented by the symbol T. For a particle of mass m, T is given by the following equation:

$$T = (1/2)\, mv^2 \tag{8.1}$$

where v is the speed of the particle. Kinetic energy of a system of particles is given by the following equation:

$$T = (1/2)\Sigma\, m^{i}\, (v^{i})^2 \tag{8.2}$$

where m_i and v_i are the mass and the speed of particle i, respectively, and the summation is over all particles in the collection.

Kinetic energy of a continuous body is given by the expression

$$T = (1/2) \int (\mathbf{v} \cdot \mathbf{v})\, dm \tag{8.3}$$

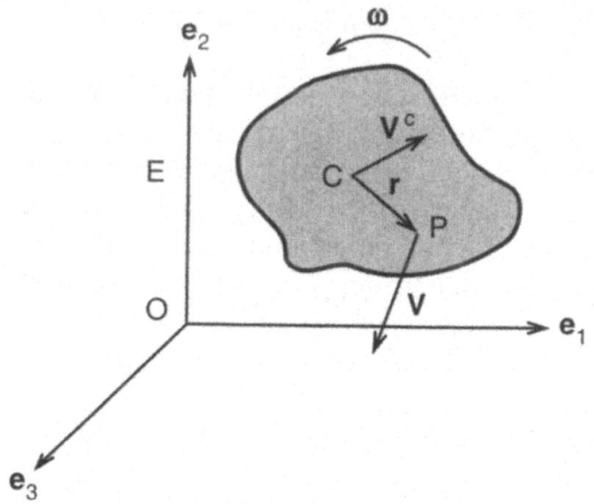

FIGURE 8.1. Velocity **v** of a point in a rigid body is equal to the velocity of the center of mass ($\mathbf{v}^c$) plus the vector product of the angular velocity ($\boldsymbol{\omega}$) and the position vector connecting the center of mass to the point under consideration (**r**).

where **v** is the velocity of a small mass element dm. Integration is over the whole body.

We have seen previously that the velocity of a point P in a rigid body can be written as

$$\mathbf{v} = \mathbf{v}^c + \omega \times \mathbf{r} \tag{8.4}$$

where **v** and $\mathbf{v}^c$ represent the velocities of point P and the center of mass C, respectively, ω is the angular velocity of the rigid body, and **r** is the position vector from C to P (Fig. 8.1).

In the case of planar motion of a rigid body, Eqns. 8.3 and 8.4 result in a simple expression for the kinetic energy:

$$T = (1/2)\, m\, \mathbf{v}^c \cdot \mathbf{v}^c + (1/2)\, \mathrm{I}^c\, \omega\, \mathbf{e}_3 \cdot \omega\, \mathbf{e}_3$$

$$T = (1/2)\, m\, (v^c)^2 + (1/2)\, \mathrm{I}^c\, \omega^2 \tag{8.5}$$

in which I^c is the moment of inertia with respect to the center of mass. The kinetic energy has two identifiable parts: the kinetic energy of the center of mass and the kinetic energy of rotation relative to the center of mass.

As an illustration of this expression, let us consider the kinetic energy of a spherical ball rolling without a slip on a planar surface (Fig. 8.2). Let $v\, \mathbf{e}_1$ denote the velocity of the center of mass of the ball and let a be its radius. The no-slip requirement means that the velocity of that point of the sphere which touches the planar surface must be equal to zero:

$$0 = v\, \mathbf{e}_1 + \omega\, \mathbf{e}_3 \times (-a\, \mathbf{e}_2) \Rightarrow \omega = -v/a$$

Noting that for a sphere of radius a and mass m, $\mathrm{I}^c = (2/5)\, ma^2$, the kinetic energy of the spherical ball rolling without slip is equal to

$$T = (1/2)\, m\, (v)^2 + (1/2)\, (2/5)\, ma^2\, (-v/a)^2 = 0.7\, m\, v^2$$

This equation shows that more than two-thirds of the kinetic energy is associated with the translational motion of the center of mass.

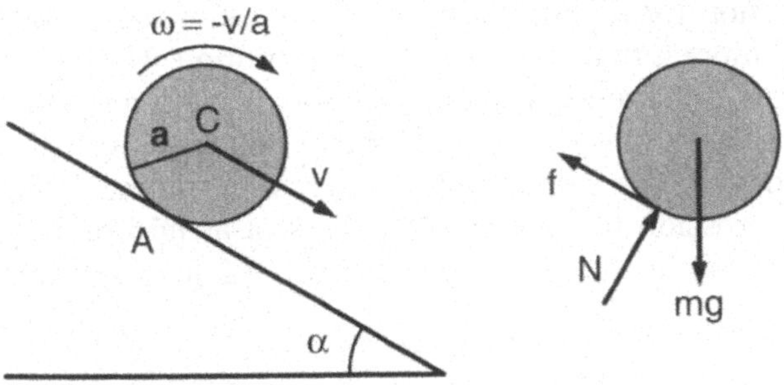

FIGURE 8.2. Rolling of a spherical ball along an inclined plane. The symbol v denotes the velocity of the center of mass of the ball. The symbol ω is as usual the angular velocity.

In some important problems of human body dynamics, the body will pivot around a point O. The kinetic energy of a rigid body that rotates around a fixed point O is given by the following equation:

$$T = (1/2)\, I^o\, \omega^2 \tag{8.6}$$

where I^o is the moment of inertia of the body with respect to an axis that passes through point O and is perpendicular to the plane of the motion. Let us illustrate the use of this equation by considering the kinetic energy of a pendulum. Let a slender rod of length L swing about a pivot O as shown in Fig. 8.3. We can compute the kinetic energy of the rod by using Eqn. 8.6:

$$T = (1/2)\,[(1/3)\, mL^2]\, \omega^2 = (1/6)\, mL^2\omega^2$$

Equation 8.5 can also be used to compute the same kinetic energy:

$$T = (1/2)\, m\, (v^c)^2 + (1/2)\, I^c\, \omega^2$$

$$T = (1/2)\, m\, (L\omega/2)^2 + (1/2)\, (mL^2/12)\, \omega^2$$

$$T = (1/6)\, mL^2\omega^2$$

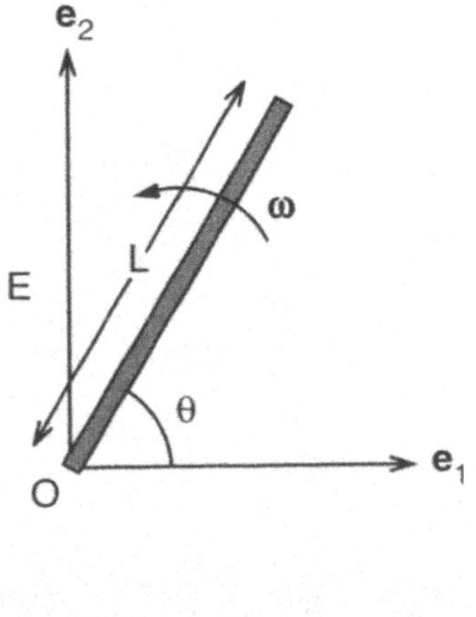

FIGURE 8.3. The rotation of a rod of length L with respect to point O, which is stationary in the reference frame E.

The expression for kinetic energy of a body in three-dimensional motion contains more terms than that of planar motion. The definition of angular velocity in three-dimensional motion is presented in Chapter 9. However, it suffices to say that Eqn. 8.4 that relates the velocity of a point in a rigid body to the velocity of another point in the same body still holds. An expression for the kinetic energy of a rigid body in three dimensions can then be obtained by substituting Eqn. 8.4 into Eqn. 8.3:

$$T = (1/2)\, m\, (v^c)^2 + (1/2)\, [I^c_{11}\, \omega_1^2 + I^c_{22}\, \omega_2^2 + I^c_{33}\, \omega_3^2 + 2I^c_{12}\, \omega_1\, \omega_2 + 2I^c_{13}\, \omega_1\, \omega_3 + 2I^c_{23}\, \omega_2\, \omega_3] \quad (8.7)$$

in which ω_i is the angular velocity in the direction of i ($i = 1, 2, 3$), and the terms I^c_{ij} ($i = 1, 2, 3$, and $j = 1, 2, 3$) represents the components of the mass moment of inertia tensor. The mass moment of inertia is a set of geometric parameters; it depends only on the shape of the object and the distribution of mass within the object. Mathematically, the mass moment of inertia tensor is defined as

$$I^c_{11} = \int (x_2^2 + x_3^2)\, \rho dV;\ I^c_{22} = \int (x_1^2 + x_3^2)\, \rho dV;\ I^c_{33} = \int (x_1^2 + x_2^2)\, \rho dV \quad (8.8)$$

$$I^c_{12} = -\int (x_1 x_2)\, \rho dV;\ I^c_{13} = -\int (x_1 x_3)\, \rho dV;\ I^c_{23} = -\int (x_2 x_3)\, \rho dV$$

in which ρ is the mass density (kg/m^3), V is volume, and x_i is the distance along the axis $\mathbf{e}_i$, as measured from the center of mass of an object. For a slender rod of mass m and length L, the components of the moment of inertia with respect to the coordinate frame shown in Fig. 8.4 can be written as

$$I^c_{11} = 0,\ I^c_{22} = I^c_{33} = (mL^2/12);\ I^c_{12} = I^c_{13} = I^c_{23} = 0$$

Components of the moment of inertia for geometrically simple homogeneous bodies are listed at the end of this volume.

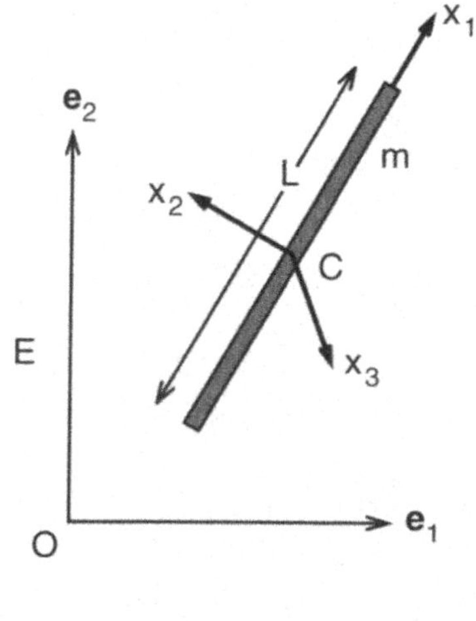

FIGURE 8.4. A Cartesian coordinate system (x_1, x_2, x_3) that is attached to the center of mass of a slender rod.

8.3 Work

The power exerted by a force on an object is

$$P = \mathbf{F} \cdot \mathbf{v} \tag{8.9}$$

in which **F** is the force and **v** is the velocity of the point of application of the force. Power is a scalar quantity with dimension N-m/s^2. Let us illustrate this definition by considering the case of a ball falling from a height h. Both the velocity and the gravitational force are in the same direction, and therefore the power produced must be positive. The velocity of a ball falling a distance d is given by the equation

$$v = (2\, g\, d)^{1/2}$$

in which v is the speed of the spherical ball and d is the distance of the fall. Thus we obtain the following equation for the power produced by gravity:

$$P = mg\, [2\, g\, (h - y)]^{1/2}$$

in which y is the height of the falling ball at time t and h is the height from which it is falling. Note that the power produced by gravity is not constant but increases with the falling distance. The power exerted by gravity is negative when a ball moves upward in the opposite direction of the gravitational acceleration.

Next consider the power produced by the ground reaction force acting on a spherical ball rolling without slip down an inclined plane (see Fig. 8.2). Because the ball does not slip, the velocity of the point of application of the ground force (point A) is zero. Therefore, the power produced by the ground force is equal to zero.

The power of external forces and couples acting on a rigid body is given by the following equation:

$$P = \Sigma(\mathbf{F} \cdot \mathbf{v}) \tag{8.10}$$

in which the summation is over all forces acting on the body. The terms **F** and **v** have their usual meanings. Let us express the velocity of the point of application of **F** in terms of the velocity of the center of mass and the angular velocity of the body:

$$\mathbf{v} = \mathbf{v}^c + \omega \times \mathbf{r}$$

Using this expression in Eqn. 8.10, we obtain:

$$\mathrm{P} = \Sigma(\mathbf{F} \cdot (\mathbf{v}^c + \boldsymbol{\omega} \times \mathbf{r}))$$

$$\mathrm{P} = (\Sigma\mathbf{F}) \cdot \mathbf{v}^c + \Sigma(\mathbf{F} \cdot \omega \times \mathbf{r})$$

$$\mathrm{P} = (\Sigma\mathbf{F}) \cdot \mathbf{v}^c + \Sigma(\mathbf{r} \times \mathbf{F}) \cdot \omega$$

$$\mathrm{P} = (\Sigma\mathbf{F}) \cdot \mathbf{v}^c + (\Sigma\mathbf{M}^c) \cdot \omega \tag{8.11}$$

in which $\Sigma\mathbf{F}$ denotes the resultant force and $(\Sigma\mathbf{M}^c)$ is the resultant moment about the center of mass. According to Newton's laws of motion, the resultant external force acting on an object is equal to the product of the mass and the acceleration of the center of mass of the object:

$$(\Sigma\mathbf{F}) \cdot \mathbf{v}^c = m\, \mathbf{a}^c \cdot \mathbf{v}^c = dT^c/dt \tag{8.12a}$$

$$T^c = (1/2)\, m\, (\mathbf{v}^c \cdot \mathbf{v}^c) = (1/2)\, m\, (v^c)^2 \tag{8.12b}$$

where the term T_c represents the kinetic energy associated with the center of mass.

The second term on the right-hand side of Eqn. 8.11 is a product of the resultant moment acting on the rigid object with the angular velocity of the object. Using the principle of conservation of angular momentum, we can express the resultant external moment in terms of the angular acceleration and angular velocity. For the planar motion studied in Chapter 4:

$$\Sigma\mathbf{M}^c = \mathrm{I}^c\, \alpha$$

in which I^c is the mass moment of inertia with respect to an axis that passes through the center of mass. The direction of the axis is perpendicular to the plane of motion. The symbol $\boldsymbol{\alpha}$ denotes angular acceleration, as earlier. Substituting this equation on the right-hand side of Eqn. 8.11, we find:

$$\Sigma\mathbf{M}^c \cdot \omega = dT^r/dt \tag{8.12c}$$

$$T^r = (1/2)\, \mathrm{I}^c\, \omega^2 \tag{8.12d}$$

where the symbol T^r represents the kinetic energy of the rigid object that is associated with the rotation around the center of mass. Although we have derived this equation for a rigid body whose plane of motion is parallel to a plane of symmetry of the body, it can be shown that even in the most general three-dimensional motion, mechanical power of external moment acting on a rigid object is equal to the time derivative of the part of the kinetic energy associated with rotation around the center of mass. Thus, we have the principle of conservation of mechanical energy:

$$(dT/dt) = \Sigma\mathbf{F} \cdot \mathbf{v}^c + \Sigma\mathbf{M}^c \cdot \omega\, \mathbf{e}_3 = P \tag{8.13}$$

According to this equation, the time rate of change of kinetic energy of a rigid body is equal to the rate of work done on it by the external forces and moments. Integrating this equation with respect to time, we arrive at the following relation:

$$T_2 = T_1 + W_{1\text{-}2} \tag{8.14}$$

in which T_1 and T_2 denote the kinetic energy of the body at times t_1 and t_2 and $W_{1\text{-}2}$ is the work done on the body by external forces and couples between times t_1 and t_2.

Example 8.1. Rolling of a Spherical Ball Down an Inclined Plane. A sphere of radius a is released from rest and rolls without sliding down an inclined plane (see Fig. 8.2). Find the velocity of the ball after its center of mass moves a distance b.

Solution: The free-body diagram of the rolling sphere is shown in Fig. 8.2. As noted, the ground reaction force produces no power. Thus, the increase in kinetic energy must be equal to the work done on the ball by the gravitational force:

$$T_2 = (^1/_2)\, m\, (v)^2 + (^1/_2)\, (2/5)\, ma^2\, (v/a)^2 = 0.7\, m\, v^2$$

$$W_{1\text{-}2} = m\, g\, b \sin\alpha$$

$$T_2 = 0 + W_{1\text{-}2}$$

$$v^2 = 1.43\, g\, b \sin\alpha$$

8.4 Potential Energy

The work done by a force on a body during a time interval $t_2 - t_1$ is equal to the integral of the power exerted by the force on the body over the same time interval:

$$W_{1\text{-}2} = \int P\, dt = \int \mathbf{F} \cdot \mathbf{v}\, dt = \int \mathbf{F} \cdot d\mathbf{r} \tag{8.15}$$

in which $W_{1\text{-}2}$ denotes the work done by the force on the body during the time interval $t_2 - t_1$. In this equation, the vector **v** represents the velocity of the point of application of force **F**. The position vector connecting the origin O of the inertial reference frame E to the point of application of the force **F** is termed **r** and is represented in terms of its projections to the axes of the coordinate system as follows:

$$\mathbf{r} = x_1\, \mathbf{e}_1 + x_2\, \mathbf{e}_2 + x_3\, \mathbf{e}_3$$

In the following, we present expressions for the work done by various types of forces that are commonly associated with human movement and motion.

Work Done by the Gravitational Force

The gravitational force acting on a body with mass m is equal to $-mg\, \mathbf{e}_2$. The work done by the gravity can be written as

$$W = \int \mathbf{F} \cdot d\mathbf{r} = -m\, g\, \mathbf{e}_3 \cdot \int d\mathbf{r}$$

$$W = -m\, g\, [(x_2)_2 - (x_2)_1] \tag{8.16}$$

in which $(x_2)_2$ and $(x_2)_1$ are the position of the center of mass at times t_2 and t_1 in the $\mathbf{e}_2$ direction. Gravity does positive work when the body moves downward and negative work when the body moves upward.

Equation 8.16 shows that the work done by the gravitational force on a body is independent of the path taken as the body moves from one configuration to another. In mechanics, such a force is said to be conservative. In the case of conservative forces, the work is expressible in terms of a scalar function V that is called the potential energy:

$$W_{1\text{-}2} = V(t_1) - V(t_2) \tag{8.17}$$

where $V(t_1)$ and $V(t_2)$ denote, respectively, the values of potential energy at times t_1 and t_2. Taking account of Eqn. 8.16, the potential energy V_g for gravitational force can be written as

$$V_g = m\, g\, y \tag{8.18a}$$

where y is the vertical distance of the center of mass from an arbitrarily chosen datum plane. The work done by the gravitational force can be expressed as the difference in potential energy between time points t_1 and t_2:

$$W_{1\text{-}2} = -V_{g2} + V_{g1} \tag{8.18b}$$

Work Done by Contact Forces

A contact force does no work if the velocity of the point of application of the force is perpendicular to the force itself. For example, when a ball rolls on a planar surface, the normal force acting on the ball creates no work and therefore does not affect the kinetic energy of the rigid body. If there is no relative movement between the interacting surfaces, the displacement is zero, and hence there is no work done by the frictional force. This is important because the frictional forces that enable us to walk or run do zero work during these activities because they act on a point of zero velocity. If, however, one body moves relative to the other at the point of contact, friction contributes to the work done by external forces. This is the case, for example, in cross-country skiing.

Work Done by the Spring Forces

Consider a spring with spring constant k and force-free length L_o. As shown in Fig. 8.5, the spring is force free at time t_o. At later times it is at first compressed (t_1) and subsequently stretched (t_2). Let x_j denote the displacement of the end of the spring marked with symbol A along the direction of the unit vector $\mathbf{e}$ at time t_j. Then the force exerted by the spring on the mass m at time t_j can be written as

$$\mathbf{F}_j = -k\, x_j\, \mathbf{e} \tag{8.19a}$$

Note that, for the spring shown in Fig. 8.5, $x_1 > 0$ at time t_1 and thus the spring exerts a force on mass m in the $-\mathbf{e}$ direction. On the other hand, the spring force is in the direction of $\mathbf{e}$ at time t_2 because $x_2 < 0$.

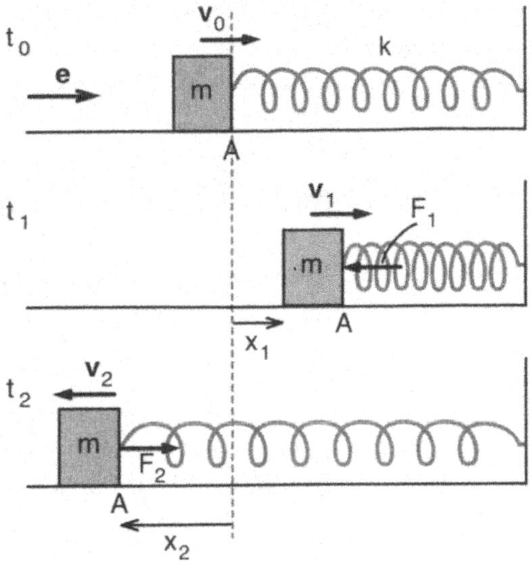

FIGURE 8.5. The force exerted by a spring on a mass m that is in contact with the spring at point A. The spring is force free at time t_o, neither stretched nor compressed. The spring is compressed at time t_1, and the force exerted by the spring is in the opposite direction of the velocity of the point A. Therefore, in this configuration, the power exerted by the spring on the mass m is negative, indicating that it causes a reduction in the kinetic energy of mass m. The spring is under tension at time t_2, and the power produced by the spring is again negative.

The power exerted by the spring shown in Fig. 8.5 on the mass m is equal to the scalar product of the spring force and velocity of the point of application of the force:

$$P = (-kx)\,(dx/dt) \Rightarrow W = -(^1/_2)\,kx_2^2 + (^1/_2)\,kx_1^2 \qquad (8.19b)$$

If both ends of a spring undergo displacements, the work done by the spring on the objects at both ends of the spring is given by the equation:

$$W_{1\text{-}2} = -(^1/_2)\,k\delta_2^2 + (^1/_2)\,k\delta_1^2 \qquad (8.19c)$$

in which δ_2 and δ_1 represent the extension (compression) of the spring at times t_2 and t_1, respectively. The work done by the spring force is independent of the path taken during compression or tension, and thus the spring force is conservative. If we call the term $(^1/_2)\,k\delta^2$ the potential energy V_s of the spring, then we have

$$W_{1\text{-}2} = -V_{s2} + V_{s1} \qquad (8.20)$$

in which V_{s2} and V_{s1} denote the potential energy of the spring at times t_2 and t_1, respectively.

Work Done by the Tensile Force in an Inextensible Cable

Displacement at one end of an inextensible cable (cord, string) is always equal to the displacement at the other end. Forces acting on the two endpoints, however, are equal in magnitude but opposite in direction. Thus, if two bodies are connected by a cable, the work done by the cable on the system of two bodies is equal to zero. Although ligaments and tendons of the human body are cable-like structures, they nonetheless undergo small stretches in response to applied force. Therefore, they behave more like an elastic spring in tension than an inextensible cable.

8.5 Conservation of Mechanical Energy

As a rigid body moves from a position at time t_1 to another position at time t_2, the change in its kinetic energy is given by the following relation:

$$T_2 = T_1 + W_{1\text{-}2} \tag{8.21}$$

in which T_2 and T_1 are the kinetic energy of the body at times t_2 and t_1, and $W_{1\text{-}2}$ is the work done during the time between t_1 and t_2. Part of the work done may be the result of conservative forces acting on the rigid body:

$$W_{1\text{-}2} = -V_{g2} + V_{g1} - V_{s2} + V_{s1} + W'_{1\text{-}2} \tag{8.22}$$

in which $W'_{1\text{-}2}$ represents the work done by forces other than gravity and springs.

If all the external forces that do work on the rigid body are conservative, then the conservation of mechanical energy holds:

$$T_2 + V_{g2} + V_{s2} = T_1 + V_{g1} + V_{s1} \tag{8.23}$$

Thus, the sum of kinetic and potential energy of a rigid body is constant unless work is done on the body by dissipative forces.

Example 8.2. Delivery Person Drops a Box Containing a Laptop from a Height h. The dynamics of a fall of a box containing a computer may be modeled as a mass of m striking a spring of stiffness k with velocity v (Fig. 8.6). When a box containing a computer is dropped from a height h, the mass–spring system is subjected to an impact. Develop an equation for the peak spring force produced when the delivery person drops the box from height h.

Solution: We assume that the box comes to rest as soon as it strikes the floor. The mass m of the computer in the box exerts zero force on the spring at the instant of impact. Nevertheless, the computer has a downward velocity v. This velocity can be found by considering the conservation of mechanical energy of mass m:

$$mgh = (^1/_2)\, mv^2 \Rightarrow v = (2gh)^{1/2}$$

When the mass–spring system comes to rest at its lowest position, the decrease of kinetic energy must equal to the increase in the potential energy of the spring:

$$mgh = (^1/_2)k(\Delta x)^2 \Rightarrow (\Delta x) = (2mgh/k)^{1/2}$$

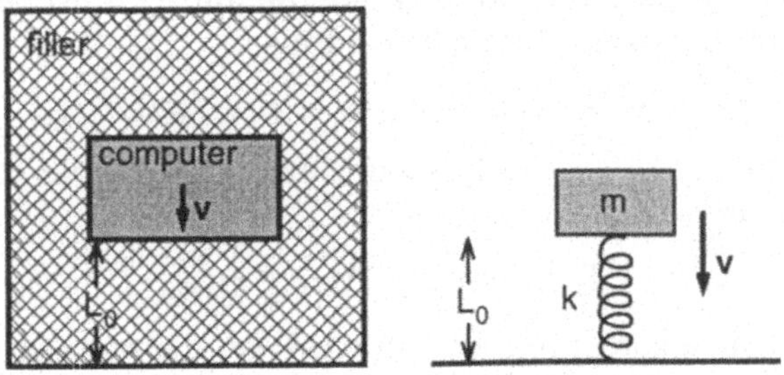

FIGURE 8.6. A schematic diagram of a box containing a computer packaged for shipment. If the box is dropped from a height h, the mechanical behavior of the computer will be similar to that of a mass m falling on a spring with downward velocity v.

The maximum spring force acting on the computer is then equal to

$$F = -k\,\Delta x = (2mghk)^{1/2}$$

The peak force during the collision increases with the falling height, the mass of the computer, and the stiffness of the spring used as cushioning. Thus, the more compliant the spring, the smaller is the peak force. However, if the spring is chosen to be very compliant, the computer might actually hit the box sitting on the floor, and then the equation for the peak force just given will not be correct.

Example 8.3. Pendulum Hit by a Bullet. A ballistic pendulum is used to determine the speed of rifle bullets. This pendulum is a rectangular block of mass m_2 that is supported by two cords (Fig. 8.7). A bullet of mass m_1 and velocity v_1 strikes the pendulum at time t_1 and becomes embedded in the block. Develop an equation that relates the velocity of the bullet to the amplitude of the pendulum swing.

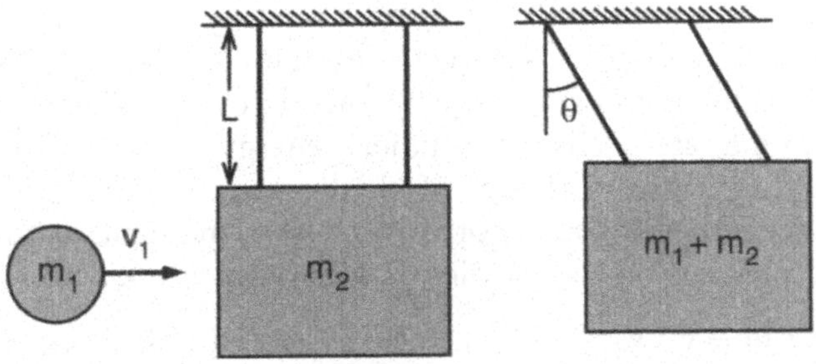

FIGURE 8.7. Schematic diagram of a rectangular block of mass m_2 that is supported on two cables. A bullet of mass m_1 and velocity v_1 strikes the mass m_2 at an arm length L.

Solution: The linear momentum before the impact must be equal to the linear momentum after the impact and thus:

$$m_1 v_1 = (m_1 + m_2)\, v_2 \Rightarrow v_2 = m_1 v_1/(m_1 + m_2)$$

The kinetic energy of the bullet–pendulum–inextensible cord complex right after the impact is

$$T_2 = (m_1 v_1)^2/[2(m_1 + m_2)] \tag{8.24a}$$

The inextensible cord produces no work on the mass m_2 because the tensile force exerted by the cord is perpendicular to the the velocity of the point to which it is attached. Conservation of mechanical energy of the system causes the kinetic energy right after the impact to transform into potential energy at the end of the pendulum swing:

$$T_2 + V_2 = T_3 + V_3 = (m_1 + m_2)\, g\, L\, (1 - \cos\theta) \tag{8.24b}$$

in which $V_2 = T_3 = 0$. Solving Eqns. 8.24a and 8.24b for $v_1{}^2$, we obtain the following equation that relates the velocity of the bullet before it strikes the pendulum to the maximum swing of the pendulum which occurs as a result of the impact:

$$v_1{}^2 = 2gL(1\text{-}\cos\theta)[(m_1 + m_2)/(m_1)]^2$$

Knowing the values of L, m_1, and m_2, one can determine the incoming speed of the bullet by measuring the maximum amplitude of the pendulum.

8.6 Multibody Systems

The time rate of change of kinetic energy of a deformable body is not necessarily equal to the work of external forces and couples acting on the body. This holds true even for a tree of rigid bodies. What is the equation governing the time rate of change of kinetic energy of serially linked rigid bodies?

Consider two rigid objects connected at point A (Fig. 8.8). We want to know whether the forces and moments acting on the objects at point A contribute to the rate of change of kinetic energy of the two objects. Let dT/dt represent the rate of change of kinetic energy of the two bodies as a result of the mechanical power of the forces and moments acting at point A. Then, according to the principle of conservation of kinetic energy:

$$dT/dt = \mathbf{F} \cdot \mathbf{v} + (-\mathbf{F}) \cdot \mathbf{v} + \mathbf{M}_{1\text{-}2} \cdot \boldsymbol{\omega}_1 + (-\mathbf{M}_{1\text{-}2} \cdot \boldsymbol{\omega}_2) \tag{8.25a}$$

in which $\mathbf{F}$ is the force exerted by body 2 on body 1 at point A, $\mathbf{v}$ is the velocity of point A, $\mathbf{M}_{1\text{-}2}$ is the moment exerted by body 2 on body 1, and $\boldsymbol{\omega}_1$ and $\boldsymbol{\omega}_2$ are, respectively, the angular velocities of the bodies 1 and 2. According to Newton's third law, action is equal to reaction, and thus the

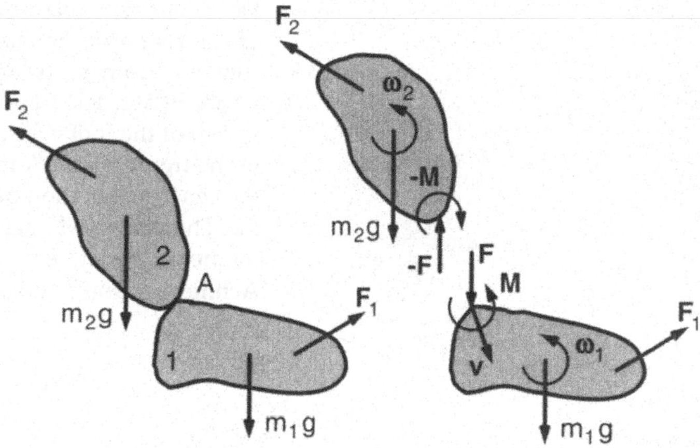

FIGURE 8.8. Two rigid bodies held together at point A by a joint force (**F**) and a joint couple (**M**). The velocity of point A is **v**. The angular velocities of bodies 1 and 2 are $\boldsymbol{\omega}_1$ and $\boldsymbol{\omega}_2$, respectively.

force (moment) exerted by body 1 on body 2 must be equal to minus the force (moment) exerted by body 2 on body 1. Thus, the reaction force at the joint connecting the two bodies cannot contribute to the total kinetic energy.

After algebraic simplifications, the time rate of change of kinetic energy of two bodies connected at a point (because of forces and moments acting at that point) can be put in the following simple form:

$$dT/dt = \mathbf{M}_{1\text{-}2} \cdot (\boldsymbol{\omega}_1 - \boldsymbol{\omega}_2) \tag{8.25b}$$

This equation illustrates what differentiates a living creature from a tree of bodies that are serially linked using hinges. As a hinge (pin) joint cannot resist a moment, the joint would move in such a way to make the resultant moment acting on the joint equal to zero. Hence, in a serially linked system of bodies, the rate of increase in the kinetic energy of the system from joint reaction forces is equal to zero. In the human body, however, the tendons exert moments of appreciable value to bone joints, and therefore the right-hand side of Eqn. 8.25 is not equal to zero.

Equation 8.25 provides insights into the mechanics of human movement. Consider, for example, the case of a skater spinning about a point on ice. When she draws her arms in, her moment of inertia decreases, resulting in an increase in her angular speed. This increase results from the chemical energy used in producing the movement.

Example 8.4. Transfer of Kinetic Energy During Running. The ground force acting on a runner during an instant in the stance phase was measured to be equal to 2,400 N in magnitude and vertically upward in direction (Fig. 8.9). At that time, the shank had an angular velocity of $\boldsymbol{\omega}_2 =$

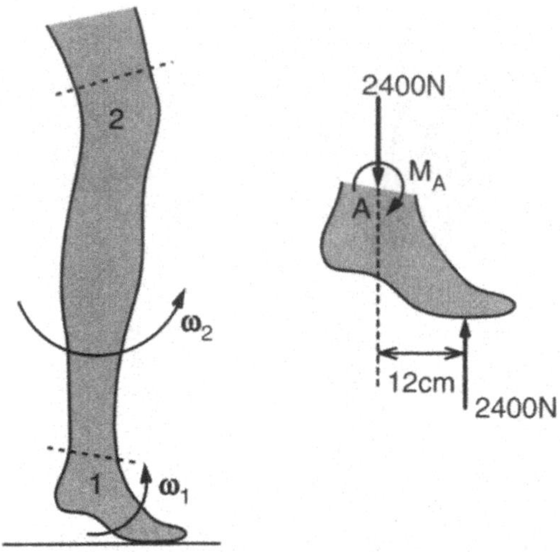

FIGURE 8.9. The relative motion of the foot and the shank of a runner during an instant in the stance phase. The free-body diagram of the foot is also shown in the figure. Point A identifies the center of rotation of the ankle. The symbol M_A denotes the moment generated by the Achilles tendon at point A.

-6 rad/s $\mathbf{e}_3$ and the foot had an angular velocity $\boldsymbol{\omega}_1 = -10.5$ rad/s $\mathbf{e}_3$. The moment arm h of the ground force with respect to the center of rotation of the ankle joint was 12 cm. Determine the rate of increase of kinetic energy of the shank and the foot at that instant. Determine also the changes in the kinetic energy of the shank and the foot separately.

Solution: Let the foot be represented as body 1 and the shank as body 2. The moment applied by the shank on the foot is equal to the moment of the force exerted by the Achilles tendon with respect to point A. The value of this force is unknown, however, the moment it generates at point A can be computed by considering the free-body diagram of the foot. Because the weight of the foot is small relative to the forces acting on it, we assume that the resultant moment acting on it must be equal to zero. Thus, the resultant moment acting on the foot at the center of rotation of the ankle is given by the following relation:

$$M_A \, \mathbf{e}_3 + 2400 \text{ N } (0.12 \text{ m}) \, \mathbf{e}_3 = 0 \Rightarrow M_A = -288 \text{ N-m}$$

in which M_A represents the moment generated by the Achilles tendon at A.

The time rate of change of kinetic energy of the foot and shank from the moment acting on joint A can be written as

$$dT/dt = M_A \, \mathbf{e}_3 \cdot (\boldsymbol{\omega}_1 - \boldsymbol{\omega}_2) = -288 \text{ N-m } \mathbf{e}_3 \cdot (-10.5 \text{ rad/s} + 6 \text{ rad/s}) \, \mathbf{e}_3$$

$$dT/dt = 1{,}296 \text{ N-m/s}$$

Thus, part of the chemical energy used by the calf muscles during running is transformed into kinetic energy at a rate of 1,296 N-m/s.

Next, let us consider the rate of change of kinetic energy of the foot and the shank separately. The force at the joint A balances the ground force F_G. This force is in the vertical direction. Thus, to calculate the power generated by this force, we need to determine the vertical component of the velocity of point A. We denote this component as v:

$$v\,\mathbf{e}_2 = -10.5\text{ rad/s }\mathbf{e}_3 \times -0.12\text{ m }\mathbf{e}_1 = 1.26\text{ m/s }\mathbf{e}_2$$

The rate of change of kinetic energy of the foot resulting from the joint forces and moments at A is given by the following relation:

$$dT^{(1)}/dt = -2{,}400\text{ N }\mathbf{e}_2 \cdot 1.26\text{ m/s }\mathbf{e}_2 + (-288\text{ N-m})\,\mathbf{e}_3 \cdot (-10.5)\text{ rad/s }\mathbf{e}_3$$

$$dT^{(1)}/dt = -3{,}024\text{ N-m/s} + 3024\text{ N-m/s} = 0$$

The time rate of change in kinetic energy of the shank from the joint forces and moments at A can be written as

$$dT^{(2)}/dt = 2{,}400\text{ N }\mathbf{e}_2 \cdot 1.26\text{ m/s }\mathbf{e}_2 + (288\text{ N-m})\,(-6)\text{ rad/s }\mathbf{e}_3$$

$$dT^{(2)}/dt = 3{,}024\text{ N-m/s} - 1728\text{ N-m/s} = 1296\text{ N-m/s}$$

Thus, the kinetic energy of the foot remains constant at the instant considered whereas the kinetic energy of the shank increases at a rate of 1,296 N-m/s.

8.7 Applications to Human Body Dynamics

Pole Vaulting

Pole vaulting is an exciting athletic event in which a vaulter clears a crossbar resting on two metal standards placed approximately 4 m apart (Fig. 8.10). On the ground in front of the crossbar is a small wedge-cut hole called the vaulting box that holds the end of the pole during the vault. Behind the standards is a landing pit that is at least 5 m wide. Back in 1877, the first championship was won with a vault of 2.92 m, but today vaulters reach the sky with much longer and flexible poles. We present two examples concerning the mechanical analysis of vaulting.

Example 8.5. Vaulting with a Rigid Pole. A vaulter of mass m grips a rigid vault of length L at a distance d from the front end. With the pole held to his side, the vaulter begins running from a distance of about 30 m from the crossbar (Fig. 8.10). Before he plants his pole in the vaulting box his running velocity is v_o and the vertical distance between his center of mass and the floor is h. Unlike the vaults presently used in athletic competitions, the one this vaulter uses is stiff; the pole does not bend or change in length in response to the ground force acting on it. Once the distal end of the vault is firmly on the ground, the vaulter pulls his hips forward and then begins rising in the air holding onto the vault. What should be his minimum running speed so that the vaulter can push the pole into a vertical configuration?

Solution: The linear momentum of the athlete and the pole before the pole hits the ground at point O is equal to (mv_o). This linear momentum is not conserved during impact because of the impulsive ground forces

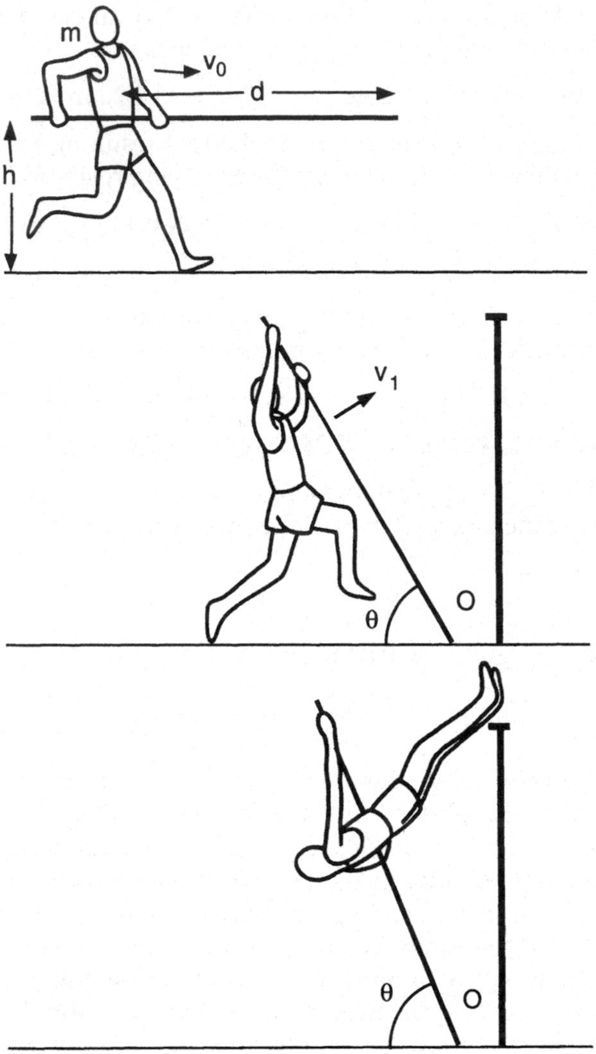

FIGURE 8.10. A vaulter clearing the crossbar using a stiff pole. The schematic diagrams show instances from the pole-vaulting events.

acting at O. However, the angular momentum with respect to O is conserved. Assuming that the mass of the athlete can be lumped at a single point on the weightless pole at a distance d from point O, the conservation of moment of momentum yields the following equation:

$$-mv_o\, h\, \mathbf{e}_3 = -mv_1\, d\, \mathbf{e}_3 \Rightarrow v_1 = v_o\, h/d \tag{8.26a}$$

in which v_1 is the speed of the athlete immediately after the impact. In deriving this equation, we took into consideration the fact that the velocity of the athlete right after the impact is no longer in the horizontal direction, but it is normal to the direction of the pole as shown in Fig. 8.10.

Equation 8.26a shows that the impact of the vault with the ground reduces the speed of the athlete from v_o to $v_o\ h/d$. Thus, the ratio of the kinetic energy after the impact to kinetic energy before the impact is equal to $(h/d)^2$. For $h = 0.9$ m and $d = 2$ m, approximately 80% of the kinetic energy of the center of mass of the vaulter is dissipated during the impact.

Let us now consider the conservation of energy between two time points, time t_1 right after the impact and time t_2 when the pole becomes vertical. Assuming that much of the kinetic energy of the vaulter is associated with his center of mass, we find

$$T_1 + V_1 = (1/2)\ m\ (v_o h/d)^2 + mgh \tag{8.26b}$$

$$T_2 + V_2 = 0 + mgd \tag{8.26c}$$

Equating the mechanical energies at times t_1 and t_2, we arrive at the following relationship between the initial speed v_o of the vaulter and the vertical distance that his center of mass can be lifted through the use of the vaulting pole:

$$v_o^2 = 2g\ (d/h)^2\ (d - h) \tag{8.26d}$$

This is the value of v_o necessary to push the pole into the vertical position. Considering that the best of the athletes run 100 m in slightly less than 10 s, it is clear that an upper bound for v_o is about 10 m/s. According to Eqn. 8.26d, the center of mass of a pole vaulter of height $2h = 1.8$ m would only rise to about 2.0 m for $v_o = 10.3$ m/s. Why is a rigid pole so inefficient in raising a human above a barrier? To answer this question, we need to consider the speed v_1 of the vaulter immediately after the impact. For $v_o = 12$ m/s and $d = 4$ m, $v_1 = 2.7$ m/s. Clearly much of the linear momentum and the kinetic energy of the athlete are wasted when the rigid pole hits the ground. If the athlete were to use a flexible pole, part of the kinetic energy of the vaulter would have been stored as elastic energy in the pole. That additional energy can push the pole into a vertical position, as illustrated in the next example.

Example 8.6. Pole Vaulting with a Flexible Pole. The athlete in the aforementioned example replaces his rigid pole with a flexible one. The new pole is such that under the application of compressive forces at both ends it assumes the shape of an arc (Fig. 8.11). Slender rods subject to thrust buckle when the compressive forces acting on them reach a certain level. If the rod can rotate freely at both ends as shown in Fig. 8.11, it will buckle when the axial force F reaches the value

$$\mathrm{F} = \pi^3 \mathrm{E}\ r_o^4/(4L^2) \tag{8.27a}$$

in which E represents Young's modulus of the circular rod, r_o is its radius, and L is its length. This equation shows that the smaller the radius and the longer the rod, the more likely it is to buckle under thrust.

The potential energy stored in a buckled pole is like the potential energy of a spring:

$$V = (1/2)\ k\ \Delta x^2 \tag{8.27b}$$

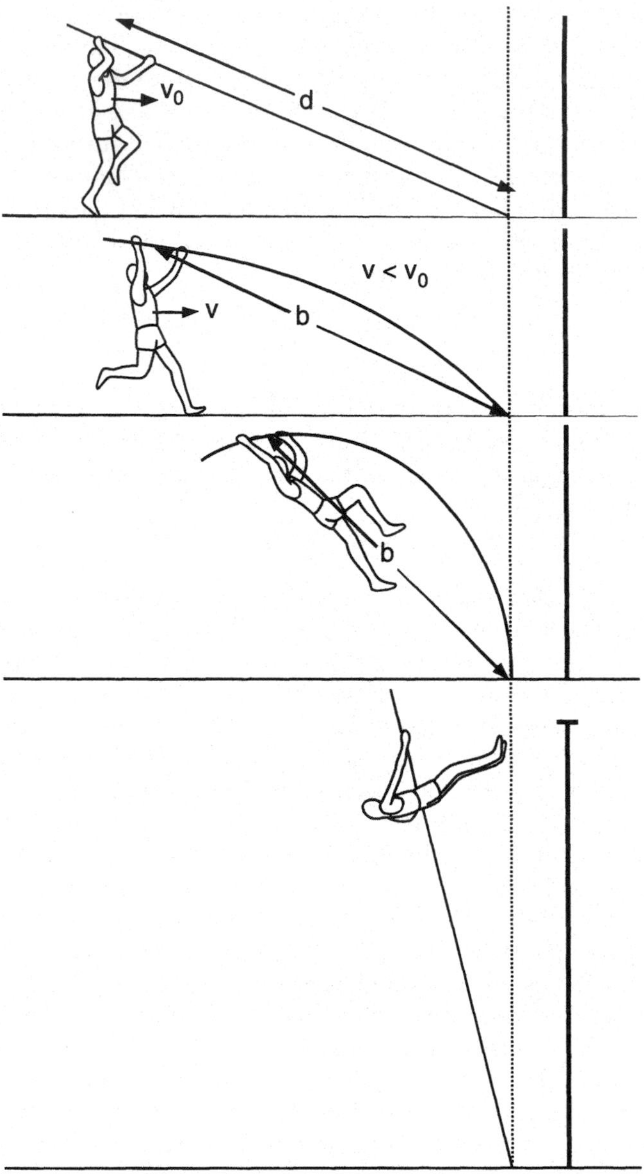

FIGURE 8.11. Pole vaulting using a flexible pole. The pole buckles in response to the thrust exerted by the vaulter as he runs toward the cross bar while keeping the front end of the pole fixed to the ground.

in which k (n/m) is the stiffness of the pole, and Δx is the difference between the length d of the curved pole between the hand grip and the distal end and the shortest distance between these points (b):

$$\Delta x = d - b \tag{8.27c}$$

When a vaulter uses a rigid pole, much of his initial kinetic energy is dissipated into heat during the brief period of impact between the pole and the ground. However, in the present case when the vaulter uses a flexible pole, as he hits the pole onto the ground the pole buckles and takes the shape of an arc. The kinetic energy lost before the takeoff is transformed into elastic energy and is stored in the pole.

Because the ground force acting on the distal end of the pole does no work, mechanical energy is conserved from the moment the vaulter places the distal end of the pole onto the ground to the moment when the vaulter lets go of the pole. Assuming again that the vaulter can be represented as a lumped mass of m, the mechanical energy of the vaulter–pole system at time t_o right before the impact is

$$T_o + V_o = (1/2)\, m v_o^2 + mgh \tag{8.27d}$$

where h denotes the vertical distance between the center of mass of the vaulter and the ground.

Initially, during the swing phase of pole vaulting when the front end of the pole is fixed on the ground, the pole will be compressed into a curved shape. The radius of curvature for the pole decreases until it reaches a minimum value. After that, the pole begins to extend, raising the center of mass of the vaulter and supplying him with additional kinetic energy. As the athlete clears the crossbar, all springlike elastic energy stored in the pole has been released ($d = b$), and the kinetic energy supplied by the pole to the athlete has been used for raising the legs above the crossbar. The mechanical energy at that time point is

$$T_2 + V_2 = mg\, d \tag{8.27e}$$

Equating the mechanical energy at time t_o given by Eqn. 8.27d to the energy at time t_2 given by Eqn. 8.27e, we arrive at the following relationship:

$$v_o^2 = 2g\,(d - h) \tag{8.27f}$$

This equation shows that, using a flexible pole, a 1.8-m-tall athlete (h = 0.9 m) can clear a bar set at 6.0 m with an initial velocity of 9.1 m/s. [For more on the mechanical analysis of pole vaulting, see Ekevad and Lundberg (1995) and other references at the end of this book.]

Example 8.7. Abdominal Wheel. Late in the 1980s, the abdominal wheel was a popular device for strengthening stomach muscles. An abdominal wheel is composed of a small rubber wheel with holders on both sides. As shown in Fig. 8.12, a person fixed their feet onto the ground and moved the wheel forward and backward in the direction of the body axis. The person begins to contract their abdominal muscles when the angle θ between the floor and the lower body is 20°. Determine the motion of the person immediately following the contraction of the abdominal muscle. In solving the problem, represent the body with two rods of equal length

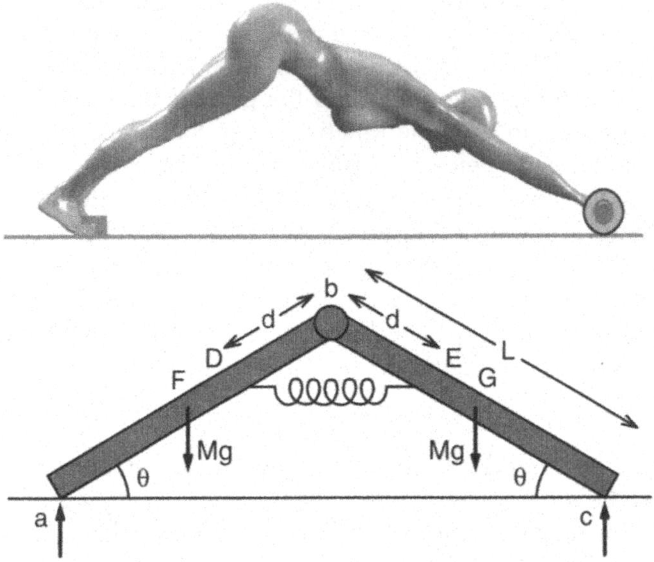

FIGURE 8.12. Schematic drawing of a woman using an abdominal wheel to strengthen abdominal muscle groups. The mechanical model of the person composed of two rigid links plus a contractile cord is shown in the lower part of the figure.

(L) and mass (M) that are hinged together at b. Model the contracting abdominal muscle as a spring connecting the two rods as shown in the figure. The muscle force is zero at $\theta = 45°$. Assume also that the floor is smooth so that the roller slides along the floor without frictional resistance.

Solution: The vertical forces acting on the rods ab and bc are shown in the figure. For each rod, the principle of conservation of mechanical energy holds: the rate of change of kinetic energy must be equal to the work done on the rod by the external forces. At time $t = t_1$ and $\theta = 20°$, both rods are stationary and therefore their kinetic energy must be equal to zero. There is potential energy, however, stored in the contracting abdominal muscle. Let us next consider the mechanical energy at a later time t_2. The reaction force at point a (feet) acting on rod ab does no work because point a remains stationary. Because the wheel is polished (frictionless), the ground force acting at point c is perpendicular to the displacement of c and thus does no work. With these considerations, the conservation of mechanical energy takes the following forms for rods ab and bc:

$$T_2^{ab} + Mg\,(L/2)\sin\theta = Mg\,(L/2)\sin 20° + W_s^{ab} + W_{1\text{-}2} \quad (8.28a)$$

$$T_2^{bc} + Mg\,(L/2)\sin\theta = Mg\,(L/2)\sin 20° + W_s^{bc} - W_{1\text{-}2} \quad (8.28b)$$

in which W_s^{ab} is the work done on rod ab by the spring DE (abdominal muscle) and $W_{1\text{-}2}$ denotes the work done by the reaction force acting on point B on rod ab. Because action equals to reaction, $W_{2\text{-}1} = -W_{1\text{-}2}$. Adding these two equations together, we obtain the following equation for the whole system:

$$T_2^{ab} + T_2^{bc} + Mg\,L \sin\theta = Mg\,L \sin 20° + W_s^{ab} + W_s^{bc} \quad (8.28c)$$

The work done by the spring ($W_s^{ab} + W_s^{bc}$) can be expressed as the change in potential energy of the spring, and therefore Eqn. 8.28c reduces to

$$T_2^{ab} + T_2^{bc} + Mg\,L \sin\theta + (1/2)\,\mathrm{k}\,[2d(\cos\theta - \cos 45°)]^2$$
$$= Mg\,L \sin 20° + (1/2)\,\mathrm{k}\,[2d(\cos 20° - \cos 45°)]^2 \quad (8.28d)$$

in which k denotes the spring stiffness of the abdominal muscle and d is defined in the figure.

Let us now compute the kinetic energy of rods *ab* and *bc*. In each case we need to evaluate the component of the kinetic energy due to the velocity of the center of mass and the remaining part due to the rotation of the rod. Each rod has the same angular speed $\|d\theta/dt\|$. The speeds of the center of masses of *ab* and *bc* are obtained from the time derivatives of their respective coordinates:

$$\|\mathbf{v}^F\| = (d\theta/dt)L/2 \quad (8.28e)$$

$$\|\mathbf{v}^G\| = (d\theta/dt)(L/2)\,(\cos^2\theta + 9\sin^2\theta)^{0.5} \quad (8.28f)$$

Because the mass moment inertia of a rod with mass M and length L about the center of mass is equal to $ML^2/12$, the kinetic energy of the two rods can be expressed as

$$T_2^{ab} = (1/8)\,M\,L^2\,[(d\theta/dt)^2 + (1/2)\,(M\,L^2/12)\,(d\theta/dt)^2 = T_2^{ab}$$
$$= (1/6)\,M\,L^2\,(d\theta/dt)^2$$

$$T_2^{bc} = (1/8)\,M\,L^2\,(d\theta/dt)^2\,(1 + 8\sin^2\theta) + (1/2)\,(M\,L^2/12)\,(d\theta/dt)^2$$

$$T_2^{bc} = (1/8)\,M\,L^2\,(d\theta/dt)^2\,[4/3 + 8\sin^2\theta] \quad (8.28g)$$

Substituting Eqn. 8.28g into Eqn. 8.28d, we determine the time rate of change of angle θ as a function of θ. In particular, we want to know the angle at which the shortening of the abdominal muscle comes to a halt. At that instant kinetic energy must be equal to zero, and thus

$$Mg\,L \sin\theta + (1/2)\,\mathrm{k}\,[2d(\cos\theta - \cos 45°)]^2 =$$
$$Mg\,L \sin 20° + (1/2)\,\mathrm{k}\,[2d(\cos 20° - \cos 45°)]^2$$

This equation can be written in the following convenient form:

$$\sin\theta - \sin 20° = \beta\{(\cos 20° - \cos 45°)^2 - (\cos\theta - \cos 45°)^2\} \quad (8.28h)$$

in which the dimensionless parameter β is defined by the following equation:

$$\beta = 2\mathrm{k}d^2/(MgL) \quad (8.28i)$$

Parameter β is a measure of the strength of the abdominal muscle relative to the weight of the body. Using the method of trial and error one can show that for $\beta = 6$, the abdominal muscle will contract approximately to the angle $\theta = 40°$.

Running

Running is a speedy form of locomotion in which the body moves in a periodic fashion. The period of steady-state running is the time duration in between consecutive landing of the same foot on the ground. The term gait cycle refers to running events (changes in the configuration of the body and various transfers of energy) that occur during that period of time. In running, all feet are in the air sometime during the gait cycle. During the rest of the cycle only one foot is on the ground at a time. In contrast, in walking there is always at least one foot on the ground. Figure 8.13 presents the configuration of a runner's body in four distinct instances during a gait cycle.

The body configuration shown in Fig. 8.13a belongs to a time toward the end of the stance phase. As shown in the diagram, frictional force acting on the runner is propulsive during this later phase of the stance

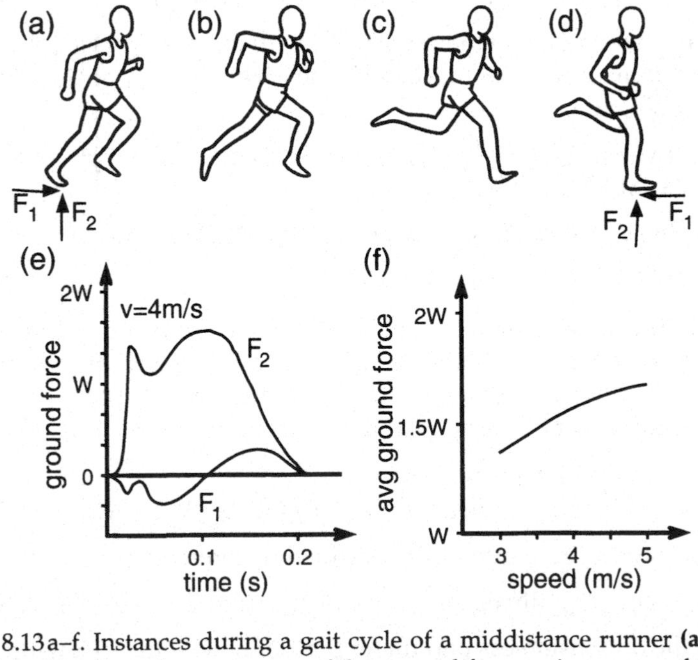

FIGURE 8.13a–f. Instances during a gait cycle of a middistance runner **(a–d)**. The vertical and horizontal components of the ground force acting on an athlete running at 4 m/s are illustrated in **(e)**. The average ground force is shown to increase with the speed of the runner **(f)**.

period. Swing phase begins once the foot leaves the ground. There are brief periods of time at the beginning and toward the end of the swing phase when both feet are off the ground. Body configurations in these phases are shown in Fig. 8.13b and 8.13c. At the initial stages of the stance phase, friction force is in the opposite direction of locomotion (Fig. 8.13d). The vertical and horizontal ground forces acting on the runner during the stance phase are shown in Fig. 8.13e as a function of time. As can be seen, the vertical ground force is much greater in magnitude than the horizontal ground force. The sum of horizontal ground force in a gait cycle during steady-state running on a horizontal surface adds up to zero. Average ground force in a gait cycle increases with the speed of running (Fig. 8.13d). Thus, forces resisted by the muscles, tendons, and bones of the lower limbs increase with increasing running speed. This is why overuse injuries become more apparent at high running speeds.

Alexander (1992) estimated that about one-third of the energy turnover during each stance of a 75-kg person running at a speed of 4.5 m/s is stored as elastic energy in the Achilles tendon. Using a free-body diagram of the foot in contact with the ground, one can compute the tension in the Achilles tendon in terms of the ground force measured during running. The elastic strain energy stored in the tendon is proportional to the square of the tensile force carried by this tendon. The Achilles tendon is not the only component of the lower limbs where strain energy is stored. The arch of the heel, the quad tendon, and the patellar ligament also have the capacity to store considerable elastic energy. As a result, when muscles act as brakes during the initial phase of stance, about half the energy used in muscle contraction is returned by the tendons during the final phase of the stance, making running an efficient activity.

Example 8.8. Minimum Time Running: An Optimal Approach. Maronski (1996) introduced an optimal control approach to address this question: How should a runner vary their speed with distance to minimize the time during which they cover a given distance? The author developed an optimal velocity profile by formulating and solving a problem in optimal control theory. In his model, the racer is regarded as a particle of mass m. The vertical displacements of the body associated with the cyclic nature of the stride pattern are neglected. The equation of motion in the direction of running is assumed to be in this form:

$$m\,(dv/dt) = m\,f_o\,\eta - m\,v/\tau \qquad (8.29a)$$

in which v denotes the speed of the runner, t is the time, and f_o and τ are constants. The parameter $mf_o\eta$ is the propulsive force, and mv/τ is the resistive force. The propulsive force is the product of the mass m, the maximum propulsive force per unit mass f_o, and the propulsive force setting η. This is a dimensionless parameter with values in the range

$$0 \le \eta \le 1 \qquad (8.29b)$$

The parameter τ appearing in Eqn. 8.29a is called the constant damping coefficient. The overall resistive force may include air resistance (assistance) on the runner as well as the steady slope of the track. For simplicity, we will focus on a horizontal track. Admittedly, the mechanical model of running adopted by this article is rather elementary. The model could not be used, for example, in investigating overuse injuries during running. However, the analysis presented here may provide insights into the decision making process on minimizing running time during a competition.

The energy transformations in the competitor's body are represented by a differential equation on the power balance:

$$de/dt = b - (v \, f_o \, \eta)/\mu \tag{8.30}$$

in which e denotes the actual reserves of chemical energy per unit mass in excess of the nonrunning metabolism, b is the recovery rate of chemical energy per unit mass, $v \, f_o\eta$ is the actual mechanical power per unit mass used by the runner, and μ is the efficency of transforming the chemical energy into the mechanical one.

The following intitial conditions are assumed:

$$x = 0, \; v = 0, \; e = e_o \text{ at } t = 0 \tag{8.31}$$

in which x denotes the distance the runner has covered at time t. The runner should adjust their speed v during a race over a given distance D_o to minimize the time t^* of the event. This is possible because of the variations of the propulsive force setting η, which may be adjusted arbitrarily in any point of the race. In mathematical terms, this is formulated as follows.

Find $v = v(x)$, $\eta = \eta(x)$, and $e(x)$ so that t^*, given by the equation

$$t^* = \int dt = \int dx/v \tag{8.32}$$

is minimized. The details of the optimization procedure may be found in Maronski (1996). His results indicate that the distance of the race may be broken into three phases:

1. The early phase of running (acceleration phase) during which the competitor moves with maximal propulsive force ($\eta = 1$).
2. The middle phase of the race (cruise phase) when the runner moves with partial propulsive force and the velocity is constant.
3. The negative kick phase where reserves of chemical energy has been depleted and the recovery rate of chemical energy is used to propel the runner at the final stage of the competition.

Let us illlustrate this model with a simple case in which the runner completes the race by going through phase 1 and ends the race at phase 2. The solution of Eqn. 8.29a for $\eta = 1$ and the initial conditions specified in Eqn. 8.31 can be shown to be equal to

$$v = f_o\tau \, [1 - \exp(-t/\tau)] \tag{8.33}$$

Integration of this equation with respect to time gives the distance covered by the runner at time t:

$$x = f_o\tau\,[t + \tau(1 - \exp(-t/\tau))]$$

The energy balance during the acceleration phase ($\eta = 1$) is given by the relation

$$de/dt = b - (f_o/\mu)\,v = b - (f_o^2\,\tau/\mu)\,[1 - \exp(-t/\tau)]$$

Integrating this equation with respect to time, we find

$$e = e_o + bt - (f_o^2\,\tau t/\mu) - (f_o^2\,\tau^2/\mu)\exp(-t/\tau)$$

Let us denote by t_1 the time duration of phase 1. Then the velocity of the runner at the end of phase 1, the distance covered by the runner during phase 1, and the remaining reserves of chemical energy at the end of phase 1 are given by the relations:

$$v_1 = f_o\tau\,[1 - \exp(-t_1/\tau)] \tag{8.34a}$$

$$x_1 = f_o\tau\,[t_1 + \tau(1 - \exp(-\,t_1/\tau))] \tag{8.34b}$$

$$e_1 = e_o + bt_1 - (f_o^2\,\tau t_1/\mu) - (f_o^2\,\tau^2/\mu)\exp(-t_1/\tau) \tag{8.34c}$$

Next, we consider the constant velocity phase. The equations governing the speed v and the rate of change of chemical energy e are then as follows:

$$f_o\eta - v_1/\tau = 0 \Rightarrow \eta = v_1/(f_o\tau) \tag{8.35a}$$

$$(de/dt) = b - v_1^2/(\mu\tau)$$

$$e = e\,(t) = e_1 + [b - v_1^2/(\mu\tau)](t - t_1) \tag{8.35b}$$

The second phase ends when e reduces to zero.

Let us consider a competitor in a 400-m race on a horizontal track. The parameters indicating the energy and force producing capacity of the athlete are

$$f_o = 12.0\ \mathrm{m/s^2},\ \tau = 0.9\ \mathrm{s},\ e_o = 2{,}400\ \mathrm{m^2/s^2},\ b = 42\ \mathrm{m^2/s^3},\ \mu = 0.3$$

The athlete would like to know with which velocity he should run so that he finishes the race in phase 2, with e_o all used up at the end of the run.

We first look at the equations of phase 1 and see how long would it take for the runner to complete phase 1. In this phase, the speed of the runner is given by the equation

$$v_1 = 10.8[1 - \exp(-t_1/0.9)]$$

This equation shows that the racer reaches the optimum speed within a fraction of a second into phase 1. For example, 2 s into the race, if phase 1 were to continue, that competitor would have a speed of 9.63 m/s. Thus, we will neglect the time and energy spent in phase 1 and assume that the competitor runs the race with constant speed. We want to determine his speed. As the chemical energy at the end of the second phase must be equal to zero, we set Eqn. 8.35b equal to zero. In this equation we set the chemical energy at the end of phase 1 (e_1) equal to the initial chemical energy e_o. This assumption is justified because the first phase takes only a few seconds. Thus, we have

$$e_o + [b - v^2/(\mu\tau)]\, t^* = 0 \qquad (8.36a)$$

The distance D of the racing event must be equal to the constant velocity times t^*:

$$D = v\, t^* \qquad (8.36b)$$

Eliminating t^* from Eqns. 8.36a and 8.36b, we obtain

$$e_o v + b\, D - D\, v^2/(\mu\tau) = 0 \qquad (8.37a)$$

Using the parameter values supplied as input this equation reduces to the following algebraic equation:

$$6v + 42 - 3.7\, v^2 = 0 \qquad (8.37b)$$

Solution of this equation yields the result that the runner completes the race in 48 s with an average speed of 8.32 m/s.

8.8 Summary

The power P exerted by a force $\mathbf{F}$ on a particle of mass m moving with velocity $\mathbf{v}$ is defined as the dot product of the force and the velocity. Power is a scalar quantity. The power of the resultant force acting on a particle can be shown to be equal to the time rate of change of kinetic energy of the particle:

$$P = \mathbf{F} \cdot \mathbf{v} = dT/dt$$

in which $T = (1/2)\, mv^2$ is the kinetic energy and v is the speed of the particle. Kinetic energy of a rigid body in planar motion is given by the expression:

$$T = (1/2)\, m\, (v^c)^2 + (1/2)\, I^c\, \omega^2$$

in which v^c is the velocity of the center of mass, I^c is the mass moment of inertia about the center of mass, and ω is the angular speed. When a rigid body rotates around a fixed point O, this equation reduces to the form:

$$T = (1/2)\, I^o\, \omega^2$$

in which I^o is the mass moment of inertia with respect to point O.

The power of a force acting on a rigid body is equal to the dot product of the force $\mathbf{F}$ and the velocity $\mathbf{v}$ of the point of its application. If the point has zero velocity or its velocity is perpendicular to the force, then the force creates no power. The power of the resultant force and the resultant couple acting on a rigid body is equal to the time rate of change of kinetic energy:

$$P = (\Sigma \mathbf{F}) \cdot \mathbf{v}^c + (\Sigma \mathbf{M}_c) \cdot \boldsymbol{\omega} = dT/dt$$

where $\Sigma\mathbf{F}$ denotes the resultant force acting on the object and $(\Sigma\mathbf{M}^c)$ is the resultant moment acting on the body about the center of mass.

The change of kinetic energy of a rigid body is equal to the work done on it by the external forces and moments:

$$T_2 = T_1 + W_{1\text{-}2}$$

in which T_1 and T_2 denote the kinetic energy of the body at times t_1 and t_2 and W is the work done on the body by external forces and couples.

The change of kinetic energy of a tree of rigid bodies is equal to the work done on it by the external forces and moments plus the work done by the joint moments.

$$dT/dt = P + \Sigma \mathbf{M}_{\text{i-j}} \cdot (\boldsymbol{\omega}_{\text{i}} - \boldsymbol{\omega}_{\text{j}})$$

in which P is as usual the power produced by the external forces and couples acting on the tree of rigid bodies, $\mathbf{M}_{\text{i-j}}$ denotes the moment exerted by body j on body i at their joint, and $\boldsymbol{\omega}_{\text{j}}$ is the angular velocity of body j.

The work done by gravity on a body B in going from configuration 1 to configuration 2 can be written as

$$W_{1\text{-}2} = -mg\,(h_2 - h_1)$$

in which h_2 and h_1 represent the vertical distance from the center of mass to some arbitrarily chosen datum plane in configurations 2 and 1, respectively.

The work done by a spring in going from configuration 1 to configuration 2 can be written as

$$W_{1\text{-}2} = -(1/2)\,k\,(x_2^2 - x_1^2)$$

in which k is the stiffness of the spring, and x_2 and x_1 are the extensions of the spring at configurations 2 and 1, respectively.

8.9 Problems

Problem 8.1. Consider two masses m_1 (15 g) and m_2 (24 g) that are attached together by a spring with stiffness coefficient $k = 120$ dyn/mm. The spring is force free at time $t = 0$ and at that instant these particles are located at the positions

$$\mathbf{r}_1 = 3 \text{ mm } \mathbf{e}_1;\ \mathbf{r}_2 = 12 \text{ mm } \mathbf{e}_1 + 2 \text{ mm } \mathbf{e}_2$$

Suppose that the particles move with constant velocities:

$$\mathbf{v}_1 = 5\ \text{mm/s}\ \mathbf{e}_1;\ \mathbf{v}_2 = 4\ \text{mm/s}\ \mathbf{e}_2$$

Determine the work done by the spring on the particle 1 and 2 at $t = 2$ s.
Answer: $W = -41$ dyn-mm.

Problem 8.2. Show that the kinetic energy of a rigid body rotating around an axis that passes through point O of the body is given by the following equation:

$$T = (1/2)\ I^o\ \omega^2$$

in which I^o is the mass moment of inertia with respect to point O and ω is the angular speed.

Problem 8.3. A Problem Concerning Geometry. Consider a two-rod system in which the rigid links are connected by a contractile cord (Fig. P.8.3). Determine the change in the angle between the two links when the contractile cord shortens by 10%.
Answer: (a) 135° → 113°

Problem 8.4. Four equal rods, each having mass of 10 kg and length 0.6 m, are hinged together to form a rhombus (Fig. P.8.4). The angle the rods make with the horizontal plane, θ, is 45° at the instant the system is let go from rest. Use the conservation of energy principle for the system of four links to determine the kinetic energy of the system when the angle θ reduces to 30°. Note that the kinetic energy of the system is the sum of kinetic energy of the four links. What is the work done by the reaction forces at B and C on the part of the system composed of the rod a and b?
Answer: $T = 48.7$ N-m, $W = 4.6$ N-m.

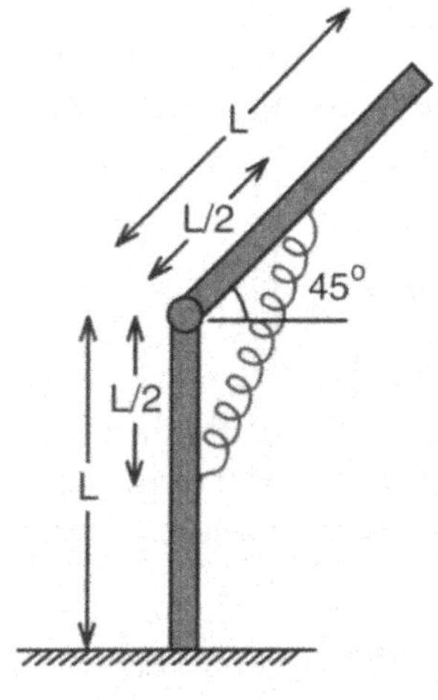

FIGURE P.8.3. Two rods connected by a hinge and a contractile cord as a simple model of bending with the use of the abdominal muscle.

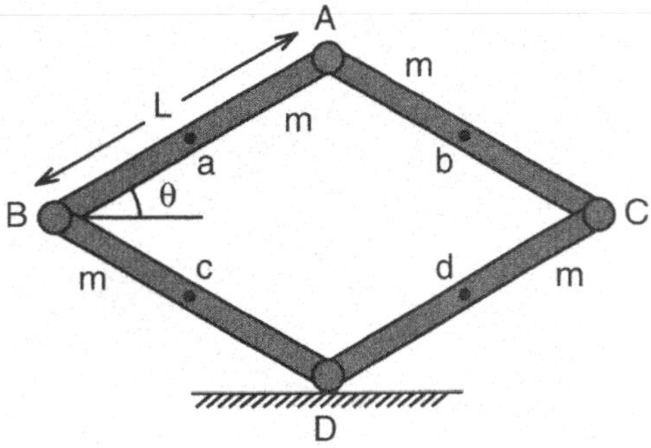

FIGURE P.8.4. A structure consisting of four equal rods that are connected to each other by hinges.

Problem 8.5. The four-rod system shown in Fig. P.8.4 is revised so that elastic springs with spring constant $k = 27$ N/cm connect the rods at midpoint as shown in Fig. P.8.5. The angle the rods make with the horizontal plane, θ, is 45° at time $t = 0$. The springs are force free when the system is let go from rest at $t = 0$. Determine the angle θ for which the system reaches static equilibrium.
Answer: $\theta = 12°$.

Problem 8.6. Consider the four-rod system shown in Fig. P.8.5. Determine the work done by the springs as θ decreases from 45° to 40°.

Problem 8.7. When a car or a railway carriage enters upon a circular path of radius r after traveling in a straight road with velocity v_o, its velocity decreases (Fig. P.8.7). Determine the velocity at the curved path.
Answer: $v^2 = v_o^2\, r^2/(r^2 + k^2)$ where k is the radius of gyration of the body.

Problem 8.8. A rod of length L and mass m is held in a vertical position on a rough horizontal surface (Fig. P.8.8). The rod is then slightly displaced from the vertical unstable equilibrium position (Fig. P.8.8a) and let go. The rod will rotate clockwise around point A without slip (Fig.

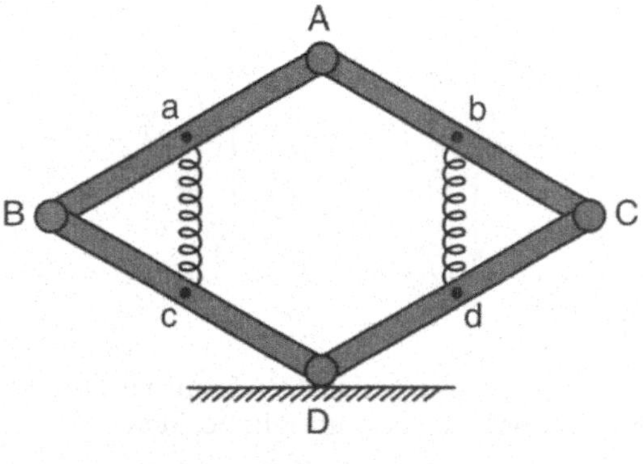

FIGURE P.8.5. The four-rod structure with compression springs attached to the rods.

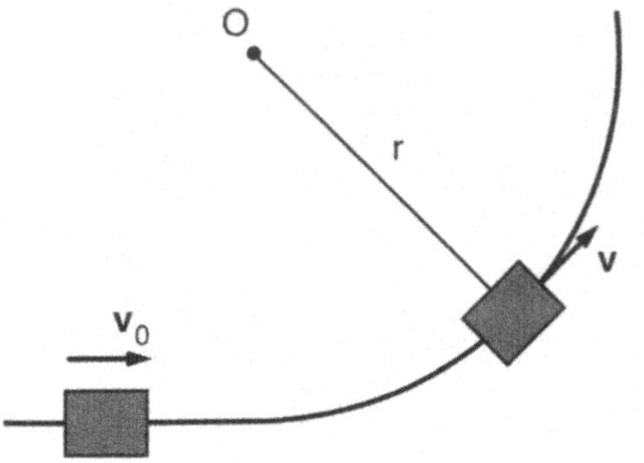

FIGURE P.8.7. The path of a train car changes from a straight line to a circular arc in the horizontal plane.

P.8.8b), but eventually the slipping must take place before the horizontal position is reached (Fig. P.8.8c). Determine the angular velocity of the rod as a function of its angle of inclination with the vertical axis before the slip takes place. At which value of θ does the slip takes place? The coefficient of static friction $\mu = 0.4$.

Hint: Conservation of mechanical energy leads to $L(d\theta/dt)^2 = 3g\,(1 - \cos\theta)$. Conservation of angular momentum with respect to point A results in $L(d^2\theta/dt^2) = -(3/2)\,g \sin\theta$.
Answer: $\theta \cong 50°$

Problem 8.9. Model the dynamics of a fall of a box containing a computer as a mass of m striking a dashpot with velocity v (Fig. P.8.9). When the box is dropped from a height h, the mass–dashpot system is subjected to an impact. Assume that the resistance force exerted by the dashpot on the computer is given by the equation

$$F = m\,v/\tau$$

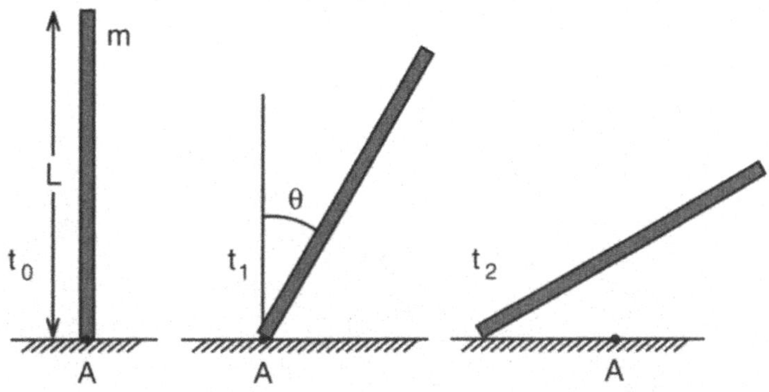

FIGURE P.8.8. Three instances from the time history of the fall of a thin rod released from a vertical upright position.

FIGURE P.8.9. A computer falling through a viscous damping material after the container hit the ground with velocity v_o.

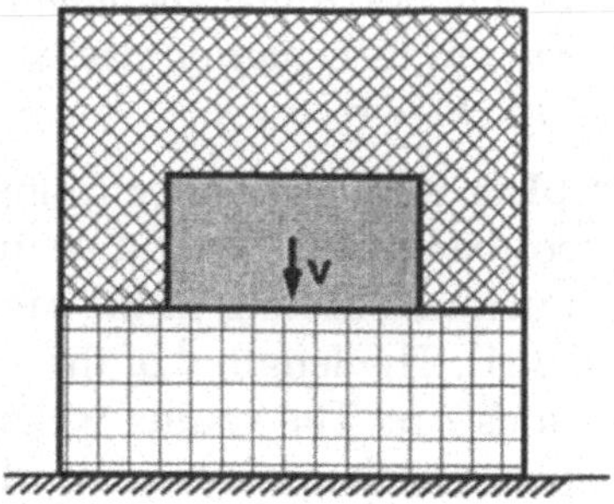

in which m is the mass of the computer, v is its speed, and τ is the dissipation time constant. Initially, before the impact, the vertical distance between the computer and the bottom of the box is d. What is the differential equation governing the motion of the computer after the box hits the ground and before the computer reaches the bottom of the box? What is the resistance force on the computer immediately after the box hits the ground? Use the following parameter values in your calculations: $m = 3$ kg, $h = 2$ m, $d = 0.3$ m, $\tau = 1.2$ s.
Answer: $F = 15.66$ N, $(d^2x/dt^2) + (1/\tau)(dx/dt) = g$ where x represents the vertical distance.

Problem 8.10. Back hyperextension is a good exercise for strengthening the back muscle (Fig. P.8.10). Supporting the heels, upper thighs, and pelvis on the padded supports, with arms folded on the chest, lower the upper torso down to the vertical position. Using the back muscle slowly pull up to the horizontal position. Once the upper body is aligned horizontally, its kinetic energy must be equal to zero. Determine the work done by the back muscle in raising the upper body from vertical down-

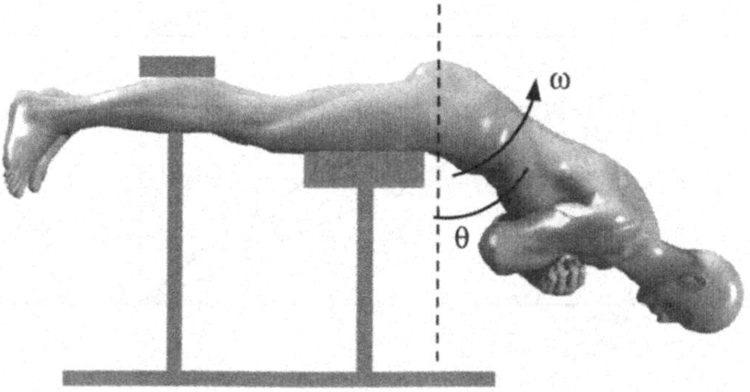

FIGURE P.8.10. A schematic diagram of a man performing back extensions.

ward to horizontal alignment. Assume that the mass of the upper body is m and its center of mass is at a distance h from the hip joint.
Answer: $W = m\,g\,h$.

Problem 8.11. A vaulter running with speed $v_o = 10$ m/s toward the crossbar places the end of the bar onto the ground at time $t = 0$ and continues running in the same direction for an additional 1.0 m (Fig. P.8.11). The length d of the part of the pole from his grip to the distal end is 7 m. The vaulter weighs 75 kg and his height is 1.91 m. The response of the pole to a thrust is given by the following relation:

$$F = k\,(b - d)$$

in which F is the compressive force acting on the pole, k = 6,000 N/m, and b is the shortest distance between the handgrip and the distal end of the pole as shown in the figure. Determine the speed v of the vaulter at the instant when he has covered a distance of 0.75 m toward the crossbar while keeping the distal end of the pole fixed on the ground.
Answer: $v = 7.7$ m/s.

Problem 8.12. McMahon and Cheng (1990) modeled hopping in place with a mass–spring system constrained to move vertically as it strikes the ground. The mass m represents the body mass, and the spring constant k of the spring represents stiffness properties of the leg (Fig. P.8.12). Let the displacement y of mass m be measured such that y is increasing when the mass is moving upward. The spring is slack, neither stretched nor compressed, when $y = 0$. The vertical velocity dy/dt

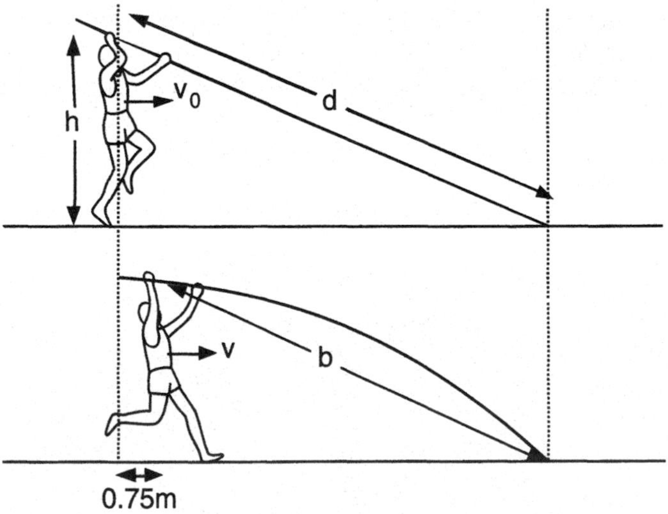

FIGURE P.8.11. A vaulter pressing the pole against the ground while running toward the crossbar.

FIGURE P.8.12. The hopping model consisting of a mass–spring system interacting with a horizontal surface.

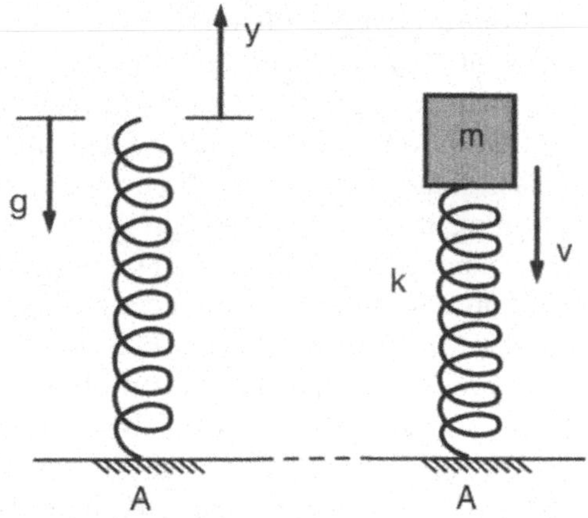

of the mass at the moment the leg-spring strikes the ground is $-v$, where v is a positive quantity. Determine the peak force the leg must bear, the time the leg spends in contact with the ground, and the stride frequency as a function of one dimensionless group, $(v\omega/g)$, where $\omega = (k/m)^{1/2}$.

Problem 8.13. McMahon and Cheng (1990) used the mass–spring model to study the mechanics of running (Fig. P.8.13). Zero-force length of the leg was represented by l_o. The leg was assumed to have a stiffness of k_{leg}. The parameter y still measures the vertical height of the body mass m, but now $y = 0$ corresponds to the ground plane. At the beginning of the rebound (the foot strikes the ground), the forward velocity dx/dt of the body mass is u and the vertical velocity dy/dt is $-v$. During the

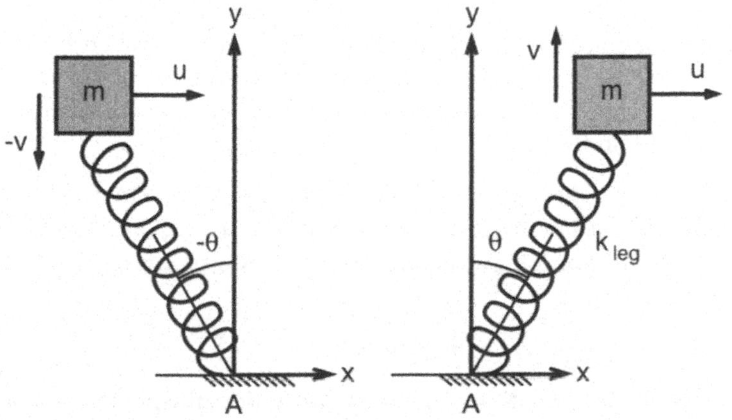

FIGURE P.8.13. A simple mass–spring system used to analyze the dependence of running speed on leg stiffness.

rebound (stance), the angle of the leg with respect to the vertical begins at $-\theta_o$ and ends at θ_o. The x velocity begins and ends with the value u, and the y velocity is reversed by the step, starting with the value $-v$ and ending with v. Determine expressions for the velocity of mass m and the ground force acting on the mass–spring system by assuming that θ_o is small ($\sin \theta_o \cong \theta_o$). For a complete solution of the problem, see McMahon and Cheng (1990).

Problem 8.14. In a review article published in August 1998 in the *Journal of Bone and Joint Surgery*, T. F. Novacheck, M.D., wrote, "Joint power is the product of the net joint moment and the joint angular velocity." How would you transform this sentence into the notation adopted in this book? Is there such a thing as a joint angular velocity?

Problem 8.15. The data presented by Roberts et al. (1997) suggest that the changes in length of the muscle belly itself are relatively slight in running turkeys. Instead, the muscles appear to function as tensioners of the tendons, meaning they act as tension generators. Discuss whether finite angular changes between body segments can occur without significant shortening of the skeletal muscle. Cite examples.

Problem 8.16. In running, both the kinetic energy and the potential energy from gravity are higher at the two ends of the stance phase than at the midstance. Because the ground force performs no work on the feet of the runner, it is clear that the sum of kinetic energy and gravitational potential energy is not constant. Speculate how the elastic (strain) energy stored in the heel pad and the tendons of the leg vary with time during the stance phase.

Problem 8.17. A subject raises the arm from the resting position ($\theta = 0$) to the full arm elevation ($\theta = \pi$) such that the angle the arm makes with the horizontal axis changes according to the relation

$$\theta = (\pi/2)[1 + \sin(\alpha \pi t - \pi/2)]$$

in which t is time in seconds. Determine the work done on the arm by the muscle–tendon systems and ligaments of the shoulder as a function of arm length L and mass m for $\alpha = 1, 2,$ and 3 and for θ from 0 to $(\pi/2)$ and from 0 to π.

Hint: The sum of kinetic plus potential energy of the arm must be equal to the work W done by forces and moments acting on the arm at the shoulder joint. For $\theta = (\pi/2)$, the conservation of energy leads to the following equation:

$$W = (mL/2)[L\alpha^2 \pi^4/3 + g]$$

According to this equation, the faster the motion, the greater is the work that must be done by the muscle–tendon systems, in this case, principally the deltoids. Deltoids will have to shorten faster with increasing speed of the motion. Even when $\alpha = 1$, most of the work done goes to increasing the kinetic energy at $\theta = (\pi/2)$. In contrast, the resultant work done by joint structures in raising the arm from $\theta = 0$ to $\theta = \pi$ is given by the expression:

$$W = m\, g\, L$$

The work done at $\theta = \pi$ is smaller than the corresponding work done at $\theta = \pi/2$ for the values of $\alpha^2 > 3g/(L\pi^4)$. Thus, in raising the arm from $\theta = \pi/2$ to $\theta = \pi$, shoulder joint structures must have supplied power in the opposite direction to that supplied by the deltoids.

Problem 8.18. A walking animal pole-vaults over its rigid limb similar to using a rigid pole to clear a hurdle. There is an alternate exchange between kinetic energy of the center of mass of the animal and the gravitational potential energy as the animal rotates over its rigid limb during walking. Experiments on dogs indicate that as much as 55% of the initial kinetic energy might be transformed into gravitational potential energy during rotation over a rigid limb. What should be the angle the limb makes with the vertical axis when it first hits the ground? Represent the animal as a point mass of m attached to one end of a light rigid pole of length d. Assume also the velocity of the center of mass to be in the horizontal direction.

9

Three-Dimensional Motion: Somersaults, Throwing, and Hitting Motions

9.1 Introduction

Three-dimensional human movement plays a fundamental role in performing arts and athletics. Some of us may have observed the three-dimensional nature of baseball pitching or noticed how a diver can induce twisting rotations in air by raising an arm to the side. Appropriately chosen shape changes are crucial in gymnastics, in the graceful turns of a ballerina, and the triple and quadruple jumps in ice skating. In this chapter, we present the principles of three-dimensional mechanics of bodies and their application to human motion. The mechanics of rigid bodies was laid out in 1760 by the Swiss scientist Leonhard Euler in his book *Thoria Motus Corporum Solidurum seu Rigidorum*. Euler is considered to be a founder of pure mathematics. He was a profilic scholar. Reading and writing late into the night under a gas lamp, night after night, he lost the sight of one eye at the age of 28. He became totally blind at 60. Blindness did not diminish his productivity, however. He had a unique ability in carrying out mental computations and continued to write papers and books for another 15 years, until he died in 1783.

Euler treated a rigid body as a system of particles in which the distance between any two particles remained constant with time. Using this property, Euler derived equations that govern the rate of rotation of a rigid body in three dimensions. These equations are much more complex than those that govern the planar motion of rigid bodies. The simple definition of angular velocity in planar motion no longer applies to three-dimensional motion. The derivation of equations of motion in three dimensions involves seemingly abstract concepts concerning vector differentiation. In the first three subsections of this chapter, we outline the application of Newton's laws of motion to the three-dimensional motion of rigid objects and multibody systems. Our presentation in this chapter does not closely follow that of Euler. Especially in the description of three-dimensional motion, we follow the presentation in Thomas R. Kane's *Dynamics*.

9.2 Time Derivatives of Vectors

In many instances in movement and motion, velocity of a point in the body or even the angular velocity of a body segment may be a function of time. If either the magnitude or the direction of a vector **v** depends on time t, then the vector is said to be a function of t. The time derivative of vectors depends on the coordinate system in which they are measured. Let **P** be a vector and let two Cartesian reference frames, E and B, be defined by their respective unit vectors ($\mathbf{e}_1$, $\mathbf{e}_2$, $\mathbf{e}_3$) and ($\mathbf{b}_1$, $\mathbf{b}_2$, $\mathbf{b}_3$). Then **P** can be expressed using the unit vectors of either E or B:

$$\mathbf{P} = {}^{E}P_1\, \mathbf{e}_1 + {}^{E}P_2\, \mathbf{e}_2 + {}^{E}P_3\, \mathbf{e}_3 \qquad (9.1a)$$

$$\mathbf{P} = {}^{B}P_1\, \mathbf{b}_1 + {}^{B}P_2\, \mathbf{b}_2 + {}^{B}P_3\, \mathbf{b}_3 \qquad (9.1b)$$

where ${}^{E}P_i$ and ${}^{B}P_i$ represent the projections of the vector **P** along the unit vectors of the coordinate systems E and B, respectively.

The time derivative of **P** in E (and in B) is defined as

$$^{E}d\mathbf{P}/dt = (d^{E}P_1/dt)\, \mathbf{e}_1 + (d^{E}P_2/dt)\, \mathbf{e}_2 + (d^{E}P_3/dt)\, \mathbf{e}_3 \qquad (9.2a)$$

$$^{B}d\mathbf{P}/dt = (d^{B}P_1/dt)\, \mathbf{b}_1 + (d^{B}P_2/dt)\, \mathbf{b}_2 + (d^{B}P_3/dt)\, \mathbf{b}_3 \qquad (9.2b)$$

As an example to illustrate the use of these equations, consider an aerobic instructor asking the class to extend their arms to the side and then rotate them around a horizontal axis that connects the two shoulder joints (Fig. 9.1). The motion involved can be represented by a slender rod rotating around a fixed axis. The position vector $\mathbf{r}^{H/S}$ from the shoulder joint to the hand can be written in the reference frame B as follows:

$$\mathbf{r}^{H/S} = L\, \mathbf{b}_1$$

where L denotes the length of the upper arm. The same position vector is a function of time in the inertial reference frame E:

$$\mathbf{r}^{H/S} = L \cos\theta\, \mathbf{e}_1 + L \sin\theta\, (\cos\phi\, \mathbf{e}_2 + \sin\phi\, \mathbf{e}_3)$$

in which the angles θ and ϕ have been defined in the figure. If the arm rotates such that the angle θ remains constant, the time derivative of **P** in B and in E can be written as

$$^{B}d\mathbf{r}^{H/S}/dt = 0$$

$$^{E}d\mathbf{r}^{H/S}/dt = L \sin\theta\, (d\phi/dt)\, (-\sin\phi\, \mathbf{e}_2 + \cos\phi\, \mathbf{e}_3)$$

Clearly in the inertial reference frame E the line segment connecting the shoulder to the hand changes orientation with time and this reflects in the time derivative in the reference frame E.

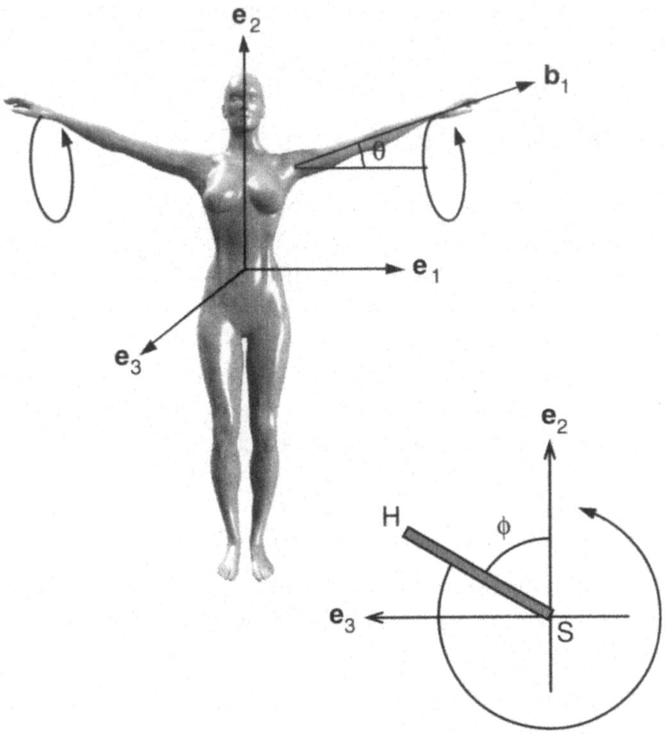

FIGURE 9.1. As part of the warm-up routine, an aerobic instructor rotates her arms around a horizontal axis that passes through her shoulder joints. The inertial coordinate system E is fixed on her trunk whereas the Cartesian coordinates B moves with the rotating arms. The unit vector along the arm length is denoted as $\mathbf{b}_1$.

9.3 Angular Velocity and Angular Acceleration

When a rigid body B moves in a reference frame E in such a way that there exists a unit vector $\mathbf{e}_3$ whose orientation in both E and B is independent of time t, B is said to have a planar motion relative to E. The angular velocity of B in E was defined in Chapter 4 as

$$^{E}\boldsymbol{\omega}^{B} = (d\theta/dt)\,\mathbf{e}_3 \tag{9.3}$$

where θ is the radian measure of the angle between a line whose orientation is fixed in E and another line whose orientation is fixed in B, both lines being perpendicular to $\mathbf{e}_3$ (Fig. 9.2). The angle θ is regarded as positive when it can be generated by a rotation of B during which a right-handed screw parallel to $\mathbf{e}_3$ and rigidly attached to B advances in the direction of $\mathbf{e}_3$.

We have also shown in Chapter 4 that, in planar motion, if $\mathbf{d}$ is any vector fixed in reference frame B then

$$(d^{E}\mathbf{d}/dt) = {}^{E}\boldsymbol{\omega}^{B} \times \mathbf{d} \tag{9.4}$$

FIGURE 9.2. Planar rotation of reference frame B with respect to reference frame E.

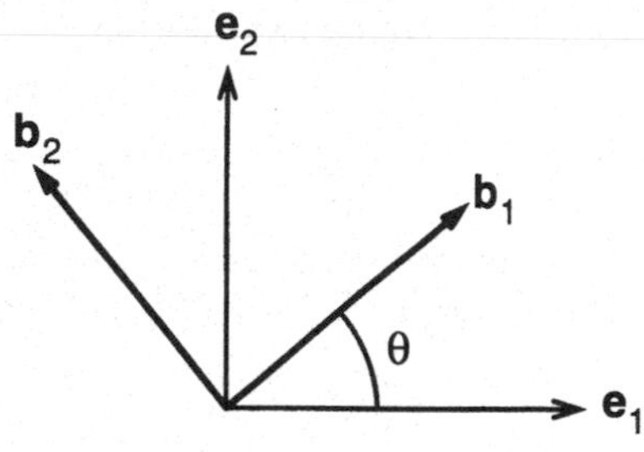

This equation illustrates why angular velocity plays such an important role in mechanics. With this operation, the time derivatives of vectors of constant length can be reduced to algebraic manipulations. Therefore, it is important to extend the concept of angular velocity to three dimensions. What is the definition of angular velocity in three-dimensional motion? The definition presented here is abstract and complex. Nevertheless this definition is at the heart of three-dimensional mechanics. It will become clear later in the section that the angular velocity thus defined satisfies Eqn. 9.4.

Let $\mathbf{b}_1$, $\mathbf{b}_2$, $\mathbf{b}_3$ be a right-handed set of mutually perpendicular unit vectors fixed in a rigid body B; the angular velocity ${}^{R}\boldsymbol{\omega}^{B}$ of B in a reference frame R is defined as

$$ {}^{R}\boldsymbol{\omega}^{B} = [({}^{R}d\mathbf{b}_1/dt) \cdot \mathbf{b}_2]\mathbf{b}_3 + [({}^{R}d\mathbf{b}_2/dt) \cdot \mathbf{b}_3]\mathbf{b}_1 + [({}^{R}d\mathbf{b}_3/dt) \cdot \mathbf{b}_1]\mathbf{b}_2 \quad (9.5) $$

where $({}^{R}d\mathbf{b}_i/dt)$ denotes the ordinary derivative of $\mathbf{b}_i$ with respect to time in reference frame R. Note that in the determination of angular velocity using this equation, we need to know only the time rate of change of two of the vectors that define the reference frame B. The time derivative of the relation $\mathbf{b}_3 = \mathbf{b}_1 \times \mathbf{b}_2$ can be used to determine the time rate of change of the third unit vector of the reference frame B.

Next, let us take the vector product of ${}^{R}\boldsymbol{\omega}^{B}$ and $\mathbf{b}_1$:

$$ {}^{R}\boldsymbol{\omega}^{B} \times \mathbf{b}_1 = [({}^{R}d\mathbf{b}_1/dt) \cdot \mathbf{b}_2]\mathbf{b}_2 - [({}^{R}d\mathbf{b}_3/dt) \cdot \mathbf{b}_1]\mathbf{b}_3 $$

On the other hand, the rule of chain differentiation requires that

$$ ({}^{R}d\mathbf{b}_3/dt) \cdot \mathbf{b}_1 + ({}^{R}d\mathbf{b}_1/dt) \cdot \mathbf{b}_3 = {}^{R}d(\mathbf{b}_3 \cdot \mathbf{b}_1)/dt = 0 $$

Thus the vector product of angular velocity and the unit vector vector $\mathbf{b}_1$ can be written as

$$ {}^{R}\boldsymbol{\omega}^{B} \times \mathbf{b}_1 = [({}^{R}d\mathbf{b}_1/dt) \cdot \mathbf{b}_2]\mathbf{b}_2 + [({}^{R}d\mathbf{b}_1/dt) \cdot \mathbf{b}_3]\mathbf{b}_3 $$

The two terms on the right-hand side of this equation are simply the projections of ${}^{R}d\mathbf{b}_1/dt$ on $\mathbf{b}_2$ and $\mathbf{b}_3$. Because ${}^{R}d\mathbf{b}_1/dt$ cannot have a projection on $\mathbf{b}_1$ (otherwise its length would not remain equal to one), it is clear that

$$ {}^{R}d\mathbf{b}_1/dt = {}^{R}\boldsymbol{\omega}^{B} \times \mathbf{b}_1 \quad (9.6a) $$

Similarly we can show that

$$^{\mathrm{R}}d\mathbf{b}_2/dt = {^{\mathrm{R}}\boldsymbol{\omega}^{\mathrm{B}}} \times \mathbf{b}_2 \tag{9.6b}$$

$$^{\mathrm{R}}d\mathbf{b}_3/dt = {^{\mathrm{R}}\boldsymbol{\omega}^{\mathrm{B}}} \times \mathbf{b}_3 \tag{9.6c}$$

Let $\mathbf{r}_i$ be unit vectors fixed in R. The time derivatives of vector $\mathbf{P}$ in reference systems R and B are related by the following equation:

$$\begin{aligned} {^{\mathrm{R}}d\mathbf{P}/dt} &= (d^{\mathrm{R}}\mathrm{P}_1/dt)\,\mathbf{r}_1 + (d^{\mathrm{R}}\mathrm{P}_2/dt)\,\mathbf{r}_2 + (d^{\mathrm{R}}\mathrm{P}_3/dt)\,\mathbf{r}_3 \\ &= (d^{\mathrm{B}}\mathrm{P}_1/dt)\,\mathbf{b}_1 + {^{\mathrm{B}}\mathrm{P}_1}(d^{\mathrm{R}}\mathbf{b}_1/dt) + (d^{\mathrm{B}}\mathrm{P}_2/dt)\,\mathbf{b}_2 + {^{\mathrm{B}}\mathrm{P}_2}(d^{\mathrm{R}}\mathbf{b}_2/dt) \\ &\quad + (d^{\mathrm{B}}\mathrm{P}_3/dt)\,\mathbf{b}_3 + {^{\mathrm{B}}\mathrm{P}_3}(d^{\mathrm{R}}\mathbf{b}_3/dt) \end{aligned}$$

Substituting Eqns. 9.2 and 9.6 into this equation, we obtain the following:

$$^{\mathrm{R}}d\mathbf{P}/dt = {^{\mathrm{B}}d\mathbf{P}/dt} + {^{\mathrm{R}}\boldsymbol{\omega}^{\mathrm{B}}} \times \mathbf{P} \tag{9.7}$$

This is one of the most important equations in mechanics. According to this equation, if $\mathbf{P}$ is constant in B, then the time derivative of $\mathbf{P}$ in R is simply the vector product of the angular velocity of B with respect to R and the vector $\mathbf{P}$.

Next let us consider the time derivative of $\mathbf{P}$ in the inertial reference frame E:

$$\begin{aligned} {^{\mathrm{E}}d\mathbf{P}/dt} &= {^{\mathrm{R}}d\mathbf{P}/dt} + {^{\mathrm{E}}\boldsymbol{\omega}^{\mathrm{R}}} \times \mathbf{P} = {^{\mathrm{B}}d\mathbf{P}/dt} + {^{\mathrm{R}}\boldsymbol{\omega}^{\mathrm{B}}} \times \mathbf{P} + {^{\mathrm{E}}\boldsymbol{\omega}^{\mathrm{R}}} \times \mathbf{P} \\ &= {^{\mathrm{B}}d\mathbf{P}/dt} + ({^{\mathrm{R}}\boldsymbol{\omega}^{\mathrm{B}}} + {^{\mathrm{E}}\boldsymbol{\omega}^{\mathrm{R}}}) \times \mathbf{P} \end{aligned}$$

Thus, one can show that angular velocities can be added according to the following equation:

$$^{\mathrm{E}}\boldsymbol{\omega}^{\mathrm{B}} = {^{\mathrm{R}}\boldsymbol{\omega}^{\mathrm{B}}} + {^{\mathrm{E}}\boldsymbol{\omega}^{\mathrm{R}}} \tag{9.8}$$

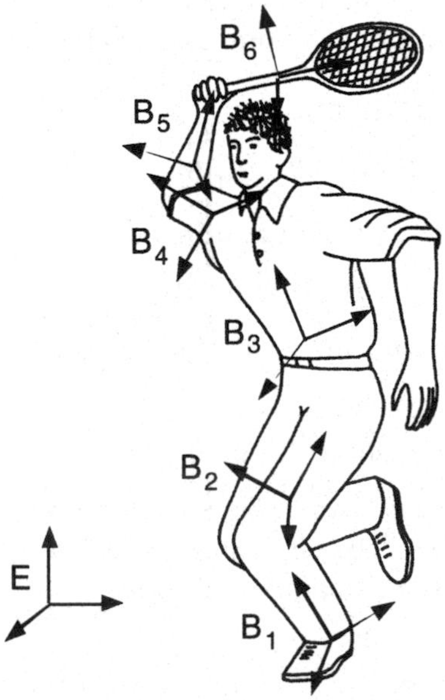

FIGURE 9.3. A tennis player preparing to hit a ball. The reference frames B_i are attached to the body segments shown and move with them.

It is straightforward to extend the additivity relationship to the case of successive reference frames in the study of motion (Fig. 9.3):

$$^{E}\boldsymbol{\omega}^{B5} = {}^{E}\boldsymbol{\omega}^{B1} + {}^{B1}\boldsymbol{\omega}^{B2} + {}^{B2}\boldsymbol{\omega}^{B3} + {}^{B3}\boldsymbol{\omega}^{B4} + {}^{B4}\boldsymbol{\omega}^{B5} \tag{9.9}$$

Example 9.1. Seated Dumbbell Press. Determine the angular velocities of the upper arm and the forearm of a weight lifter during seated dumbbell press (Fig. 9.4). The rotational angles θ and ϕ shown in Fig. 9.4 were determined with the use of two cameras. Both angles varied with time:

$$\theta = 20t$$

$$\phi = 30t + 100t^2$$

If B and D denote the reference frames fixed in the upper arm and forearm, respectively, determine the angular velocities $^{E}\boldsymbol{\omega}^{B}$, $^{E}\boldsymbol{\omega}^{D}$, and $^{B}\boldsymbol{\omega}^{D}$.

Solution: Using Eqn. 9.3, it is straightforward to show that the following holds:

$$^{E}\boldsymbol{\omega}^{B} = 20\mathbf{e}_3 = 20\mathbf{b}_3$$

$$^{E}\boldsymbol{\omega}^{D} = (30 + 200t)\mathbf{b}_3$$

$$^{B}\boldsymbol{\omega}^{D} = {}^{E}\boldsymbol{\omega}^{D} - {}^{E}\boldsymbol{\omega}^{B} = (10 + 200t)\mathbf{b}_3$$

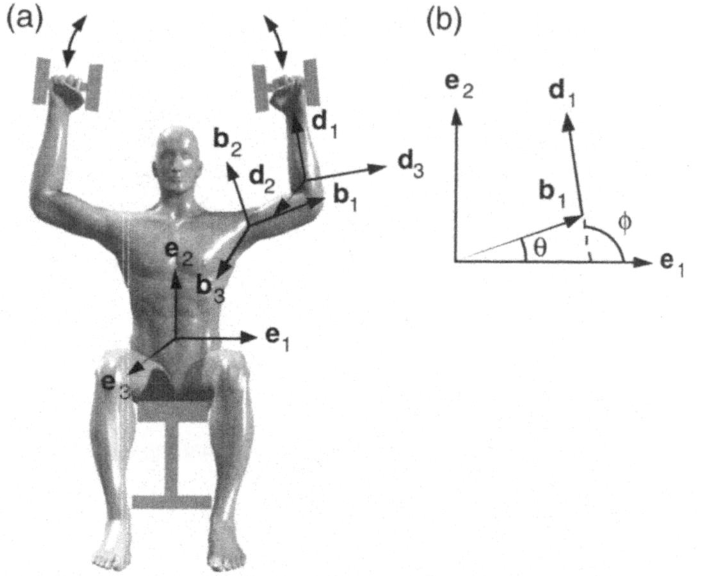

FIGURE 9.4a,b. Planar rotation of upper arm and forearm of a weight lifter during seated dumbbell press **(a)**. The symbols θ and ϕ indicate the angles made by the unit vectors $\mathbf{b}_1$ and $\mathbf{d}_1$, respectively, with the $\mathbf{e}_1$ axis **(b)**.

Velocity and Acceleration in a Rigid Body

To relate the velocity of two points in a rigid body, one takes the time derivative of the position vector connecting the two points:

$$ {}^{E}\mathbf{v}^{P} = {}^{E}\mathbf{v}^{Q} + {}^{E}\boldsymbol{\omega}^{B} \times \mathbf{r}^{P/Q} \tag{9.10a} $$

Subsequent differentiation of this relation with respect to time yields the following equation relating accelerations of points P and Q:

$$ {}^{E}\mathbf{a}^{P} = {}^{E}\mathbf{a}^{Q} + {}^{E}\boldsymbol{\alpha}^{B} \times \mathbf{r}^{P/Q} + {}^{E}\boldsymbol{\omega}^{B} \times ({}^{E}\boldsymbol{\omega}^{B} \times \mathbf{r}^{P/Q}) \tag{9.10b} $$

where ${}^{E}\boldsymbol{\alpha}^{B}$ is the angular acceleration of the rigid body B. The angular acceleration is defined as the time derivative of angular velocity:

$$ {}^{E}\boldsymbol{\alpha}^{B} = d^{E}\boldsymbol{\omega}^{B}/dt \tag{9.11} $$

Equations 9.10 and 9.11 are identical to those presented earlier for planar motion. It is just that the definition of angular velocity is more complex in three-dimensional motion than in planar motion.

Example 9.2. Leg-Lifting Turns of a Dancer. A dancer rotates his body around an axis normal to the stage at a constant rate of 2 rad/s, counterclockwise (Fig. 9.5). At the same time, his one leg moves away from his body axis (extends) at a uniform rate of 10 rad/s. Determine the an-

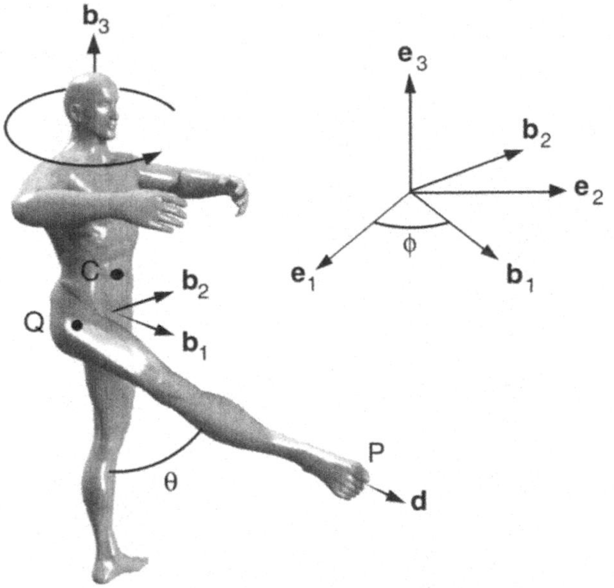

FIGURE 9.5. Leg-lifting turns of a dancer. The body rotates around an axis perpendicular to the stage. At the same time the one leg moves away from the trunk in the sagittal plane.

gular velocity of the dancer's trunk as well as the angular velocity of his leg with respect to his trunk.

Solution: Let $\mathbf{b}_i$ and $\mathbf{e}_i$ represent the unit vectors attached to the trunk of the dancer and to the floor (Fig. 9.5). The angular velocity of the trunk with respect to the inertial reference frame E is given by the following equation:

$$^E\boldsymbol{\omega}^B = 2\mathbf{e}_3$$

Similarly, we can show that

$$^B\boldsymbol{\omega}^D = -10\mathbf{b}_2 = -10\,(-\sin\phi\,\mathbf{e}_1 + \cos\phi\,\mathbf{e}_2)$$

where D denotes the reference frame attached to the extended leg of the dancer. The angular velocity of the leg of the dancer with respect to earth is given by the equation:

$$^E\boldsymbol{\omega}^D = {}^B\boldsymbol{\omega}^D + {}^E\boldsymbol{\omega}^B = -10\,(-\sin\phi\,\mathbf{e}_1 + \cos\phi\,\mathbf{e}_2) + 2\mathbf{e}_3 = -10\mathbf{b}_2 + 2\mathbf{e}_3$$

Example 9.3. Velocity and Acceleration During a Leg-Lifting Turn. Compute the velocity and acceleration of the hip and the ankle of the dancer performing a leg-lifting turn. The velocity of the center of mass C at the moment considered was equal to zero. The distance between the two hip joints of the dancer was measured to be 22 cm. The total length of lower limb was 88 cm. At the instant considered the leg was aligned with the negative $\mathbf{e}_3$ axis. All other parameters were the same as in Example 9.2.

Solution: Let E, B, and D represent the reference frames fixed on the stage, on his trunk, and on his leg, respectively. Let Q denote the rotation center of the hip joint. Using the angular velocity that was determined in the previous example, we find that:

$$^E\mathbf{v}^Q = {}^E\mathbf{v}^C + {}^E\boldsymbol{\omega}^B \times \mathbf{r}^{Q/C} = 0 + 2\mathbf{e}_3 \times (0.11\mathbf{b}_2) = -0.22\mathbf{b}_1$$

$$\begin{aligned} ^E\mathbf{a}^Q &= {}^E\mathbf{a}^C + {}^E\boldsymbol{\alpha}^B \times \mathbf{r}^{Q/C} + {}^E\boldsymbol{\omega}^B \times ({}^E\boldsymbol{\omega}^B \times \mathbf{r}^{Q/C}) \\ &= 0 + 0 \times \mathbf{r}^{P/Q} + 2\mathbf{e}_3 \times (2\mathbf{e}_3 \times 0.11\mathbf{b}_2) = -0.44\mathbf{b}_2 \end{aligned}$$

Let us now use Eqn. 9.10 between the hip joint (Q) and the center of rotation of the ankle (P) to determine the velocity and acceleration of his ankle.

$$\begin{aligned} ^E\mathbf{v}^P = {}^E\mathbf{v}^Q + {}^E\boldsymbol{\omega}^D \times \mathbf{r}^{P/Q} &= -0.22\,\mathbf{b}_1 \\ &+ (-10\mathbf{b}_2 + 2\mathbf{e}_3) \times (-0.88\mathbf{e}_3) = 8.58\mathbf{b}_1 \end{aligned}$$

To compute the acceleration of P, we need to determine the angular acceleration $^E\boldsymbol{\alpha}^D$ of the leg with respect to E.

$$\begin{aligned} ^E\boldsymbol{\alpha}^D &= d({}^E\boldsymbol{\omega}^D)/dt = d(-10\,\mathbf{b}_2 + 2\mathbf{e}_3)/dt \\ &= -10\,({}^E\boldsymbol{\omega}^D \times \mathbf{b}_2) = -10\,(2\mathbf{e}_3 \times \mathbf{b}_2) = 20\mathbf{b}_1 \end{aligned}$$

Now we can use Eqn. 9.10b to compute $^E\mathbf{a}^P$:

$$^E\mathbf{a}^P = {}^E\mathbf{a}^Q + {}^E\boldsymbol{\alpha}^D \times \mathbf{r}^{P/Q} + {}^E\boldsymbol{\omega}^D \times ({}^E\boldsymbol{\omega}^D \times \mathbf{r}^{P/Q})$$

Because the leg is oriented in the direction of $\mathbf{e}_3$, we find:

$$^{E}\mathbf{a}^{P} = -0.44\mathbf{b}_2 + 20\mathbf{b}_1 \times 0.88\mathbf{e}_3 + (-10\mathbf{b}_2 + 2\mathbf{e}_3) \times [(-10\mathbf{b}_2 + 2\mathbf{e}_3) \times 0.88\mathbf{e}_3] = 34.76\mathbf{b}_2 + 88.00\mathbf{e}_3$$

9.4 Conservation of Angular Momentum

The angular momentum of a rigid object with respect to point O is defined as

$$\mathbf{H}^{o} = \int (\mathbf{r}^{P/O} \times {}^{E}\mathbf{v}^{P})\, dm \tag{9.12}$$

where $\mathbf{r}^{P/O}$ is the position vector from point O to a point P in the body B, ${}^{E}\mathbf{v}^{P}$ is the velocity of point P in reference frame E, and dm is a small mass element surrounding point P (Fig. 9.6). According to the parallelogram law

$$\mathbf{r}^{P/O} = \mathbf{r}^{C/O} + \boldsymbol{\rho}$$

where $\boldsymbol{\rho}$ denotes the position vector from C to P. Note that from the definition of the center of mass $\int \boldsymbol{\rho} dm = 0$. The integration of velocity over the mass elements of the body is by definition equal to the mass of the body times the velocity of the center of mass. Using these two relationships in Eqn. 9.12 we obtain:

$$\mathbf{H}^{o} = \int (\mathbf{r}^{C/O} + \boldsymbol{\rho}) \times ({}^{E}\mathbf{v}^{P})\, dm = m\, \mathbf{r}^{C/O} \times {}^{E}\mathbf{v}^{C} + \int \boldsymbol{\rho} \times {}^{E}\mathbf{v}^{P}\, dm$$

in which m denotes the mass of the body B.

The velocity of a point P in body B can be expressed as a function of the velocity of the center of mass and the angular velocity of the body B:

$${}^{E}\mathbf{v}^{P} = {}^{E}\mathbf{v}^{C} + {}^{E}\boldsymbol{\omega}^{B} \times \boldsymbol{\rho}$$

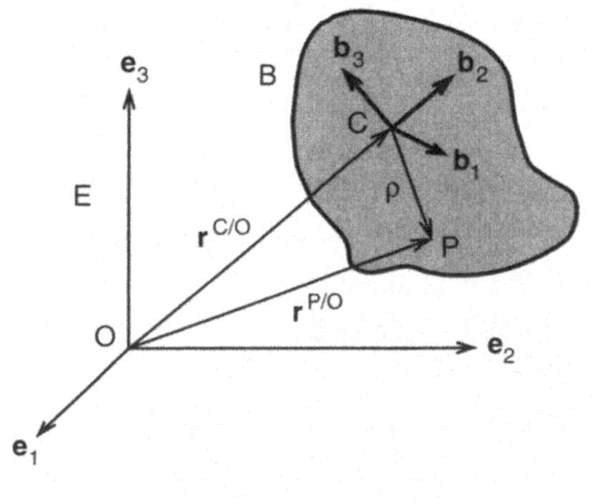

FIGURE 9.6. A rigid body undergoing three-dimensional motion with respect to the reference frame E that is fixed on earth. The point O is the origin of the Cartesian coordinate E. The rigid body and the reference frame embedded onto it are denoted by B. The symbol P denotes an ordinary point on B whereas C refers to the center of mass.

The angular momentum about point O can then be written as

$$\mathbf{H}^o = m\,\mathbf{r}^{C/O} \times {}^E\mathbf{v}^C + \int \boldsymbol{\rho} \times {}^E\mathbf{v}^P\, dm$$
$$= m\,\mathbf{r}^{C/O} \times {}^E\mathbf{v}^C + \left(\int \boldsymbol{\rho} dm\right) \times {}^E\mathbf{v}^C + \int \boldsymbol{\rho} \times ({}^E\boldsymbol{\omega}^B \times \boldsymbol{\rho})\, dm$$

$$\mathbf{H}^o = m\,\mathbf{r}^{C/O} \times {}^E\mathbf{v}^C + \int \boldsymbol{\rho} \times ({}^E\boldsymbol{\omega}^B \times \boldsymbol{\rho})\, dm$$

$$\mathbf{H}^o = m\,\mathbf{r}^{C/O} \times {}^E\mathbf{v}^C + \mathbf{H}^c \tag{9.13a}$$

where $\mathbf{H}^c$ is the angular momentum about the center of mass:

$$\mathbf{H}^c = \int \boldsymbol{\rho} \times ({}^E\boldsymbol{\omega}^B \times \boldsymbol{\rho})\, dm \tag{9.13b}$$

Expressing $\boldsymbol{\rho}$ and ${}^E\boldsymbol{\omega}^B$ in terms of unit vectors $\mathbf{b}_i$ associated with rigid body B, the angular momentum with respect to the center of mass can be written as

$$\mathbf{H}^c = \mathrm{H}^c{}_1\, \mathbf{b}_1 + \mathrm{H}^c{}_1\, \mathbf{b}_2 + \mathrm{H}^c{}_1\, \mathbf{b}_3 \tag{9.14a}$$

where $\mathbf{b}_1$, $\mathbf{b}_2$, and $\mathbf{b}_3$ are a set of orthogonal unit vectors defined in B. The scalar components of the angular momentum are given by the following relationships:

$$\mathrm{H}^c{}_1 = (\mathrm{I}^c{}_{11}\, \omega_1 + \mathrm{I}^c{}_{12}\, \omega_2 + \mathrm{I}^c{}_{13}\, \omega_3) \tag{9.14b}$$

$$\mathrm{H}^c{}_2 = (\mathrm{I}^c{}_{21}\, \omega_1 + \mathrm{I}^c{}_{22}\, \omega_2 + \mathrm{I}^c{}_{23}\, \omega_3) \tag{9.14c}$$

$$\mathrm{H}^c{}_3 = (\mathrm{I}^c{}_{31}\, \omega_1 + \mathrm{I}^c{}_{32}\, \omega_2 + \mathrm{I}^c{}_{33}\, \omega_3) \tag{9.14d}$$

where

$${}^E\boldsymbol{\omega}^B = \omega_1\, \mathbf{b}_1 + \omega_2\, \mathbf{b}_2 + \omega_3\, \mathbf{b}_3 \tag{9.14e}$$

and the symbols such as $\mathrm{I}^c{}_{13}$ refer to the components of mass moment of inertia. This matrix is defined as follows:

$$\mathrm{I}^c{}_{ij} = \int [(x_i x_j)\, \delta_{ij} - (x_i x_j)]\, dm \tag{9.15a}$$

The integration in this equation is over the mass of the body. The symbol x_i denotes the distance from the center of mass of the body to the center of mass of the small mass element dm along the $\mathbf{b}_i$ direction. The entity δ_{ij} is equal to one when $i = j$, and is equal to zero otherwise.

$$\delta_{ij} = 0 \text{ if } i \neq j \text{ and } \delta_{ij} = 1 \text{ if } i = j \tag{9.15b}$$

Also, summation over $i = 1, 2,$ and 3 is implied in Eqn. 9.15a when an index such as i or j is repeated twice in a term. Thus, for example:

$$\mathrm{I}^c{}_{11} = \int [x_1 x_1 + x_2 x_2 + x_3 x_3 - x_1 x_1]\, dm = \int [x_2{}^2 + x_3{}^2] dm$$

$$\mathrm{I}^c{}_{12} = -\int [x_1 x_2] dm$$

Overall, there are nine elements of the inertia matrix, but because the matrix is symmetric, only six independent elements need to be determined. The components of mass moment of inertia are functions of geometry of the body and the mass distribution within the body. They do not depend on the velocity and acceleration of the body. Mass moment of inertia is a measure of a body's resistance to a change in its rate of rotation. The components of mass moment of inertia for simple shapes have been tabulated in the back of this book.

If a point in a rigid body B, say point O, is fixed in the inertial reference frame E, using Eqns. 9.10a and 9.13a we can express the angular momentum about O as follows:

$$H^o_1 = (I^o_{11}\,\omega_1 + I^o_{12}\,\omega_2 + I^o_{13}\,\omega_3) \tag{9.16a}$$

$$H^o_2 = (I^o_{21}\,\omega_1 + I^o_{22}\,\omega_2 + I^o_{23}\,\omega_3) \tag{9.16b}$$

$$H^o_3 = (I^o_{31}\,\omega_1 + I^o_{32}\,\omega_2 + I^o_{33}\,\omega_3) \tag{9.16c}$$

in which H^o_i and ω_i are the components of angular momentum and angular velocity along the unit vectors associated with reference frame B.

Once the inertia matrix I^c_{ij} is determined, I^o_{ij} can be obtained from it using the following equation:

$$I^o_{ij} = I^c_{ij} + m\,(h_i\,h_j\,\delta_{ij} - h_i h_j) \tag{9.17}$$

in which m denotes the mass of the body, and h_i are the coordinates of the center of mass of the body relative to point O.

Example 9.4. The Mass Moment of Inertia Matrix of Two Slender Rods. Determine the mass moment of inertia matrix of the body shown in Fig. 9.7 about its center of mass C. The body shown in the figure is composed of two slender rods of mass m and length L.

Solution: The mass moment of inertia of the structure shown in the figure is the sum of contributions from each slender rod. The mass moment of inertia of a rod about its center of mass with respect to an axis perpendicular to the long axis of the rod is equal to $mL^2/12$. Its moment of

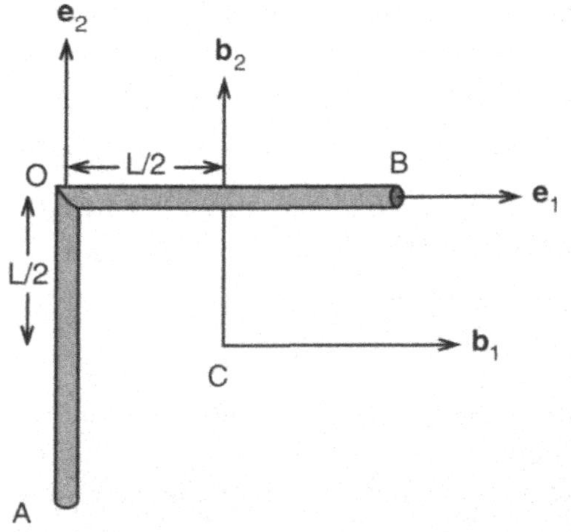

FIGURE 9.7. A rigid object composed of two slender rods of length L and mass m. The mass moment of inertia matrix was computed with respect to the reference frame shown in the figure.

inertia with respect to point C can be computed by using Eqn. 9.17. Let us illustrate this by computing I^c_{11}:

$$I^c_{11} = I^{c1}_{11} + m\,(h_2^2 + h_3^2) + I^{c2}_{11} + m\,(d_2^2 + d_3^2)$$
$$= mL^2/12 + m\,(0 + 0) + 0 + m\,(L^2/4 + 0) = mL^2/3$$

where the symbols h_i and d_i denote the coordinates of the center of mass of OA and OB relative to point C. The first two terms on the right-hand side represent the contribution of rod OA, and the last two terms are the contribution of OB.

Similarly, we can show that

$$I^c_{22} = mL^2/3;\; I^c_{33} = 2\,mL^2/3$$

$$I^c_{12} = I^c_{21} = 0$$

$$I^c_{13} = I^c_{31} = I^c_{23} = I^c_{32} = 0$$

Moment of Momentum of a Tree of Rigid Bodies

In the analysis of some movements, the human body can be considered as a tree of rigid body segments. What is the angular momentum of a tree of rigid objects about the center of mass of the system?

We can address this question by computing the angular momentum ${}^i\mathbf{H}^c$ of each rigid body B_i in the system about the center of mass of the system by using Eqn. 9.13a:

$${}^i\mathbf{H}^c = m_i\,\mathbf{r}^{Ci/C} \times {}^E\mathbf{v}^{Ci} + \mathbf{H}^{ci}$$

where m_i and Ci represent the mass and the center of mass of body B_i. Summing over all bodies in the multibody system, we obtain

$$\mathbf{H}^c = \Sigma m_i\,\mathbf{r}^{Ci/C} \times {}^E\mathbf{v}^{Ci} + \Sigma\mathbf{H}^{ci} \tag{9.18a}$$

$$\mathbf{H}^c = \mathbf{H}^{*c} + \Sigma\mathbf{H}^c_i \tag{9.18b}$$

The first term on the right-hand side, $\mathbf{H}^{*c}$, is the moment of momentum about point C of the system of lumped masses where the mass of each body in the system is lumped in the center of mass of the body. The second term is the sum of angular momentums of each rigid body in the system about its own center of mass.

If the velocity of center of mass of each of these bodies are the same, then the first term will be equal to zero because $\Sigma m_i\,\mathbf{r}^{Ci/C}$ defines position of the center of mass of the system. In this case the moment of momentum will be the sum of angular momentum of each body about its own center of mass.

In Chapter 3, we have seen that the laws of motion for a system of particles can be written as follows:

$$\Sigma\mathbf{F} = m\mathbf{a}^c \tag{9.19}$$

$$\Sigma\mathbf{M}^c = {}^E d\mathbf{H}^c/dt \tag{9.20a}$$

where $\Sigma\mathbf{F}$ is the resultant external force acting on the system, $\mathbf{a}^c$ is the acceleration of the center of mass, and $\Sigma\mathbf{M}^c$ is the resultant external moment with respect to the center of mass of the system. These laws apply for a rigid body because it is composed of a large number of particles. When the resultant external moment acting on a system is zero, then $\mathbf{H}^c$ must be equal to a constant. This is the case for airborne heavy bodies where the only significant external force is the force of gravity and it passes through the center of mass of the body.

If a point within the body is fixed in the inertial reference system, Eqn. 9.20a can be replaced with the following:

$$\mathbf{M}^o{}_{ext} = {}^E d\mathbf{H}^o/dt \tag{9.20b}$$

where $\mathbf{M}^o{}_{ext}$ is the resultant external moment with respect to the fixed point O. In Eqns. 9.20 we have specified the frame in which a time derivative must be taken. If the frame is not specified, it is understood to be the inertial reference frame E, that is:

$$d\mathbf{H}^c/dt = {}^E d\mathbf{H}^c/dt$$

Let the angular momentum of a body B expressed in the following form:

$$\mathbf{H}^c = H^c{}_1\mathbf{d}_1 + H^c{}_2\,\mathbf{d}_2 + H^c{}_3\,\mathbf{d}_3$$

where $\mathbf{d}_i$ represents the unit vectors in a reference frame D.

Let $\boldsymbol{\Omega} = \Omega_1\,\mathbf{d}_1 + \Omega_2\,\mathbf{d}_2 + \Omega_3\,\mathbf{d}_3$ be the angular velocity of D in E. Note that if D were a reference frame embedded in body B then the angular velocity $\boldsymbol{\Omega}$ would be equal to the angular velocity $\boldsymbol{\omega}$ of body B in reference frame E.

Using Eqn. 9.7, the time derivative of $\mathbf{H}^c$ in the inertial reference frame E can be expressed as follows:

$${}^E d\mathbf{H}^c/dt = {}^D d\mathbf{H}^c/dt + \boldsymbol{\Omega} \times \mathbf{H}^c \tag{9.21}$$

Substituting Eqns. 9.14 into this equation, we derive three differential equations governing the time rate of change of angular velocity of B. These equations are called the equations of angular momentum:

$$\begin{aligned}
M_1 &= I^c{}_{11}\,(d\omega_1/dt) + I^c{}_{12}\,(d\omega_2/dt) + I^c{}_{13}\,(d\omega_3/dt) \\
&\quad - \Omega_3\,(I^c{}_{21}\omega_1 + I^c{}_{22}\omega_2 + I^c{}_{23}\,\omega_3) + \Omega_2\,(I^c{}_{31}\omega_1 + I^c{}_{32}\,\omega_2 + I^c{}_{33}\,\omega_3) \\
M_2 &= I^c{}_{21}\,(d\omega_1/dt) + I^c{}_{22}\,(d\omega_2/dt) + I^c{}_{23}\,(d\omega_3/dt) + \\
&\quad + \Omega_3\,(I^c{}_{12}\,\omega_2 + I^c{}_{11}\,\omega_1 + I^c{}_{13}\,\omega_3) - \Omega_1\,(I^c{}_{31}\omega_1 + I^c{}_{32}\,\omega_2 + I^c{}_{33}\,\omega_3) \\
M_3 &= I^c{}_{31}\,(d\omega_1/dt) + I^c{}_{32}\,(d\omega_2/dt) + I^c{}_{33}\,(d\omega_3/dt) - \\
&\quad - \Omega_2\,(I^c{}_{11}\omega_1 + I^c{}_{12}\,\omega_2 + I^c{}_{13}\,\omega_3) + \Omega_1\,(I^c{}_{21}\omega_1 + I^c{}_{22}\omega_2 + I^c{}_{23}\,\omega_3)
\end{aligned} \tag{9.22}$$

Taken together with the primary equation for angular velocity:

$${}^D d\mathbf{b}_i/dt = {}^D\boldsymbol{\omega}^B \times \mathbf{b}_i$$

Equation 9.22 determines both the angular velocity and the orientation of the body in space, provided that the orientation and the angular velocity are known at an initial time and that the resultant external moment acting on the body is specified as a function of time. If an object rotates about a fixed point O, the equations of angular momentum with respect to point O are identical to that of Eqns. 9.22 with I^c_{ij} replaced by I^o_{ij}.

Integration of Euler's equations is not trivial even for simple-shaped bodies. Often, however, the angular velocity of a body may be a function of only one or two variables. In other cases, as in the study of airborne motion, the rotational velocity is found by setting the angular momentum about the center of mass equal to a constant. Additionally, commercially available programs such as *Working Model 3D* can be used to determine both the motion of the center of mass and the rotation of the body around the center of mass for a rigid body or a tree of rigid bodies. Yet another option is to quantify the motion using photogrammetric means, and then use the laws of motion to evaluate the resultant force and the resultant moment acting on a body segment.

Example 9.5. Precession of a Football. When a top or an American football is set to spinning about the vertical axis, the long axis of the top (football) may initially remain vertical (Fig. 9.8). As friction reduces the spin rate, the spin axis begins to lean over and rotate about the vertical axis. This phase of the top's motion approximates steady precession. Thus it is reasonable to assume the rate of rotation to be constant. Determine an equation that relates the spin rates to the mass moment of inertia and the angle of inclination of the top with the vertical axis.

Solution: Let $\mathbf{e}_1$, $\mathbf{e}_2$, and $\mathbf{e}_3$ denote the unit vectors of reference frame E. We choose origin O of E to coincide with the point of the top and the $\mathbf{e}_3$ vertical upward. The reference frame D has also its origins at O.

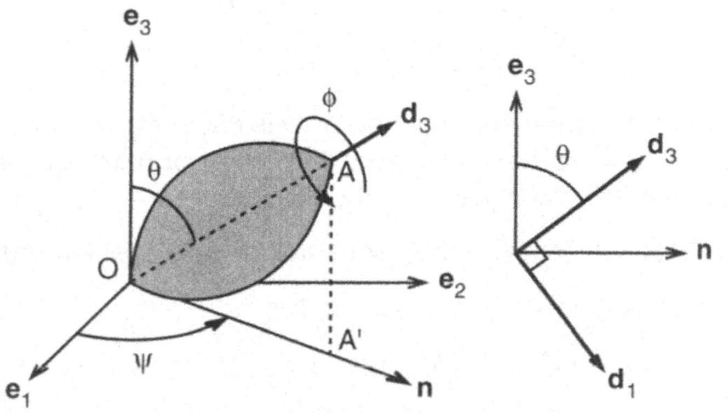

FIGURE 9.8. A spinning top (American football) seems to defy gravity. The orientation of the spin axis $\mathbf{d}_3$ is defined by the angles θ and Ψ.

Let $\mathbf{d}_3$ denote the unit vector along the axis of the top as shown in Fig. 9.8. The $\mathbf{d}_1$ axis is taken to be in the plane of $\mathbf{e}_3$ and $\mathbf{d}_3$. The unit vector $\mathbf{d}_2$ is equal to the vector product of $\mathbf{d}_1$ and $\mathbf{d}_3$. The nutation angle between $\mathbf{d}_3$ and $\mathbf{e}_3$ is denoted by the symbol θ. This angle remains constant during the motion. The top spins about the $\mathbf{d}_3$ axis with a constant spin rate equal to $(d\phi/dt)$. The center of mass of the top traverses a circle about the vertical axis $\mathbf{e}_3$ at a constant precession rate equal to $(d\Psi/dt)$.

The unit vector $\mathbf{e}_3$ remains constant in both reference frames D and E during the spinning of football, indicating that the motion is planar. The angular velocity of D in E is then given by the following expression:

$$^{E}\boldsymbol{\omega}^{D} = \boldsymbol{\Omega} = (d\Psi/dt)\ \mathbf{e}_3 = (d\Psi/dt)\ (-\sin\theta\ \mathbf{d}_1 + \cos\theta\ \mathbf{d}_3)$$

$$\Omega_1 = -(d\Psi/dt)\sin\theta,\ \Omega_2 = 0\ ,\ \Omega_3 = (d\Psi/dt)\cos\theta \qquad (9.23a)$$

The angular velocity of the top with respect to D is

$$^{D}\boldsymbol{\omega}^{B} = (d\phi/dt)\ \mathbf{d}_3$$

where B denotes a reference frame that is fixed in B. Angular velocity of the top with respect to E is obtained using the relation:

$$^{E}\boldsymbol{\omega}^{B} = \boldsymbol{\omega} = {}^{E}\boldsymbol{\omega}^{D} + {}^{D}\boldsymbol{\omega}^{B} = (d\Psi/dt)\ (-\sin\theta\ \mathbf{d}_1 + \cos\theta\ \mathbf{d}_3) + (d\phi/dt)\ \mathbf{d}_3$$
$$-(d\Psi/dt)\sin\theta\ \mathbf{d}_1 + [(d\Psi/dt)\cos\theta + (d\phi/dt)]\ \mathbf{d}_3$$

$$\omega_1 = -(d\Psi/dt)\sin\theta,\ \omega_2 = 0,\ \omega_3 = (d\Psi/dt)\cos\theta + (d\phi/dt) \qquad (9.23b)$$

Because θ remains constant during motion and the angles Ψ and ϕ vary with time at constant rates, angular acceleration is equal to zero:

$$d\omega_1/dt = 0,\ d\omega_2/dt = 0,\ d\omega_3/dt = 0 \qquad (9.23c)$$

As the top is symmetric in the reference frame D, the mass moment of inertia in D can be written as

$$I^{o}_{11} = I^{o}_{22} = I_o;\ I^{o}_{22} = I_3$$

$$I^{o}_{ij} = 0 \text{ when } i \neq j. \qquad (9.23d)$$

The only external moment about point O is the moment of the force of gravity. Let h be the distance between the center of mass and the point O. The moment $\mathbf{M}$ is then given by the equation:

$$\mathbf{M} = h\ \mathbf{d}_3 \times (-mg)\ \mathbf{e}_3 = h\ \mathbf{d}_3 \times (-mg)\ (-\sin\theta\ \mathbf{d}_1 + \cos\theta\ \mathbf{d}_3)$$

$$\mathbf{M} = mgh\sin\theta\ \mathbf{d}_2$$

$$M_1 = 0,\ M_2 = mgh\sin\theta,\ M_3 = 0 \qquad (9.23e)$$

Now we can use Euler's equations for rotation about the stationary point O. Substituting Eqns. 9.23a–e into Eqns. 9.22 and replacing I^{c}_{ij} by I^{o}_{ij}, we find:

$$M_2 = mgh \sin\theta = I^o_{21}\,(d\omega_1/dt) + I^o_{22}\,(d\omega_2/dt) + I^o_{23}\,(d\omega_3/dt)$$
$$+ \Omega_3\,(I^o_{12}\,\omega_2 + I^o_{11}\,\omega_1 + I^o_{13}\,\omega_3) - \Omega_1\,(I^o_{31}\omega_1 + I^o_{32}\,\omega_2 + I^o_{33}\,\omega_3)$$

$$mgh \sin\theta = -(d\Psi/dt)^2 \cos\theta \sin\theta\, I_o + (d\Psi/dt)^2 \sin\theta \cos\theta\, I_3$$
$$+ (d\Psi/dt) \sin\theta\, (d\phi/dt)\, I_3$$

$$mgh = (d\Psi/dt)^2 \cos\theta\,(I_3 - I_o) + (d\Psi/dt)\,(d\phi/dt)\,I_3 \qquad (9.24a)$$

The components of Euler's equations in $\mathbf{d}_1$ and $\mathbf{d}_3$ directions are identically satisfied. If we know the spin rate $(d\phi/dt)$ and the nutation angle θ, we can solve Eqn. 9.24a for the top's precession rate $(d\Psi/dt)$. Note that the top must spin fast for this motion to occur. This is because $(I_3 - I_o) <$ 0, and thus Eqn. 9.24a can be put in the form:

$$(d\phi/dt) = mgh/[(d\Psi/dt)I_3] + (d\Psi/dt)\cos\theta\,(I_o - I_3)/I_3 \qquad (9.24b)$$

A spinning top seems to defy gravity. The solution presented indicates that a top could remain inclined to the vertical axis with only the tip touching a horizontal plane so long as it spins fast enough about its own long axis. As the spin rate decreases, the top leans and the center of mass moves downward toward the horizontal plane. If the body under consideration is slender ($I_3 = 0$), then the motion described is not possible. The example of a spinning top has implications for human movement and motion. If an aging ballet dancer spins faster on stage in between leaps than he did years before, it is because it is easier to stay in certain postures at higher spin rates.

9.5 Dancing Holding on to a Pole

An example of planar motion of a body that has no symmetry with respect to a plane parallel to the plane of motion is considered. We find that angular momentum is not perpendicular to the plane of motion. Furthermore, contact forces that support the planar motion are found to be three-dimensional.

Example 9.6. Dancing Around a Pole. Consider a dancer rotating around a pole while holding onto it (Fig. 9.9). The dancer's feet are placed next to the pole and she skips in place as she rotates around the pole at a constant rate ω_o. The angle of inclination θ of the dancer with the vertical axis remains constant at all times. Our aim is to determine the reaction forces exerted on the dancer as a function of angular velocity, height, and weight. These forces consist of the ground forces exerted on the feet and the force exerted by the pole on the hands of the dancer.

Solution: We represent the dancer as a slender rod with length L and mass m (Fig. 9.9b). We seek to determine the angular velocity of the slender rod. We denote the unit vector along the line of the rod $\mathbf{b}_3$. Let $\mathbf{n}$ and $\mathbf{t}$ be auxiliary unit vectors in the horizontal plane ($\mathbf{e}_1$, $\mathbf{e}_2$) defined in Fig.

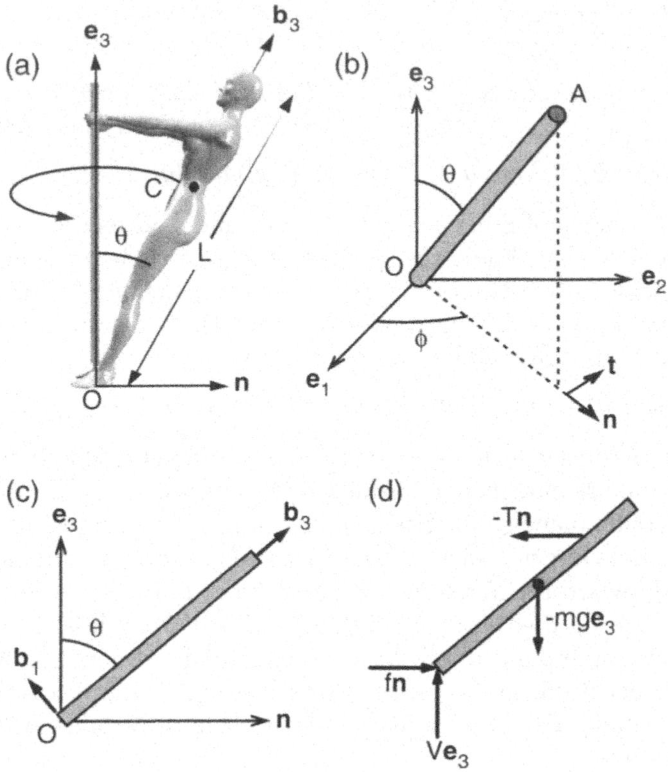

FIGURE 9.9a–d. A woman dancing around a pole **(a)**. The dancer is represented by a slender rod of mass m and length L in **(b–d)**. The reference frames E and B are defined in **(b)** and **(c)**, respectively. The free-body diagram of the dancer is presented in **(d)**.

9.9b. Note that $\mathbf{t}$ is in the direction of the velocity of the center of mass. Let $\mathbf{b}_1$ be the unit vector normal to $\mathbf{b}_3$ in the plane of $\mathbf{b}_3$ and $\mathbf{n}$. The unit vector $\mathbf{b}_2$ is then perpendicular to both $\mathbf{b}_1$ and $\mathbf{b}_3$:

$$\begin{aligned}
\mathbf{n} &= \cos\phi\, \mathbf{e}_1 + \sin\phi\, \mathbf{e}_2 \\
\mathbf{t} &= -\sin\phi\, \mathbf{e}_1 + \cos\phi\, \mathbf{e}_2 \\
\mathbf{b}_1 &= \cos\theta\,(-\cos\phi\, \mathbf{e}_1 - \sin\phi\, \mathbf{e}_2) + \sin\theta\, \mathbf{e}_3 \qquad (9.25a) \\
\mathbf{b}_2 &= -\mathbf{t} = \sin\phi\, \mathbf{e}_1 - \cos\phi\, \mathbf{e}_2 \\
\mathbf{b}_3 &= \sin\theta\,(\cos\phi\, \mathbf{e}_1 + \sin\phi\, \mathbf{e}_2) + \cos\theta\, \mathbf{e}_3
\end{aligned}$$

The unit vectors $\mathbf{b}_1$, $\mathbf{b}_2$, and $\mathbf{b}_3$ comprise a set of orthogonal unit vectors embedded in the slender bar B. We call this the reference system B. The angular velocity of the rod in the inertial frame E can then be obtained by using the three-dimensional definition of angular velocity given in Eqn. 9.5. Note that the angle θ remains constant during motion and therefore its time derivative is zero. Note also that the unit vectors $\mathbf{e}_1$, $\mathbf{e}_2$, and $\mathbf{e}_3$ are also constant in E. Thus we obtain:

$$d\mathbf{b}_1/dt = \omega_o \cos\theta\,(\sin\phi\,\mathbf{e}_1 - \cos\phi\,\mathbf{e}_2)$$

$$d\mathbf{b}_2/dt = \omega_o\,(\cos\phi\,\mathbf{e}_1 + \sin\phi\,\mathbf{e}_2)$$

$$d\mathbf{b}_3/dt = \omega_o \sin\theta\,(-\sin\phi\,\mathbf{e}_1 + \cos\phi\,\mathbf{e}_2)$$

in which $\omega_o = (d\phi/dt)$.

Substituting these expressions into Eqn. 9.5 yields the result:

$$\boldsymbol{\omega} = \omega_o\,\mathbf{e}_3 \tag{9.25b}$$

We could have arrived at this expression without the use of Eqn. 9.5 by recognizing that a line drawn in the vertical direction onto a dancer never changes its orientation during the dancer's motion. Thus the planar definition of angular velocity, Eqn. 9.3, holds and once again $\boldsymbol{\omega} = \omega_o\,\mathbf{e}_3$.

Computation of Reaction Forces

To determine the external forces acting on the dancer, we need to consider the motion of her center of mass. The position of the center of mass of the rod is given by the expression:

$$\mathbf{r}^c = (L/2)\sin\theta\,(\cos\phi\,\mathbf{e}_1 + \sin\phi\,\mathbf{e}_2) + (L/2)\cos\theta\,\mathbf{e}_3 \tag{9.26a}$$

Because the angle of inclination of the dancer with the vertical axis (θ) remains constant during the rotation, $(d\theta/dt) = 0$. Taking the time derivative of Eqn. 9.26a, the velocity of the center of mass can then be written in the form:

$$\mathbf{v}^c = (L/2)\,\omega_o \sin\theta\,(-\sin\phi\,\mathbf{e}_1 + \cos\phi\,\mathbf{e}_2) = (-L/2)\,\omega_o \sin\theta\,\mathbf{b}_2 \tag{9.26b}$$

where $\omega_o = (d\phi/dt)$.

As the rate of rotation ω_o was specified as constant, $d^2\phi/dt^2 = 0$. The acceleration of the center of mass is obtained by taking the time derivative of Eqn. 9.26b:

$$\mathbf{a}^c = (L/2)\,\omega_o^2 \sin\theta\,(-\cos\phi\,\mathbf{e}_1 - \sin\phi\,\mathbf{e}_2) \tag{9.26c}$$

Substituting Eqn. 9.26c into Newton's law of motion for the center of mass (Eqn. 9.19), and using the free-body diagram shown in Fig. 9.9d, we find:

$$-mg\,\mathbf{e}_3 + T\,(-\cos\phi\,\mathbf{e}_1 - \sin\phi\,\mathbf{e}_2) + V\,\mathbf{e}_3 + \mathrm{f}\,(\cos\phi\,\mathbf{e}_1 + \sin\phi\,\mathbf{e}_2)$$
$$= (1/2)\,mL\omega_o^2 \sin\theta\,(-\cos\phi\,\mathbf{e}_1 - \sin\phi\,\mathbf{e}_2) \tag{9.27}$$

in which T represents the magnitude of the tensile force exerted by the pole on the dancer's hands and V is the magnitude of the vertical force exerted by the ground. The frictional force exerted by the ground is represented by the last term in the left-hand side of Eqn. 9.27.

Writing this vectorial equation in scalar components in $-\mathbf{n}$ and $\mathbf{e}_3$ directions, we find:

$$T - f = (1/2)\, mL\omega_o^2 \sin\theta \tag{9.28a}$$

$$-mg + V = 0 \tag{9.28b}$$

These two equations are not sufficient to solve for the three unknowns: T, V, and f. Thus, we need to consider the principle of conservation of angular momentum to arrive at an additional equation. Because the dancer rotates around a fixed point O, the time rate of change of her angular momentum about O must be equal to the resultant moment with respect to O. Assuming that the arms of the dancer are positioned at a distance $(3L/4)$ from the feet, this resultant moment $\mathbf{M}_{ext}$ is given by the expression:

$$\begin{aligned}\mathbf{M}_{ext} &= (L/2)\, \mathbf{b}_3 \times (-mg\, \mathbf{e}_3) + (3L/4)\, \mathbf{b}_3 \times T\, (-\mathbf{n}) \\ &= [-mg\, (L/2) \sin\theta + (3L/4)\, T \cos\theta]\, \mathbf{b}_2 \end{aligned} \tag{9.29}$$

The resultant moment about O must be equal to the time rate of change of angular momentum $\mathbf{H}^o$ about the same point. Angular momentum $\mathbf{H}^o$ was defined as follows:

$$\mathbf{H}^o = \int \mathbf{r} \times \mathbf{v}\, dm \tag{9.30}$$

where $\mathbf{r}$ is the position vector from the base of the rod to any point on the rod, $\mathbf{v}$ is the velocity of the point with respect to the inertial reference frame E, and dm is a small mass element surrounding the point.

Let h denote distance from point O along the bar $(0 < h < L)$; then, a small mass element dm can be created such that

$$dm = (m/L)\, dh$$

where dh is a small length element. Expressing $\mathbf{r}$ and $\mathbf{v}$ in Eqn. 9.30 as a function of h, θ, and ϕ, one obtains:

$$\mathbf{H}^o = (m/L)\int h\mathbf{b}_3 \times (-h\, \omega_o \sin\theta\, \mathbf{b}_2)\, dh = (mL^2/3)\, \omega_o \sin\theta\, \mathbf{b}_1 \tag{9.31a}$$

$$= (mL^2/3)\omega_o \sin\theta\, [\cos\theta\, (-\cos\phi\, \mathbf{e}_1 - \sin\phi\, \mathbf{e}_2) + \sin\theta\, \mathbf{e}_3] \tag{9.31b}$$

We can check the validity of this expression by using the mathematical formulation presented in Section 9.3. Let us illustrate this by going over a few mathematical steps. The angular velocity of the slender rod given by Eqn. 9.25b can be expressed in terms of the unit vectors associated with the reference frame B as

$$\omega_o\, \mathbf{e}_3 = \omega_o \sin\theta\, \mathbf{b}_1 + \omega_o \cos\theta\, \mathbf{b}_3$$

Therefore, in reference frame B, $\omega_1 = \omega_o \sin\theta$ and $\omega_3 = \omega_o \cos\theta$.

The elements of mass moment of inertia of a slender bar with respect to an endpoint are given as

$$\mathrm{I}^o_{11} = \mathrm{I}^o_{22} = mL^2/3 \text{ and all other } \mathrm{I}^o_{ij} = 0$$

Equation 9.16 then dictates that

$$\mathbf{H}^{o} = (mL^2/3)\ \omega_o \sin\theta\ \mathbf{b}_1$$

which is what we had found using the direct approach. As can be seen from this equation, $\mathbf{H}^{o}$ is not parallel to the angular velocity $\omega_o \sin\theta\ \mathbf{b}_1 + \omega_o \cos\theta\ \mathbf{b}_3$. This is in contrast with the examples of planar motion presented earlier in this book. The discrepancy stems from the fact that the slender rod (dancer) has no horizontal plane of symmetry in this particular movement. If the rod had swept a plane rather than the surface of a cone during motion, the asymmetry would vanish and once again angular momentum and angular velocity would be colinear.

To compute the reaction forces let us take the time derivative of $\mathbf{H}^{o}$ in the inertial reference frame E. In E the unit vectors $\mathbf{e}_1$, $\mathbf{e}_2$, and $\mathbf{e}_3$ remain constant. Using the expressions given in Eqn. 9.25a for the unit vectors $\mathbf{b}_i$, one can show that

$$d\mathbf{H}^{o}/dt = (mL^2/3)\ \omega_o^2 \sin\theta \cos\theta\ \mathbf{b}_2 \tag{9.32}$$

According to the conservation of angular momentum this time derivative must be equal to the resultant external moment given by Eqn (9.29):

$$-mg\ (L/2)\sin\theta + (3L/4)\ T\cos\theta = (mL^2/3)\ \omega_o^2 \sin\theta\cos\theta$$

Thus we arrive at an expression for the tensile force T:

$$T = (2/3)\ mg\tan\theta + (4/9)\ (mL)\ \omega_o^2 \sin\theta \tag{9.33a}$$

To find the ground reaction forces, we combine Eqn. 9.33a with the equation of motion of the center of mass (Eqn. 9.28). The results are

$$V = mg \tag{9.33b}$$

$$f = (2/3)\ mg\tan\theta - (1/18)\ (mL)\ \omega_o^2 \sin\theta \tag{9.33c}$$

These equations show that the vertical ground force acting on the dancer is equal to the weight of the dancer. The force with which the dancer pulls on the pole increases and the ground friction force decreases with increasing rate of rotation.

9.6 Rolling of an Abdominal Wheel on a Horizontal Plane

In this section, we consider the three-dimensional motion of an abdominal wheel on a flat surface. The solution is obtained by using the definition of angular momentum directly. Later we explain how the problem could be solved by expressing angular momentum in terms of the mass moment of inertia and angular velocity.

Example 9.7. Abdominal Wheel. A simple mechanical device called an abdominal wheel was promoted in the 1980s as the miracle tool for strengthening abdominal muscles. It is composed of a circular disk with two short cylindrical bars on both sides (holders) (Fig. 9.10). The radius of the disk is R and the length of each of the holders is L. When placed on a flat floor, the disk of the abdominal wheel rolls without slip in such a way that the center C of the disk travels on a horizontal circle at constant velocity v_O. The center of this circle is the distal end of one of the holders. The disk has a mass of m and the weight of the holders is negligible. Determine the angular velocity of the rotating disk in the inertial reference frame E. Determine also the ground forces acting on the abdominal wheel.

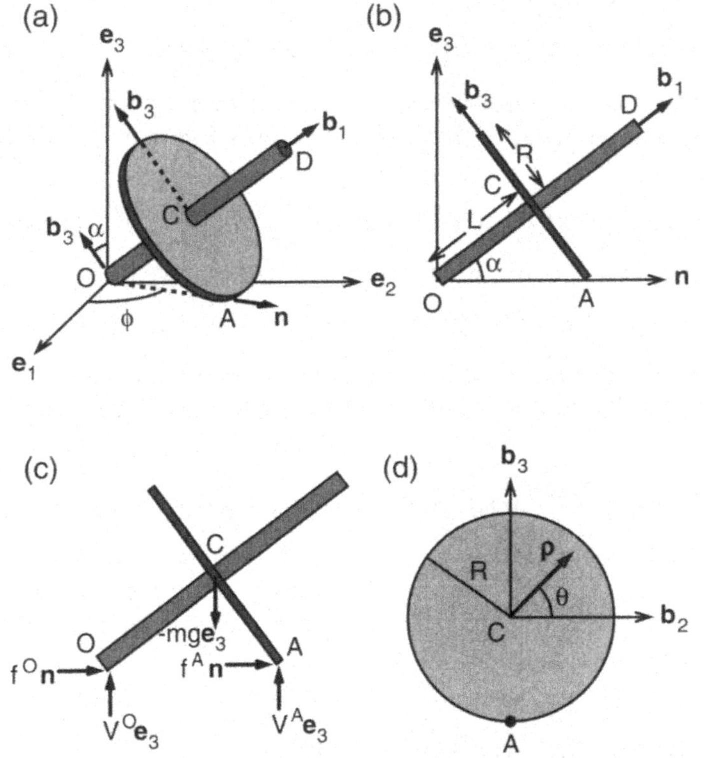

FIGURE 9.10a–d. The motion of an abdominal wheel on a horizontal plane **(a)**. The abdominal wheel is composed of a disk of mass m and two rods attached to it. These rods are used to hold the wheel and push it out on the floor and away from the body. The weights of these rods are small compared to the weight of the disk. When set on the floor with a velocity, the abdominal wheel tends to rotate around one end of one of the holders. The reference frames used in the analysis are identified in **(b)** and **(c)**. The free-body diagram of the abdominal wheel is given in **(d)**.

Determination of the Angular Velocity

When the abdominal wheel rotates on the horizontal plane, two points of the wheel have zero velocity: the point O representing one end of one of the holders and the point A on the outer edge of the disk that is in immediate contact with the floor. As both O and A belong to the same rigid object:

$$\mathbf{v}^{A} = \mathbf{v}^{O} + \omega \times \mathbf{r}$$

In the coordinate system E defined in Fig. 9.2, the displacement vector $\mathbf{r}$ can be written as

$$\mathbf{r} = (L^2 + R^2)^{(1/2)} (\cos\phi\ \mathbf{e}_1 + \sin\phi\ \mathbf{e}_2) = (L^2 + R^2)^{(1/2)}\ \mathbf{n}$$

in which the unit vector $\mathbf{n}$ is in the horizantal plane and in radial direction.

Thus, we have

$$(\omega_1\ \mathbf{e}_1 + \omega_2\ \mathbf{e}_2 + \omega_3\ \mathbf{e}_3) \times (\cos\phi\ \mathbf{e}_1 + \sin\phi\ \mathbf{e}_2) = 0$$

That means that either the angular velocity is zero or it is in the direction of $\mathbf{n}$:

$$\boldsymbol{\omega} = \omega_o\ \mathbf{n}$$

where the scalar quantity ω_o is yet to be determined.

Let us next relate the velocity of center of mass (point C) to the velocity of point A and to the angular velocity:

$$\begin{aligned}\mathbf{v}^{c} &= v_o\,(-\sin\phi\ \mathbf{e}_1 + \cos\phi\ \mathbf{e}_2)\\ &= \omega_o\,(\cos\phi\ \mathbf{e}_1 + \sin\phi\ \mathbf{e}_2)\\ &\quad \times R\,[-\sin\alpha\,(\cos\phi\ \mathbf{e}_1 + \sin\phi\ \mathbf{e}_2) + \cos\alpha\ \mathbf{e}_3]\end{aligned}$$

in which $\tan\alpha = (R/L)$.

Solving this equation for ω_o we find:

$$\boldsymbol{\omega} = -[(v_o/(R\cos\alpha)]\,(\cos\phi\ \mathbf{e}_1 + \sin\phi\ \mathbf{e}_2)$$

$$\omega_o = -[(v_o/(R\cos\alpha)]$$

We could also express $\boldsymbol{\omega}$ in the coordinate system B that is associated with the abdominal wheel. Let $\mathbf{b}_1$ be the unit vector along the vector connecting point O to point C and let $\mathbf{b}_3$ be perpendicular to it in the plane created by $\mathbf{e}_3$ and $\mathbf{n}$. The unit vector $\mathbf{b}_2$ is then given by the equality $\mathbf{b}_2 = \mathbf{b}_3 \times \mathbf{b}_1$. It can be shown that

$$\begin{aligned}\mathbf{b}_1 &= \cos\alpha\,(\cos\phi\ \mathbf{e}_1 + \sin\phi\ \mathbf{e}_2) + \sin\alpha\ \mathbf{e}_3\\ \mathbf{b}_2 &= -\sin\phi\ \mathbf{e}_1 + \cos\phi\ \mathbf{e}_2 \qquad (9.34)\\ \mathbf{b}_3 &= -\sin\alpha\,(\cos\phi\ \mathbf{e}_1 + \sin\phi\ \mathbf{e}_2) + \cos\alpha\ \mathbf{e}_3\end{aligned}$$

Note that the reference frame B moves with the axis of the disk but does not rotate with the disk. In terms of these unit vectors, angular velocity $\boldsymbol{\omega}$ is expressed as

$$\boldsymbol{\omega} = (v_o/R)\,(-\mathbf{b}_1 + \tan\alpha\,\mathbf{b}_3) \tag{9.35}$$

Determination of Reaction Forces on the Abdominal Wheel

The center of mass draws a circle of radius ($L\cos\alpha$) with speed v_o. Therefore the acceleration of the center of mass is pointed toward the center of the circle and has a magnitude of $v_o^2/(L\cos\alpha)$

$$\mathbf{a}^c = -[v_o^2/(L\cos\alpha)]\,(\cos\phi\,\mathbf{e}_1 + \sin\phi\,\mathbf{e}_2)$$

Newton's second law written in vertical ($\mathbf{e}_3$) and horizontal directions ($\mathbf{n}$) for the free-body diagram shown in Fig. 9.10 is as follows:

$$V^O + V^A - mg = 0 \tag{9.36a}$$

$$f^O + f^A = -m\,(v_o^2\sin\alpha/L)]\,(\cos\phi\,\mathbf{e}_1 + \sin\phi\,\mathbf{e}_2) \tag{9.36b}$$

Because the center of mass traverses a circle with constant velocity, both the acceleration and the contact force tangential to the circle must be equal to zero. Thus, the resultant moment with respect to point O is given by

$$\mathbf{M}^o = (m\,g\,L\cos\alpha - V^A\,L/\cos\alpha)\,\mathbf{b}_2 \tag{9.37}$$

The resultant moment with respect to O must be equal to the time rate of change of angular momentum $\mathbf{H}^o$. The angular momentum is given by the expression

$$\mathbf{H}^o = \int(\mathbf{r}\times\mathbf{v})\,dm \tag{9.38}$$

where $\mathbf{r}$ is the position vector from point O to any point P on the disk, $\mathbf{v}$ is the velocity of point P, and dm is a small mass element surrounding point P. Note that the holders of the abdominal wheel were assumed to be weightless. Therefore they do not contribute to the angular momentum of the wheel. The following relationship is a direct consequence of the parallelogram addition of vectors:

$$\begin{aligned}\mathbf{r} &= L\,\mathbf{b}_1 + \boldsymbol{\rho}\\ &= L\,\mathbf{b}_1 + \rho\,(\cos\theta\,\mathbf{b}_2 + \sin\theta\,\mathbf{b}_3)\end{aligned} \tag{9.39a}$$

in which ρ and θ are the polar coordinates of the point P as shown in Fig. 9.10d.

Using the relationship between velocities of two points in a rigid body, the velocity of point P can be written as

$$\begin{aligned}\mathbf{v} &= v_o\,\mathbf{b}_2 + \omega\times\boldsymbol{\rho}\\ &= v_o\,\mathbf{b}_2 + (v_o/R)\,(-\mathbf{b}_1 + \tan\alpha\,\mathbf{b}_3)\times\rho\,(\cos\theta\,\mathbf{b}_2 + \sin\theta\,\mathbf{b}_3)\end{aligned} \tag{9.39b}$$

The first term on the right-hand side is the velocity of point C. The second term on the right-hand side of the equation is simply the vector product of angular velocity and the position vector connecting point P to point C.

Substituting Eqns. 9.39a and 9.39b into the angular momentum equation (Eqn. 9.38), we express angular momentum as the sum of four terms:

$$\mathbf{H}^o = (\mathbf{H}^o)^1 + (\mathbf{H}^o)^2 + (\mathbf{H}^o)^3 + (\mathbf{H}^o)^4$$

$$(\mathbf{H}^o)^1 = \int (L\mathbf{b}_1 \times v_o\mathbf{b}_2)\, dm = m\, L\, v_o\mathbf{b}_3$$

$$(\mathbf{H}^o)^2 = \int [(L\mathbf{b}_1 \times (\boldsymbol{\omega} \times \boldsymbol{\rho})]\, dm = L\, \mathbf{b}_1 \times (\boldsymbol{\omega} \times \int \boldsymbol{\rho} dm) = 0$$

These results follow because both the terms $L\ \mathbf{b}_1$ and $\boldsymbol{\omega}$ are constant with respect to the variable of integration and can be taken outside the integral sign. Because C is the center of mass, the integral $\int \boldsymbol{\rho} dm = 0$.

$$(\mathbf{H}^o)^3 = \int (\boldsymbol{\rho} \times v_o\mathbf{b}_2)\, dm = (\int \boldsymbol{\rho}\, dm) \times v_o\mathbf{b}_2 = 0$$

$$(\mathbf{H}^o)^4 = \int [(\boldsymbol{\rho} \times (\boldsymbol{\omega} \times \boldsymbol{\rho})] dm = (mR^2/2)\,(v_o/R)\,[-\mathbf{b}_1 + (1/2) \tan \alpha\ \mathbf{b}_3]$$

Summing all four terms we obtain the angular momentum of the rotating disk:

$$\mathbf{H}^o = m\, L\, v_o\, \mathbf{b}_3 + (mR^2/2)\,(v_o/R)\,[-\mathbf{b}_1 + (1/2) \tan \alpha\ \mathbf{b}_3] \qquad (9.40)$$

Note that this equation can be written in the form:

$$\mathbf{H}^o = \mathrm{I}^o{}_{11}\omega_1\, \mathbf{b}_1 + \mathrm{I}^o{}_{33}\, \omega_3\, \mathbf{b}_3$$

where

$$\mathrm{I}^o{}_{11} = (mR^2/2)$$

$$\mathrm{I}^o{}_{33} = (mR^2/4) + mL^2$$

$$\omega_1 = -(v_o/R)$$

$$\omega_3 = (v_o/R) \tan \alpha$$

We could have arrived at this result by using the formulation presented in Section 9.3 (Eqns. 9.16 and 9.17) and the tables of moment of inertia presented in the back of this book.

To use the principle of conservation of angular momentum, let us take the time derivative of $\mathbf{H}^o$ in the inertial reference frame E:

$$(d\mathbf{H}^o/dt) = [d(\mathrm{I}^o{}_{11}\omega_1)/dt]\ \mathbf{b}_1 + [d(\mathrm{I}^o{}_{33}\ \omega_3)/dt]\ \mathbf{b}_3 + \mathrm{I}^o{}_{11}\omega_1\ (d\mathbf{b}_1/dt) + \mathrm{I}^o{}_{33}\ \omega_3\ (d\mathbf{b}_3/dt)$$

The first two terms are zero because the expressions in the brackets are independent of time t. The time derivatives of the unit vectors can be obtained by taking the time derivative of Eqn. 9.34, yielding the result:

$$(d\mathbf{H}^o/dt) = -(v_o/R)^2 \tan \alpha\ \{mR^2/2 + (mR^2/4 + mL^2) \tan^2 \alpha\}\mathbf{b}_2 \qquad (9.41)$$

Note that we could have obtained the same result by using Eqn. 9.22 for the time derivative of $\mathbf{H}^o$:

$$^E d\mathbf{H}^o/dt = {^B d\mathbf{H}^o/dt} + {^E\boldsymbol{\omega}^B} \times \mathbf{H}^o \tag{9.42}$$

Using Eqns. 9.5 and 9.34, one can show that

$$^E\boldsymbol{\omega}^B = (v_o/R)\ [\tan^2 \alpha\ \mathbf{b}_1 + \tan \alpha\ \mathbf{b}_3]$$

Using this expression and Eqn. 9.40 in Eqn. 9.42, one again arrives at Eqn. 9.41.

Setting the resultant external moment equal to the time rate of change of angular momentum, one finds

$$V^A = mg \cos^2 \alpha + m\ (v_o{}^2/R)\ [1.5 \sin^2 \alpha/\cos \alpha + 0.25 \sin^4 \alpha/\cos^3 \alpha]$$

Note that V^A, the vertical force exerted by the ground, cannot be larger in magnitude than the weight mg of the abdominal wheel; otherwise, to assure the force balance in the vertical direction, f^A would have to be less than zero. That means, however, that the ground would have to pull point O toward the earth. This is impossible because the ground can only push back but not pull in. The maximum value of v_o, the velocity with which the disk rotates, can be found by equating the vertical ground force exerted at point A to the weight of the disk:

$$g \sin^2 \alpha = (v_o{}^2/R)\ [1.5 \sin^2 \alpha/\cos \alpha + 0.25 \sin^4 \alpha/\cos^3 \alpha]$$

For $\alpha = 45°$ and $R = 0.2$ cm, then $v_o = 0.89$ m/s.

This means that if the center of mass rotates with a velocity greater than or equal to 0.89 m/s, the motion is unstable, that is, it could rapidly transform into another mode of rotation, such as that of a disk rolling with none of the holders touching the ground.

9.7 Biomechanics of Twisting Somersaults

In this section we present an example from the airborne movement of a human. The only significant external force in this case is the force of gravity. The gravitational force passes through the center of mass and therefore the angular momentum about the center of mass must be constant. Equation 9.14 relating the angular momentum to the mass moment of inertia and the angular velocity is used to determine the changes in the rate of rotation of the body as a result of a sudden change of body shape.

Example 9.8. Three-Dimensional Diving. A diver dives off a platform with an angular velocity $^R\boldsymbol{\omega}^B = \omega_o\ \mathbf{e}_2$ (Fig. 9.11). His plane of motion immediately after the jump is the plane containing the unit vectors $\mathbf{e}_1$ and

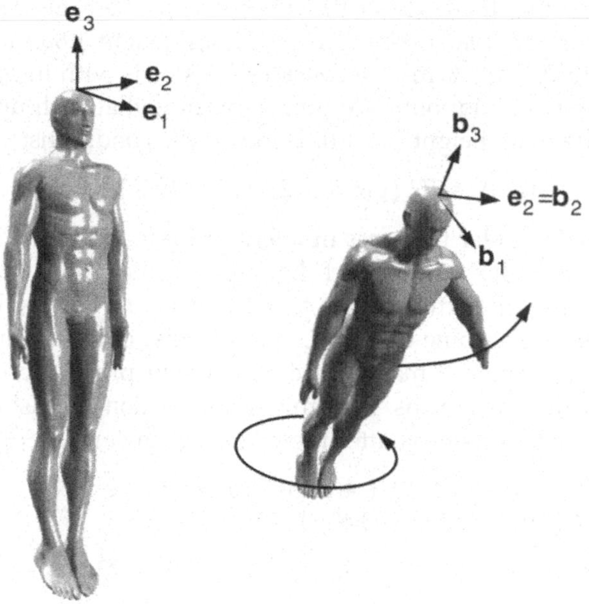

FIGURE 9.11. As soon as a diver dives off a platform, he moves one of his arms to the side. The reference frames E and B are attached to the platform and to the diver, respectively.

$\mathbf{e}_3$. As soon as the diver pushes off the platform, he moves his left arm to the side. Determine his angular velocity immediately after raising his arm to the horizontal configuration. The diver weighs 80 kg, and each of his upper limbs weigh 4 kg. His height (H), his width (W), and his depth (D) are 1.80 m, 0.35 m, and 0.20 m, respectively.

Solution: As the diver is on air in free fall, the only force acting on him is gravity and this force always passes through the center of mass. Thus there is no external moment acting on the diver and his angular momentum before he raises his arm must be equal to his angular momentum after he raises his arm. Because his motion is planar before he raises his arm, from Eqn. 9.14 we obtain

$$(H^c{}_2)_b = (\mathrm{I}^c{}_{22})_b\, \omega_o \tag{9.43a}$$

$$(H^c{}_1)_b = (H^c{}_3)_b = 0 \tag{9.43b}$$

in which the subscript b refers to time immediately before the raising of the arm. Because the diver is considered to be a symmetric body before he raises his left arm, the following holds:

$$(\mathrm{I}^c{}_{21})_b = (\mathrm{I}^c{}_{23})_b = 0$$

We assume that the body of the diver except the arms can be represented by a rectangular prism of mass 72 kg, height 1.80 m, width 0.35 m, and depth 0.20 m. Arms are considered as rods with lump masses of 4 kg positioned at midpoint. We further assume that each upper limb is equal half the body height (0.9 m). Under these conditions:

$$(I^c_{22})_b = 72\,(1.8^2 + 0.2^2)/12 = 19.7 \text{ kg-m}^2 \qquad (9.44)$$

Let us next consider the mass moment of inertia after the diver raises one arm. Because his arm is much lighter than the rest of his body, this change in shape will perturbate the center of mass only by a small amount. For simplicity we assume that the center of mass of the diver remains at all times at the center of mass of the rectangular prism representing the diver's body minus his arms. With this simplification and using Eqn. 9.17, the mass moment of inertia after the raising of the arm, $(I^c_{ij})_a$, becomes

$$(I^c_{11})_a = 72(1.8^2 + 0.35^2)/12 + 4(0.45^2 + 0.175^2) + 4\,[0.9^2 + (0.175 + 0.45)^2] = 25.3 \text{ kg-m}^2$$

$$(I^c_{22})_a = 72(1.8^2 + 0.20^2)/12 + 4(0.45)^2 + 4(0.9)^2 = 23.1 \text{ kg-m}^2$$

$$(I^c_{33})_a = 72(0.35^2 + 0.20^2)/12 + 4(0.175)^2 + 4(0.175 + 0.45)^2 = 2.7 \text{ kg-m}^2$$

$$(I^c_{23})_a = -4(-0.175)(0.45) - 4(0.175 + 0.45)0.9 = -1.9 \text{ kg-m}^2$$

$$(I^c_{32})_a = (I^c_{23})_a$$

$$(I^c_{12})_a = (I^c_{21})_a = (I^c_{13})_a = (I^c_{31})_a = 0$$

Therefore the angular momentum immediately after the raising of the arm is equal to

$$(H^c_1)_a = 25.3\,(\omega_1)_a$$

$$(H^c_2)_a = 23.1\,(\omega_2)_a - 1.9\,(\omega_3)_a$$

$$(H^c_3)_a = 2.7\,(\omega_3)_a - 1.9\,(\omega_2)_a$$

Equating the angular momentums before and after the shape change, we find:

$$25.3(\omega_1)_a = 0$$

$$23.1\,(\omega_2)_a - 1.9\,(\omega_3)_a = 19.7\,\omega_o$$

$$-1.9\,(\omega_2)_a + 2.7(\omega_3)_a = 0$$

Thus,

$$(\omega_1)_a = 0,\ (\omega_2)_a = 0.9\,\omega_o,\ (\omega_3)_a = 0.64\,\omega_o$$

This exercise shows that just by moving one arm, a diver performing plain somersaults could begin performing twisting rotations too. Note that the angular velocity component $(\omega_3)_a$ would have changed sign if the diver were to raise his right arm instead of his left arm.

9.8 Throwing and Hitting Motions

The Method of Inverse Dynamics

The method of inverse dynamics is increasingly being used to analyze the sequential ordering of body segment movements during an athletic event. The method is also useful to compute the joint moment that is resisted by muscle action in various modes of movement. Inverse dynamics is based on the experimental determination of velocity and acceleration terms that appear in the laws of motion. These laws are then used to evaluate the unknown forces and moments acting on parts of the body. As we illustrate in the following examples, the position and velocity of a point as well as the angular velocity of a body segment can be determined with reasonable accuracy by recording human movement with the use of digital cameras. Computational errors may be markedly larger in the evaluation of acceleration from data. Researchers use various numerical algorithms to enhance the accuracy of the inverse dynamics method. The sports mechanics literature is full of interesting articles that estimate muscle forces and muscle torques involved in baseball pitching and golf swing. Reader is referred to the references at the end of the book for an introduction to this rich area of biomechanics.

Example 9.9. Determination of Cartesian Coordinates of a Spatial Point Using Two Cameras. Two synchronized 500 frames/s cameras were used to capture the position of various markers on the throwing arms and shoulder of a baseball pitcher. For marker P placed near the elbow of the pitcher, the camera located at point O (0,0,0) and facing the positive $\mathbf{e}_2$ direction produced digital images which at time $t = 5$ s gave the following coordinates along the coordinate axes: (1.34 m, u, 2.15 m). Note that this camera could not yield information on the distance between the camera and the point in the $\mathbf{e}_2$ direction and therefore u is not known. The other camera, positioned at point A (−3.00 m, 2.00 m, 1.00 m) and facing the positive $\mathbf{e}_1$ direction, produced the following coordinates for point P at time $t = 5$ s: P = (v, 0.89, 1.11) where v is unknown. Determine the position of P with respect to the reference frame E positioned at point O. Evaluate the accuracy of the results.

Solution: According to the parallelogram law of vectors:

$$\mathbf{r}^{P/O} = \mathbf{r}^{A/O} + \mathbf{r}^{P/A}$$

This equation can also be written in terms of its components in the E reference frame

$$x^{P/O} = x^{A/O} + x^{P/A} \Rightarrow 1.34 = -3 + x^{P/A} \Rightarrow x^{P/A} = 1.66 \text{ m}$$

$$y^{P/O} = y^{A/O} + y^{P/A} \Rightarrow y^{P/O} = 2 + 0.89 = 2.89 \text{ m}$$

$$z^{P/O} = z^{A/O} + z^{P/A} = 2.15 \neq 1 + 1.11 = 2.11$$

Thus, with respect to the frame E at point O:

$$\mathbf{r}^P = 1.34\ \mathbf{e}_1 + 2.89\ \mathbf{e}_2 + 2.15\ \mathbf{e}_3$$

The percentage of error $\epsilon = [(2.15 - 2.11)/2.15] \times 100 = 1.9\%$.

Example 9.10. Orientation of the Throwing Arm. The three-dimensional coordinate values for two landmarks (reflective hemispheres of 20 mm in diameter), one positioned next to the shoulder joint and the other on the elbow, were determined using three cameras. The marker next to the shoulder joint was identified by the symbol SH and the marker next to the elbow by EL. Positions of SH and EL at time $t = 5$ s were as follows:

$$\mathbf{r}^{SH} = 2.31\ \mathbf{e}_1 + 1.76\ \mathbf{e}_2 + 0.28\ \mathbf{e}_3$$

$$\mathbf{r}^{EL} = 2.48\ \mathbf{e}_1 + 2.01\ \mathbf{e}_2 + 0.47\ \mathbf{e}_3$$

These positions were determined again 0.1 s later:

$$\mathbf{r}^{SH} = 2.38\ \mathbf{e}_1 + 1.85\ \mathbf{e}_2 + 0.20\ \mathbf{e}_3$$

$$\mathbf{r}^{EL} = 2.65\ \mathbf{e}_1 + 2.01\ \mathbf{e}_2 + 0.27\ \mathbf{e}_3$$

Determine the unit vector $\mathbf{b}$ along the line segment from SH to EL. What is the time rate of change of $\mathbf{b}$ at time $t = 5$ s?

Solution: We compute $\mathbf{b}$ at $t = 5$ s and at $t = 5.1$ s, and identify these vectors as $\mathbf{b}^-$ and $\mathbf{b}^+$, respectively. The time rate of change of $\mathbf{b}$ will be approximated by using the following finite-difference formula:

$$d\mathbf{b}/dt = (\mathbf{b}^+ - \mathbf{b}^-)/0.1$$

$$\begin{aligned}\mathbf{b}^- &= [0.07\ \mathbf{e}_1 + 0.25\ \mathbf{e}_2 + 0.19\ \mathbf{e}_3]/(0.0049 + 0.0625 + 0.0361)^{0.5} \\ &= 0.22\ \mathbf{e}_1 + 0.78\ \mathbf{e}_2 + 0.59\ \mathbf{e}_3\end{aligned}$$

$$\begin{aligned}\mathbf{b}^+ &= [0.27\ \mathbf{e}_1 + 0.16\ \mathbf{e}_2 + 0.07\ \mathbf{e}_3]/(0.0729 + 0.0256 + 0.0049)^{0.5} \\ &= 0.84\ \mathbf{e}_1 + 0.50\ \mathbf{e}_2 + 0.22\ \mathbf{e}_3\end{aligned}$$

$$d\mathbf{b}/\mathrm{dt} = (0.62\ \mathbf{e}_1 - 0.28\ \mathbf{e}_2 - 0.37\ \mathbf{e}_3)/0.1 = 6.2\ \mathbf{e}_1 - 2.8\ \mathbf{e}_2 - 3.7\mathbf{e}_3$$

Example 9.11. Angular Velocity of the Throwing Arm. In an investigation of the mechanics of arm swing, reflective markers were used to construct an orthogonal unit vector axis system with its origin at the glenohumeral joint. The first unit vector $\mathbf{b}_1$ was chosen along the longitudinal axis of the upper arm. The second unit vector $\mathbf{b}_2$ was constructed perpendicular to $\mathbf{b}_1$ and was taken along the direction of the rotation axis for upper arm adduction/abduction. A third unit vector, $\mathbf{b}_3$, was constructed perpendicular to $\mathbf{b}_1$ and $\mathbf{b}_2$. This vector is in the direction of the rotation axis for upper arm flexion and rotation. Using the method of the previous example, the following vector quantities were determined at time $t = 5$ s:

$$\mathbf{b}_1 = 0.22\ \mathbf{e}_1 + 0.78\ \mathbf{e}_2 + 0.59\ \mathbf{e}_3$$

$$d\mathbf{b}_1/dt = 6.2\ \mathbf{e}_1 - 2.8\ \mathbf{e}_2 + 3.7\ \mathbf{e}_3$$

$$\mathbf{b}_2 = -0.96\ \mathbf{e}_1 + 0.27\ \mathbf{e}_2$$

$$d\mathbf{b}_2/dt = 15.2\ \mathbf{e}_1 - 7.6\ \mathbf{e}_2 + 4.9\ \mathbf{e}_3$$

where $\mathbf{e}_1$ are fixed on earth.

Determine the third unit vector $\mathbf{b}_3$ and its time derivative $d\mathbf{b}_3/dt$. Determine the angular velocity of B in reference frame E and express it in terms of unit vectors in B.

Solution:

$$\mathbf{b}_3 = \mathbf{b}_1 \times \mathbf{b}_2 = -0.16\ \mathbf{e}_1 - 0.58\ \mathbf{e}_2 + 0.80\ \mathbf{e}_3$$

$$d\mathbf{b}_3/dt = d\mathbf{b}_1/dt \times \mathbf{b}_2 + \mathbf{b}_1 \times d\mathbf{b}_2/dt = (-\mathbf{e}_1 - 3.6\ \mathbf{e}_2 - \mathbf{e}_3) + (\ 8.3\mathbf{e}_1 + 8.0\ \mathbf{e}_2 - 13.5\mathbf{e}_3) = 7.3\mathbf{e}_1 + 4.4\mathbf{e}_2 - 14.5\mathbf{e}_3$$

From the definition of angular velocity given in Eqn. 9.5:

$$^{E}\boldsymbol{\omega}^{B} = [(^{E}d\mathbf{b}_2/dt) \cdot \mathbf{b}_3]\mathbf{b}_1 + [(^{E}d\mathbf{b}_3/dt) \cdot \mathbf{b}_1]\mathbf{b}_2 + [(^{E}d\mathbf{b}_1/dt) \cdot \mathbf{b}_2]\mathbf{b}_3 = \omega_1\mathbf{b}_1 + \omega_2\mathbf{b}_2 + \omega_3\mathbf{b}_3$$

$$\omega_1 = (^{E}d\mathbf{b}_2/dt) \cdot \mathbf{b}_3 = 15.2 \times (-0.16) + (-7.6) \times (-0.68) + 4.9 \times 0.80 = 6.7 \text{ rad/s}$$

$$\omega_2 = (^{E}\mathrm{d}\mathbf{b}_3/\mathrm{dt}) \cdot \mathbf{b}_1 = 7.3 \times 0.22 + 4.4 \times 0.78 - 14.5 \times 0.59 = -3.5 \text{ rad/s}$$

$$\omega_3 = (^{E}d\mathbf{b}_1/dt) \cdot \mathbf{b}_2 = 6.2 \times (-0.96) - 2.8 \times 0.27 + 3.7 \times 0 = -6.7 \text{ rad/s}$$

$$^{E}\boldsymbol{\omega}^{B} = 6.7\ \mathbf{b}_1 - 3.5\mathbf{b}_2 - 6.7\ \mathbf{b}_3$$

The component of the angular velocity along $\mathbf{b}_1$ direction may be associated with the twisting moment along the longitudinal axis of the upper arm. The component along the $\mathbf{b}_2$ direction may be altered by muscles that cause abduction/adduction in the frontal plane. The angular velocity components in the $\mathbf{b}_3$ direction, on the other hand, are dependent on the muscles that flex/extend the upper arm. The projections of angular velocity along the unit vectors $\mathbf{b}_i$ are called anatomical angular velocity components.

Contribution of Body Segments to the Velocity of the Endpoint

An important goal of sports mechanics is to determine the relative contributions of body segments to the velocity of the endpoint (the midpoint of a tennis racket or the head of a golf club). The endpoint speed displayed just before the impact results from a series of upper limb segment rotations generated by muscle torques.

Example 9.12. Tennis Player and the Velocity of the Racket. A tennis player wants to improve his serves. The task is to develop an equation that expresses racket-head speed in terms of the anatomical angular velocities of the trunk (t), upper arm (u), forearm (f), and hand (h). It is assumed that the racket is being held with a firm grip so that no rotation occurs between the racket handle and the hand.

Solution: The velocity of the elbow can be expressed in terms of the velocity of the shoulder and the angular velocity of the upper arm by using Eqn. 9.10a:

$$^{E}\mathbf{v}^{Elb} = {}^{E}\mathbf{v}^{Shl} + {}^{E}\boldsymbol{\omega}^{Bu} \times \mathbf{r}^{Elb/Shl} \tag{9.45a}$$

where Bu denotes the reference frame embedded into the upper arm (all other symbols are self-explanatory). Similarly, we could relate the velocity of the wrist to the velocity of the elbow using the same equation:

$$^{E}\mathbf{v}^{Wrs} = {}^{E}\mathbf{v}^{Elb} + {}^{E}\boldsymbol{\omega}^{Bf} \times \mathbf{r}^{Wrs/Elb} \tag{9.45b}$$

Here Bf denotes the reference frame fixed in the forearm. Substituting Eqn. 9.45a into Eqn. 9.45b, we can express the velocity of the wrist in terms of the velocity of the shoulder and the angular velocities of the upper arm and the forearm. Successive use of this methodology allows us to express the velocity of the midpoint of the racket as follows:

$$^{E}\mathbf{v}^{R} = {}^{E}\mathbf{v}^{Shl} + {}^{E}\boldsymbol{\omega}^{Bu} \times \mathbf{r}^{Elb/Shl} + {}^{E}\boldsymbol{\omega}^{Bf} \times \mathbf{r}^{Wrs/Elb} + {}^{E}\boldsymbol{\omega}^{H} \times \mathbf{r}^{R/Wrs} \tag{9.46}$$

where H denotes the reference frame that moves with the hand. We have already seen that the angular velocity of two reference frames, say Bu and Bf, are related by the following equation:

$$^{E}\boldsymbol{\omega}^{Bf} = {}^{E}\boldsymbol{\omega}^{Bu} + {}^{Bu}\boldsymbol{\omega}^{Bf} \tag{9.47a}$$

The same equation can be written for the angular velocity of the hand–racket complex:

$$^{E}\boldsymbol{\omega}^{H} = {}^{E}\boldsymbol{\omega}^{Bu} + {}^{Bu}\boldsymbol{\omega}^{Bf} + {}^{Bf}\boldsymbol{\omega}^{H} \tag{9.47b}$$

Next, we use the parallelogram law to express position of the endpoint in the following form:

$$\mathbf{r}^{R/Shl} = (\mathbf{r}^{Elb/Shl} + \mathbf{r}^{Wrs/Elb} + \mathbf{r}^{R/Wrs}) \tag{9.48}$$

Combining Eqns. 9.46 to 9.48, we arrive at the fundamental equation describing the relationships between the individual segment rotations and the linear velocity of the racket head:

$$^{E}\mathbf{v}^{R} = {}^{E}\mathbf{v}^{Shl} + {}^{E}\boldsymbol{\omega}^{Bu} \times \mathbf{r}^{R/Shl} + {}^{Bu}\boldsymbol{\omega}^{Bf} \times \mathbf{r}^{R/Elb} + {}^{Bf}\boldsymbol{\omega}^{H} \times \mathbf{r}^{R/Wrs} \tag{9.49}$$

Contribution of each of the separate anatomical angular velocity components to the racket–hand speed can be determined using this expression. Videotaping of forward swing events for a competitive male tennis player show that the highest overall rotational velocity is typically obtained by the hand segment (40 rad/s). The greatest contributor to the racket head's forward speed at ball contact is upper arm internal rotation (8 m/s), followed by wrist flexion (7 m/s) and upper arm horizontal adduction (6.5 m/s). The measured speed at the center of the racket is about 27 m/s.

The inverse dynamics method outlined here can also be applied to other sports such as baseball and golf. In these cases, one should also incorporate the rotational velocity of the trunk into the expression for the end-point velocity. The angular velocity of the trunk provides the shoulder with a velocity and it contributes additionally to the velocity of the end-point through its long moment arm. The young golf pro Tiger Woods has one of the highest rotational speeds in the game. His hips are quick through the ball. Pro golf players rotate their shoulders and hips extensively during the backswing phase of hitting a golf ball. The larger the gap between the turns of the shoulder and the hip, the greater the levels of torsional deformation imposed on the upper body. The more extensively the large muscles of the trunk are stretched in the backswing phase, the faster the body unwinds during the downswing.

9.9 Summary

Time Derivatives

Let **P** be a vector and let the two Cartesian reference frames, E and B, be defined by their respective unit vectors ($\mathbf{e}_1$, $\mathbf{e}_2$, $\mathbf{e}_3$) and ($\mathbf{b}_1$, $\mathbf{b}_2$, $\mathbf{b}_3$). Then **P** can be expressed using the unit vectors of E or B:

$$\mathbf{P} = {}^{E}P_1\ \mathbf{e}_1 + {}^{E}P_2\ \mathbf{e}_2 + {}^{E}P_3\ \mathbf{e}_3$$

$$\mathbf{P} = {}^{B}P_1\ \mathbf{b}_1 + {}^{B}P_2\ \mathbf{b}_2 + {}^{B}P_3\ \mathbf{b}_3$$

The time derivatives of **P** in E and in B are defined as

$${}^{E}d\mathbf{P}/dt = (d\,{}^{E}P_1/dt)\ \mathbf{e}_1 + (d\,{}^{E}P_2/dt)\ \mathbf{e}_2 + (d\,{}^{E}P_3/dt)\ \mathbf{e}_3$$

$${}^{B}d\mathbf{P}/dt = (d\,{}^{B}P_1/dt)\ \mathbf{b}_1 + (d\,{}^{B}P_2/dt)\ \mathbf{b}_2 + (d\,{}^{B}P_3/dt)\ \mathbf{b}_3$$

Angular Velocity

Let $\mathbf{b}_1$, $\mathbf{b}_2$, $\mathbf{b}_3$ be a right-handed set of mutually perpendicular unit vectors fixed in a rigid body B. The angular velocity ${}^{R}\boldsymbol{\omega}^{B}$ in a reference frame R is defined as

$${}^{R}\boldsymbol{\omega}^{B} = [({}^{R}d\mathbf{b}_1/dt)\cdot\mathbf{b}_2]\mathbf{b}_3 + [({}^{R}d\mathbf{b}_2/dt)\cdot\mathbf{b}_3]\mathbf{b}_1 + [({}^{R}d\mathbf{b}_3/dt)\cdot\mathbf{b}_1]\mathbf{b}_2$$

where (${}^{R}d\mathbf{b}_i/dt$) denotes the ordinary derivative of $\mathbf{b}_i$ with respect to time in reference frame R. This definition of angular velocity leads to the following results:

$${}^{R}d\mathbf{b}_1/dt = {}^{R}\omega^{B} \times \mathbf{b}_1$$

$${}^{R}d\mathbf{b}_2/dt = {}^{R}\omega^{B} \times \mathbf{b}_2$$

$${}^{R}d\mathbf{b}_3/dt = {}^{R}\omega^{B} \times \mathbf{b}_3$$

Thus, for any vector **P** defined in the reference frames R and B:

$$^{R}d\mathbf{P}/dt = {}^{B}d\mathbf{P}/dt + {}^{R}\boldsymbol{\omega}^{B} \times \mathbf{P}$$

If a series of succesive reference frames is used in the study of motion, the following equation holds:

$$^{E}\boldsymbol{\omega}^{B5} = {}^{E}\boldsymbol{\omega}^{B1} + {}^{B1}\boldsymbol{\omega}^{B2} + {}^{B2}\boldsymbol{\omega}^{B3} + {}^{B3}\boldsymbol{\omega}^{B4} + {}^{B4}\boldsymbol{\omega}^{B5}$$

Velocity and acceleration of two points in a rigid body are related by the following set of equations:

$$^{E}\mathbf{v}^{P} = {}^{E}\mathbf{v}^{Q} + {}^{E}\boldsymbol{\omega}^{B} \times \mathbf{r}^{P/Q}$$

$$^{E}\mathbf{a}^{P} = {}^{E}\mathbf{a}^{Q} + {}^{E}\boldsymbol{\alpha}^{B} \times \mathbf{r}^{P/Q} + {}^{E}\boldsymbol{\omega}^{B} \times ({}^{E}\boldsymbol{\omega}^{B} \times \mathbf{r}^{P/Q})$$

where ${}^{E}\alpha^{B}$ is the angular acceleration of the rigid body B. The angular acceleration is defined by the relation

$$^{E}\boldsymbol{\alpha}^{B} = d^{E}\boldsymbol{\omega}^{B}/dt$$

Conservation of Linear and Angular Momentum

Laws of motion for a body in three-dimensional motion are as follows:

$$\Sigma\mathbf{F} = m\ \mathbf{a}^{c}$$

$$\Sigma\mathbf{M}^{c} = {}^{E}d\mathbf{H}^{c}/dt$$

where $\Sigma\mathbf{F}$ is the resultant external force acting on the body, $\Sigma\mathbf{M}^{c}$ is the resultant external moment with respect to the center of mass, and $\mathbf{a}^{c}$ is the acceleration of the center of mass measured in the inertial reference frame E. The term $\mathbf{H}^{c}$ represents the moment of momentum about the center of mass.

If a point within the body is fixed in the inertial reference frame E, then the following equation also holds

$$\Sigma\mathbf{M}^{o} = {}^{E}d\mathbf{H}^{o}/dt$$

where $\Sigma\mathbf{M}^{o}$ and $\mathbf{H}^{o}$ represent the resultant moment and the moment of momentum about O.

The moment of momentum of a rigid body is called angular momentum. For three-dimensional motion, one obtains the following expression for angular momentum:

$$H^{c}_{1} = (I^{c}_{11}\ \omega_1 + I^{c}_{12}\ \omega_2 + I^{c}_{13}\ \omega_3)$$

$$H^{c}_{2} = (I^{c}_{21}\ \omega_1 + I^{c}_{22}\ \omega_2 + I^{c}_{23}\ \omega_3)$$

$$H^{c}_{3} = (I^{c}_{31}\ \omega_1 + I^{c}_{32}\ \omega_2 + I^{c}_{33}\ \omega_3)$$

in which H^{c}_{i} and ω_{i} are the components of angular momentum and angular velocity of the rigid body in reference frame E written in some reference frame B. Terms I^c_{ij} are elements of mass moment of inertia. These elements depend only on the geometry and mass density distribution of the rigid body.

The angular momentum about the fixed point O has a similar expression:

$$H^o_1 = (I^o_{11}\,\omega_1 + I^o_{12}\,\omega_2 + I^o_{13}\,\omega_3)$$

$$H^o_2 = (I^o_{21}\,\omega_1 + I^o_{22}\,\omega_2 + I^o_{23}\,\omega_3)$$

$$H^o_3 = (I^o_{31}\,\omega_1 + I^o_{32}\,\omega_2 + I^o_{33}\,\omega_3)$$

As shown in the chapter, once the inertia matrix I^c_{ij} is derived, the matrix I^o_{ij} can be obtained from the matrix I^c_{ij} by using a transformation equation (Eqn. 9.17).

9.10 Problems

Problem 9.1. A figure skater spins about her longitudinal axis $\mathbf{b}_2$ with constant angular speed of 5 rad/s (Fig. P.9.1). She then begins to raise her arms over her head at a rate of 2 rad/s. Determine the angular velocity of her arms (A) with respect to the inertial reference frame in E (${}^E\boldsymbol{\omega}^A$). Express this angular velocity in the auxiliary coordinate system B shown in the figure.
Answer: $\boldsymbol{\omega}^A = 5\ (\text{rad/s})\ \mathbf{b}_2 + 2\ (\text{rad/s})\ \mathbf{b}_3$.

Problem 9.2. The following equation relates the acceleration of two points in a rigid object:

$${}^E\mathbf{a}^P = {}^E\mathbf{a}^Q + {}^E\boldsymbol{\alpha}^B \times \mathbf{r}^{P/Q} + {}^E\boldsymbol{\omega}^B \times ({}^E\boldsymbol{\omega}^B \times \mathbf{r}^{P/Q})$$

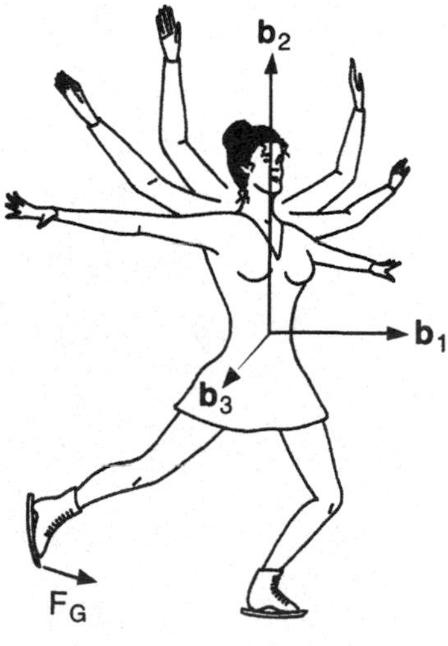

FIGURE P.9.1. A figure skater spins about her longitudinal axis with constant angular speed of 5 rad/s.

Show that it can be written in the following form for the planar motion:

$$^{E}\mathbf{a}^{P} = {^{E}\mathbf{a}^{Q}} + {^{E}\boldsymbol{\alpha}^{B}} \times \mathbf{r}^{P/Q} - ({^{E}\boldsymbol{\omega}^{B}})^{2}\, \mathbf{r}^{P/Q}$$

Problem 9.3. A gyroscope is a wheel mounted in a successive series of rings so that its axis is free to turn in any direction. When the wheel is spun rapidly, it will keep its original plane of rotation no matter which way the rings are turned. Gyroscopes are used to keep moving ships and planes level. The gyroscope shown in Fig. P.9.3 is composed of an outer gimbal ring B, an inner gimbal ring C, and a rotor D. The symbol S denotes the ship to which the gyroscope is attached. The unit vectors **b**, **c**, and **d** are parallel to the axes of B, C, and D, respectively. The angles of rotation shown in the figure were determined to obey the following equations:

$$\theta = -2t^{3}$$

$$\phi = 15 - 5t$$

$$\Psi = 67t^{2} \text{ (all in radians)}$$

Determine $^{S}\boldsymbol{\omega}^{B}$, $^{C}\boldsymbol{\omega}^{B}$, $^{S}\boldsymbol{\omega}^{D}$ as a function of time t.
Answer: $^{S}\boldsymbol{\omega}^{D} = -6t^{2}\,\mathbf{b} - 5\mathbf{c} - 134t\,\mathbf{d}$.

Problem 9.4. Determine the mass moment of inertia matrix about the center of mass of the shapes shown in Fig. P.9.4. All rod segments have a mass m and length L. Rings have a radius R and mass M.

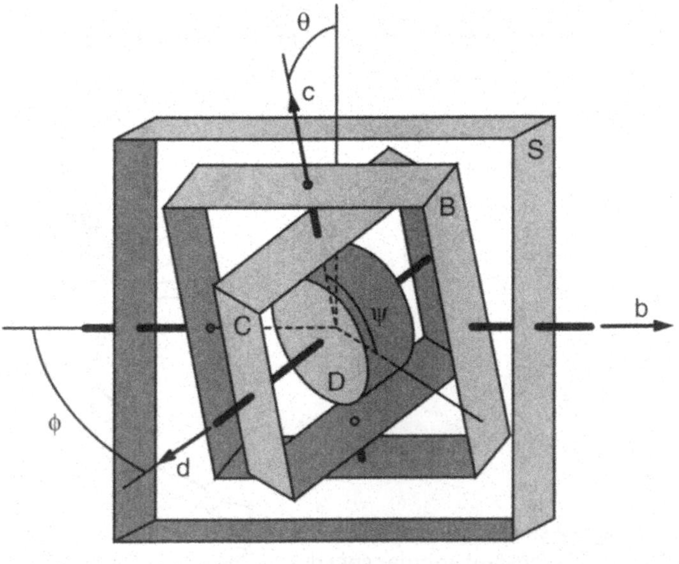

FIGURE P.9.3. A gyroscope attached to a ship.

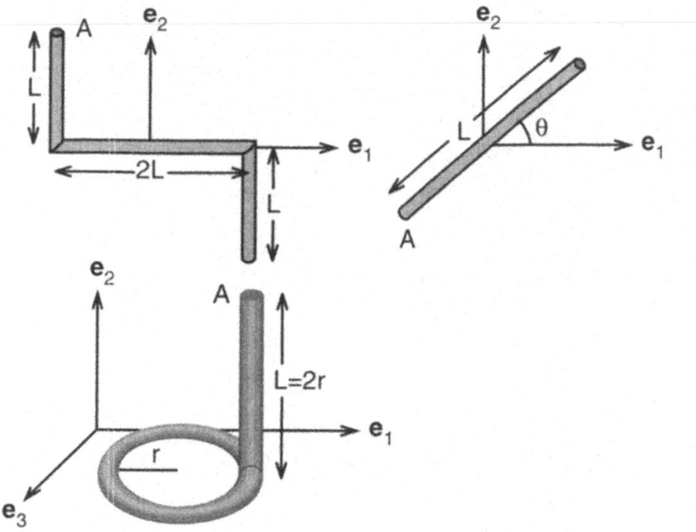

FIGURE P.9.4. Structures constructed by bending of a thin circular rod.

Answer: For the bent rod of length of $4L$ and mass $4m$ shown in the figure, the mass moment of inertia matrix is equal to

$$I^c_{11} = (2/3)\, mL^2,\ I^c_{22} = (8/3)\, mL^2,\ I^c_{33} = (10/3)\, mL^2$$

$$I^c_{12} = {}^{I}c_{21} = mL^2,\ I^c_{13} = {}^{I}c_{31} = I^c_{23} = I^c_{32} = 0$$

Problem 9.5. Determine the mass moment of inertia of the same shapes (Fig. P.9.4) with respect to the point A.

Problem 9.6. A man stands still on a turntable that isolates his body from external torques around the vertical $\mathbf{b}_3$ axis (Fig. P.9.6). He then begins to swing his arms from one side to the other, as shown in the figure. The angular momentum of the swinging arms about the center of mass of the man is equal to 8 (kg-m^2/s) $\mathbf{b}_3$. Determine the angular velocity of the trunk of the man after he begins swinging his arms. The man is 1.80 m tall and weighs 65 kg. His average width is 34 cm and average depth 19 cm.
Answer: $\boldsymbol{\omega} = -9.7$ rad/s $\mathbf{b}_3$.

Problem 9.7. A man standing on a frictionless turntable begins swinging a pendulum of 10 kg (Fig. P.9.7). The length of the pendulum is 0.5 m. Swinging of the pendulum generates angular momentum in $\mathbf{b}_1$direction. The man keeps his hands fixed at the center of the body while swinging the pendulum. Which direction does his trunk rotate to balance the angular momentum produced by the swinging of the pendulum?

Problem 9.8. A thin disk D of mass m rolls on a table while at the same time it rotates around a vertical axis that passes through point O (Fig.

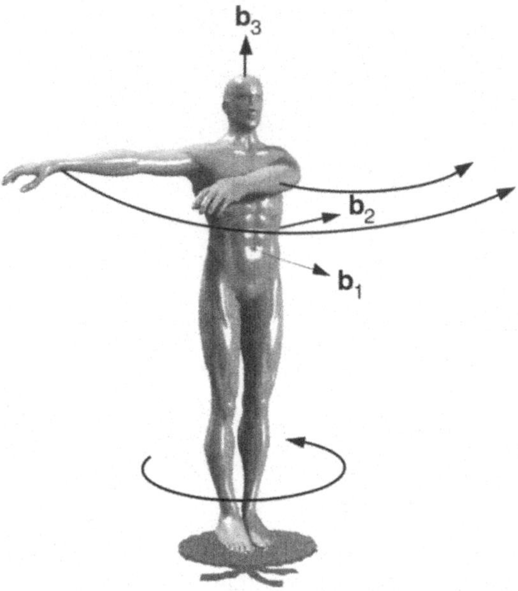

FIGURE P.9.6. A man is swinging his arms while standing on a turntable.

P.9.8). The system is set up so that the cable that connects the center of the disk to point O is horizontal. The angle the rope makes with the $\mathbf{e}_1$ direction varies with time according to the equation:

$$\phi = \pi/4 + 0.36\, t$$

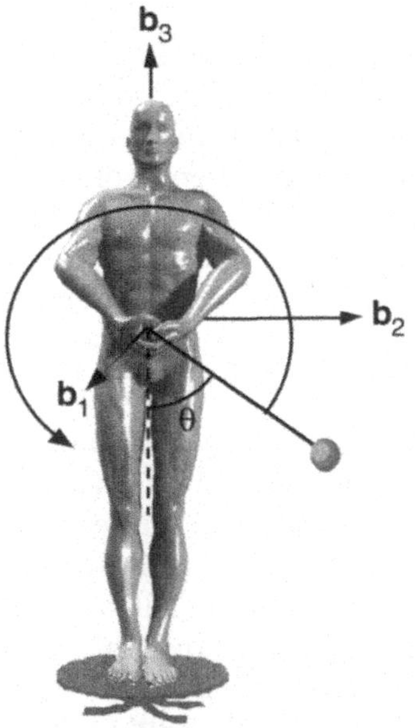

FIGURE P.9.7. A man is swinging a steel ball while standing on a turntable.

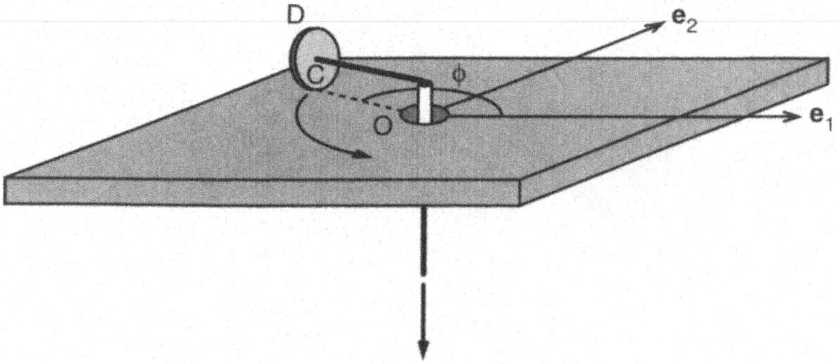

FIGURE P.9.8. A disk rolling on a horizontal table. The center of the disk is connected to a cable that passes through point O.

The length of the cable above the table is 48 cm, and the diameter of the disk is 14 cm. Determine the position, velocity, and the acceleration of the center of mass of the disk at $t = 2$s.

Answer: $\mathbf{r} = 0.48\ [\cos 86°\ \mathbf{e}_1 + \sin 86°\ \mathbf{e}_2]$, $\mathbf{a} = -0.06\ [\cos 86°\ \mathbf{e}_1 + \sin 86°\ \mathbf{e}_2]$

Problem 9.9. A diver wants to add spin (twist) to his somersaults during the aerial movement. The desired twist is in the counterclockwise direction with respect to an axis that is directed from his foot toward his head. Propose a number of shape changes involving upper and lower limbs that would cause the desired twist.

Problem 9.10. A metal coin of radius R and mass m rolls along a horizontal circle (Fig. P.9.10). The angle θ between the disk's axis and the vertical remains constant. Show that the magnitude v of the velocity of the center of mass of the coin is related to the angle θ by the following equation:

$$v^2 = (2/3)\ g \cot \theta\ (r - R\cos\theta)^2/[r - (5/6)\ R\cos\theta]$$

Hint: Let E denote the reference frame fixed on earth such that $\mathbf{e}_3$ points vertical upward. Let D be a coordinate system attached to the center of mass of the coin. The unit vector $\mathbf{d}_1$ remains always in the horizontal plane while the coin rolls without slip. The angle between the $\mathbf{e}_1$ and $\mathbf{d}_1$ is denoted as Ψ. The unit vector $\mathbf{d}_3$ is chosen along the axis of the coin as shown in Fig. P.9.10. The angle θ between $\mathbf{e}_3$ and $\mathbf{d}_3$ remains constant during the motion of the coin. The angle ϕ measures the counterclockwise rotation of the coin about $\mathbf{d}_3$. Using these angles and their time derivatives, one can derive expressions for the angular velocity of D in E ($\Omega_1, \Omega_2, \Omega_3$) and the angular velocity of the coin in E ($\omega_1, \omega_2, \omega_3$).

The rate at which the unit vector $\mathbf{d}_1$ rotates in the horizontal plane is given by the following equation

$$d\Psi/dt = v/(r - R\cos\theta)$$

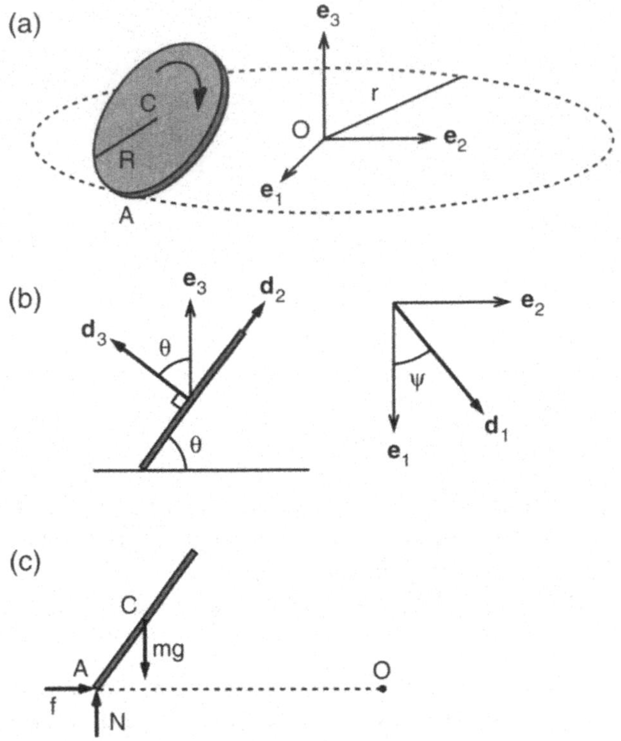

FIGURE P.9.10a–c. A coin rolls along a horizontal circular path **(a)** of radius r. The reference frame D is shown in **(b)**. The free-body diagram of the coin is shown in **(c)**.

where the term in the denominator is equal to the radius of the circular path the center of mass moves.

Note also that the no-slip condition essentially determines the spin rate $(d\phi/dt)$. Show that the condition that the velocity of the point of contact with the flat surface is zero leads to the relation

$$(d\phi/dt) = -(v/R) - \cos\theta \, v/(r - R\cos\theta)$$

The free-body diagram of the coin is shown in the figure (Fig. P.9.10). Use the equation of motion of the center of mass to determine the reaction forces. Then use Euler's equations about the center of mass to express v as a function of R, r, g, and θ.

Problem 9.11. Just before a pitcher throws a baseball, the spatial positions of the ball, ankle, and shoulder were determined experimentally (Fig. P.9.11). The data on the time course of the position of markers were then curve fit by an equation of the form:

$$\mathbf{r}^{P/O} = L\,[\cos\theta\,\mathbf{e}_1 + \cos\alpha\,\mathbf{e}_2 + \cos\phi\,\mathbf{e}_3]$$

in which $\mathbf{r}^{P/O}$ denotes the position vector from O to P, and θ, α, and ϕ are the angles $\mathbf{r}^{P/O}$ make with the unit vectors $\mathbf{e}_1$,$\mathbf{e}_2$, and $\mathbf{e}_3$, fixed on earth. The position vectors of the markers are given as follows:

$$\mathbf{r}^{Shl} = (1.6 + 10\,t^2)\,\mathbf{e}_1 + 2.10\,\mathbf{e}_3$$

$$\mathbf{r}^{Elb/Shl} = 0.41\,[\cos(1.1 + 3\,t^2)\,\mathbf{e}_1 + \cos\,(0.2 + 4t)\,\mathbf{e}_2 + \cos\,(0.2 + 7.2t^2)\,\mathbf{e}_3]$$

$$\mathbf{r}^{Hnd/Elb} = 0.47\,[\cos\,(5\,t^2)\,\mathbf{e}_1 + \cos\,(0.5 - t^2)\,\mathbf{e}_2 + \cos\,(0.4 - 0.2t^2)\,\mathbf{e}_3]$$

Assume that the upper arm weighs 4.1 kg and the lower arm (forearm, hand, and the ball) 2.9 kg. Assume further that each of these limb segments can be represented as a uniform cylindrical rod. Determine the average radii of the upper and lower arms. Assume that the mass density of the upper limb is 1 g/cm^3.

Determine the velocity and acceleration of the markers at the shoulder joint, elbow, and hand by using the parallelogram law and taking the time derivatives in the inertial reference frame E.

Determine the acceleration of the center of mass of the upper arm and the lower arm using the same method.

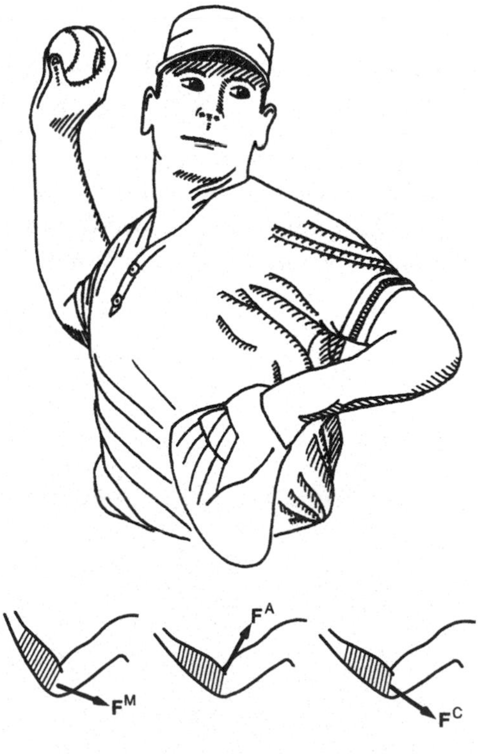

FIGURE P.9.11. A pitcher in preparation for a throw.

Determine the force applied to the forearm by the upper arm at the elbow and the force applied to the upper arm by the shoulder at time $t = 0$ by using Newton's laws of motion for the center of mass of an object.

Determine the angular velocity of the lower arm at time $t = 0$ by either of the following two methods:

Method i: Use the equation that relates the velocity of two points in a rigid body:

$$^{E}\mathbf{v}^{Elb} = {}^{E}\mathbf{v}^{Shl} + {}^{E}\boldsymbol{\omega}^{Bu} \times \mathbf{r}^{Elb/Shl}$$

The only unknown in this equation is $^{E}\boldsymbol{\omega}^{Bu}$.

Method ii: Construct a Cartesian coordinate system B embedded into the upper arm. Let $\mathbf{b}_1$ be the unit vector:

$$\mathbf{b}_1 = [\cos(1.1 + 3\,t^2)\,\mathbf{e}_1 + \cos\,(0.2 + 4t)\,\mathbf{e}_2 + \cos\,(0.2 + 7.2t^2)\,\mathbf{e}_3]$$

Let us choose $\mathbf{b}_2$ along the axis of flexion/extension of the forearm:

$$\mathbf{b}_2 = \mathbf{b}_1 \times [\cos(5t^2)\,\mathbf{e}_1 + \cos(0.5 - t^2)\,\mathbf{e}_2 + \cos\,(0.4 - 0.2t^2)\,\mathbf{e}_3]$$

This equation and its time derivative can be used to evaluate $\mathbf{b}_2$ and its time derivative $d\mathbf{b}_2/dt$.

The third unit vector $\mathbf{b}_3$ and its time derivative can be found by using the expression:

$$\mathbf{b}_3 = \mathbf{b}_1 \times \mathbf{b}_2$$

Angular velocity is then determined using the expression:

$$^{R}\boldsymbol{\omega}^{B} = [(^{R}d\mathbf{b}_1/dt) \cdot \mathbf{b}_2]\mathbf{b}_3 + [(^{R}d\mathbf{b}_2/dt) \cdot \mathbf{b}_3]\mathbf{b}_1 + [(^{R}d\mathbf{b}_3/dt) \cdot \mathbf{b}_1]\mathbf{b}_2$$

Note that one has to determine not only angular velocity but also angular acceleration to determine the net moment exerted at the elbow and the shoulder joints. Write down a "road map" for computing angular acceleration and the resultant joint moments.

Appendix 1: Units and Conversion Factors

The fundamental units in mechanics are the units of length, mass, and time. The other units may be derived from these three. The connection between force and the fundamental units is provided by Newton's second law of motion. The units of other mechanical variables are connected by definition with the fundamental units. For example, if a particle traverses a unit length in unit time, its velocity is unity. Likewise, unit acceleration is the acceleration of a body that gains unit velocity in unit time.

A. Units of Variables Often Encountered in Mechanics

Quantity	Standard International Units (SIU)	United States Units (USU)
Length	meter (m)	foot (ft)
	centimeter (cm) = 10^{-2} m	inch (in.) = 1/12 ft
	milimeter (mm) = 10^{-3} m	
	micrometer (μm) = 10^{-6} m	
Time	second (s)	second (s)
	millisecond (ms) = 10^{-3} s	millisecond (ms) = 10^{-3} s
Velocity	m/s	ft/s
	cm/s	in./s
Acceleration	m/s^2	ft/s^2
	cm/s^2	in./s^2
Mass	kilogram (kg)	slug
	gram (g) = 10^{-3} kg	
Force	newton (N)	pound (lb)
	dyne (dyn) = 10^{-5} N	
Stress	N/m^2 = pascal (Pa)	lb/in.^{2}
	dyn/cm^2 = 0.1 Pa	

Note: A force of 1 N can accelerate a body of 1 kg to 1 m/s^2. A force of 1 dyn can accelerate a body of mass 1 g to 1 cm/s^2.

Standard Prefixes Used in the SI System of Units

mega	M	10^6
kilo	k	10^3
centi	c	10^{-2}
milli	m	10^{-3}
micro	μ	10^{-6}

B. Conversion Factors

To Convert from USU	To SIU	Multiply by	Reciprocal
Length			
foot (ft)	meter (m)	0.30480	3.2808
inch (in.)	centimeter (cm)	2.5400	0.3937
Mass			
slug (lb-s^2/ft)	kilogram (kg)	14.594	0.068522
Force			
pound (lb)	newton (N)	4.4482	0.22481
Density			
slug/ft^3	kg/m^3	515.38	0.0019403
Energy, Work, Moment			
(lb-ft) or (ft-lb)	N-m = joule (J)	1.3558	0.73757
Power			
(ft-lb/s)	N-m/s = watt (W)	1.3558	0.73757
Stress			
lb/in.^{2} or psi	N/m^2 (Pa)	6894.8	1.4504×10^{-4}
Moment of Inertia			
(lb-ft-s^2)	kg-m^2	1.3558	0.73756
(lb-in.-s^2)	kg-m^2	0.11298	8.8507
Linear Momentum			
(slug-ft/s)	kg-m/s	4.4482	0.22481
Moment of Momentum			
(slug-ft^2/s)	kg-m^2/s	1.3558	0.73756

Appendix 2: Geometric Properties of the Human Body

A. Whole-Body Measurements

Body Coordinate System (Table A.2.1)

Let an orthogonal coordinate system B be defined by the intersection of the three principal planes of the body passing through the center of gravity in the standing position as shown in Fig. A.2.1. The location of the center of mass (gravity) of the body is measured along the **b**3 axis from top of the head (L_3), along the $\mathbf{b}_1$ axis from the back plane (L_1), and along the $\mathbf{b}_2$ axis from the anterior superior spine of the ilium (L_2) as shown in Fig. A.2.2. The body positions identified with symbols 1, 2, 3, and 4 are illustrated in Fig. A.2.3. The parameters I_1, I_2, and I_3 denote the mass moment of inertia elements along the $\mathbf{b}_1$, $\mathbf{b}_2$, and $\mathbf{b}_3$, respectively.

TABLE A.2.1. Mass moments of inertia and centers of gravity of young male adults: samples from data presented by Santschi et al. (1963)

Position	L_1	L_2	L_3	I_1	I_2	I_3
Subject 1: Age, 29; height, 72.2 in.; weight 173.5 lb						
1	3.62	5.31	32.2	132	119	13.1
2	3.52	5.31	29.9	174	159	12.5
3	3.34	5.31	29.5	172	135	41.9
4	7.70	5.31	23.9	41.4	37.3	29.9
Subject 2: Age, 23; height, 67.4 in.; weight, 147.3 lb						
1	3.50	4.25	29.8	95.6	81.9	10.3
2	3.49	4.25	27.5	126	110	10.2
3	3.18	4.25	27.2	125	90.3	30.0
4	7.03	4.25	21.7	32.9	30.4	21.5
Subject 3: Age, 33; height, 73.3 in.; weight, 203 lb						
1	3.81	5.61	31.8	144	129	14.2
2	3.67	5.61	29.4	188	172	13.8
3	3.56	5.61	29.5	187	142	51.2
4	7.75	5.61	23.4	48.9	41.9	36.7

Note: All lengths (L_1, L_2, L_3) are inches (in.; multiply by 2.54 to convert to centimeters, cm) and moment of inertia (I_1, I_2, I_3) in lb-in.-s^2 (multiply by 0.11298 to convert to kg-m^2).

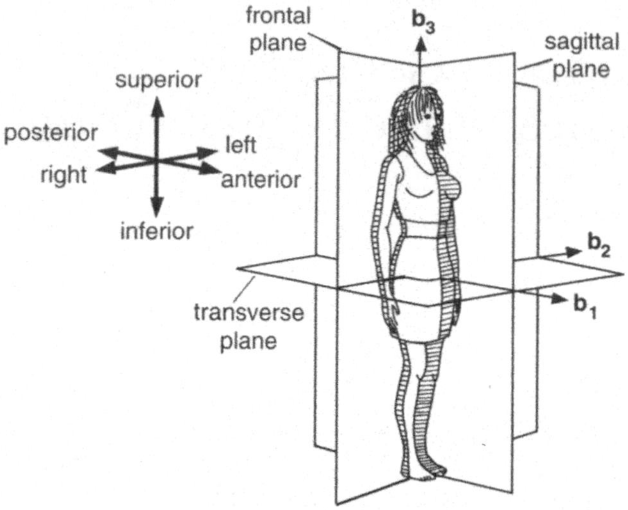

FIGURE A.2.1. Body coordinate system.

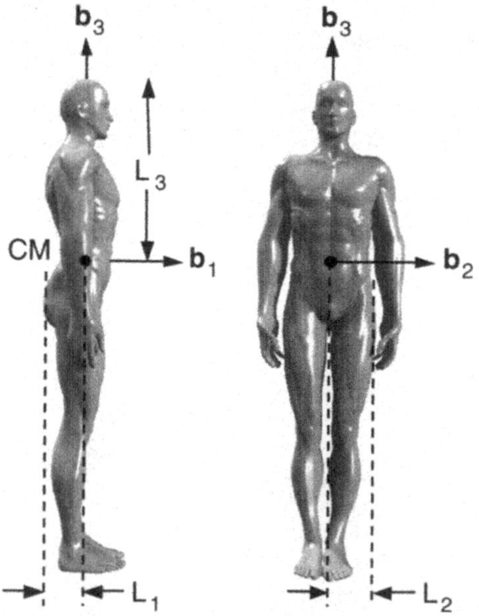

FIGURE A.2.2. Reference landmarks for the location of center of gravity.

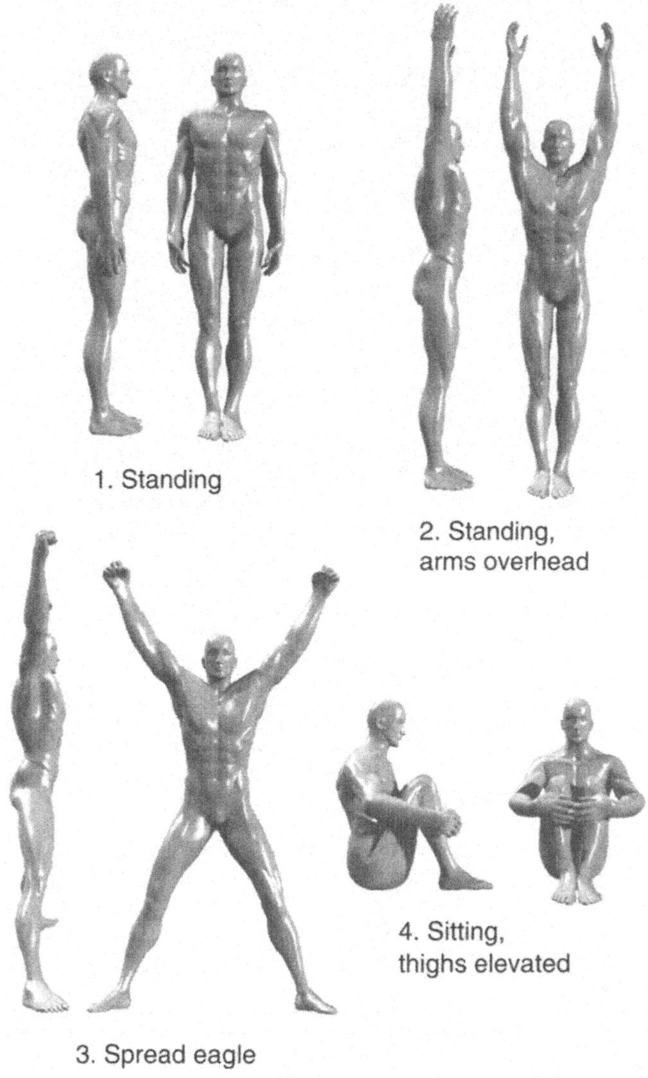

FIGURE A.2.3. Body positions.

Matsuo et al. (1995) proposed the following equations for relating the moment of inertia I_1 and I_2 of adolescent boys to their height H (in m) and weight W (kg). These equations may provide reasonably accurate results for other subpopulations.

$$I_1 = 3.44H^2 + 0.144W - 8.04 \text{ (kg-m}^2\text{)}$$

$$I_2 = 3.52H^2 + 0.125W - 7.78 \text{ (kg-m}^2\text{)}$$

B. Segment Properties (Tables A.2.2 and A.2.3)

TABLE A.2.2. Body segment parameters (Dempster 1955)

Segment	*p* (%)	*m* (%)	I_1 (kg-m^2)	I_3 (kg-m^2)
Head	0.5358	0.0730	0.0248	
Upper arm	0.4360	0.0270	0.0213	
Forearm	0.4300	0.0160	0.076	
Hand	0.5060	0.0066	0.0005	
Trunk	0.4383	0.5080	1.3080	0.3119
Thigh	0.4330	0.0988	0.1502	
Lower leg	0.4330	0.0465	0.0505	
Foot	0.4290	0.0145	0.0038	

p, the distance from the center of gravity of the segment to the proximal endpoint expressed as a fraction of the segment length.
m, segment weight as a percentage of whole body weight.
I_1 and I_3, mass moments of inertia with respect to the center of mass of a body segment about the transverse and longitudinal axis, respectively, for a subject with mass of 74.2 kg and standing height of 1.755 m. For more detailed information on segment properties, see de Leva (1996).

TABLE A.2.3. Relative weight and length of body segments for adult men and women

	Men		Women	
Body segment	Weight	Length	Weight	Length
Whole body	100	100	100	100
Trunk	48.3	30.0	50.8	30.0
Head and neck	7.1	13.8	9.4	
Thigh	10.5	23.2	8.3	24.7
Shank	4.5	24.7	5.5	25.6
Foot	1.5	4.2[a]	1.2	
Upper arm	3.3	17.2	2.7	19.3
Forearm	1.9	15.7	1.6	16.6
Hand	0.6	10.4	0.5	10.4

Weight and length measurements reported are percentages of body weight and body length, respectively.
[a]The number is associated with height of the foot not its length.
The data were gathered from de Leva (1996) and the skeletal anthropometric measurements by Santschi et al. (1963).

C. Moment Arms

Typical moment arms (*d*) of knee and ankle muscle–tendon systems
At the knee:
Hamstring Muscle Group
Semitendinosus 40 mm $< d <$ 55 mm
Semimembraneous 20 mm $< d <$ 40 mm
Biceps femoris d = 20 mm

Calf Muscles

Gastrocnemius lateralis $12\ \text{mm} < d < 20\ \text{mm}$

Gastrocnemius medialis $12\ \text{mm} < d < 20\ \text{mm}$

Quads

Rectus femoris $20\ \text{mm} < d < 40\ \text{mm}$

At the ankle:

Achilles tendon $40\ \text{mm} < d < 52\ \text{mm}$

Moment arms depend on the angles between the articulating bones, as discussed in Chapter 5. Consult Spoor et al. (1990) and Spoor and van Leeuwen (1992) for more detailed depiction of moment arms as a function of the flexion angle of the knee and the ankle.

Selected References

A. Books

Adrian, M.J., and Cooper, J.M. 1995. *Biomechanics of Human Movement*. New York: Benchmark Press.

Agur, A.M.R. 1991. *Grant's Atlas of Anatomy*. Baltimore: Williams & Wilkins.

Alexander, R.M. 1992. *The Human Machine*. New York: Columbia University Press.

Alexander, R.M. 1990. *Animals*. Cambridge: Cambridge University Press.

Alexander, R.M. 1988. *Elastic Mechanisms in Animal Movement*. Cambridge: Cambridge University Press.

Alter, M.J. 1996. *Science of Flexibility*. Champaign: Human Kinetics.

Borelli, G.A. 1670. *De Motionibus Naturalibus a Gravitate Pendentibus*. Regio Iulio: In Officina Domenici Ferri.

Borelli, G.A. 1989. *On the Movement of Animals*. Berlin; New York: Springer-Verlag.

Bronzino, J.D. 1995. *The Biomedical Engineering Handbook*. Boca Raton: CRC Press.

Chaffin, D.B., and Anderson, G.B.J. 1991. *Occupational Biomechanics*. New York: Wiley.

Dampier, Sir W.C. 1943. *A History of Science*. New York: Macmillan.

Euler, L. 1790. *Theoria Motys Corporm Solidorm sev Rigidorvm*. Gryphiswaldie: Litteris et Impensis A.F. Rose.

Frankel, V.H., and Burstein, A.H. 1970. *Orthopaedic Biomechanics*. Baltimore: Lea & Febiger.

Fung, Y.C. 1984. *Biomechanics*. New York: Springer-Verlag.

Guyton, A.C. 1997. *Human Physiology and Mechanisms of Disease*. Philadelphia: Saunders.

Hamill, J., and Knutzen, K.M. 1995. *Biomechanical Basis of Human Movement*. Baltimore: Williams & Wilkins.

Hay, J.G. 1993. *The Biomechanics of Sports Techniques*. Englewood Cliffs: Prentice-Hall.

Hole, J.W. 1990. *Human Anatomy and Physiology*. New York: William.

Johnson, A.T. 1991. *Biomechanics and Exercise Physiology*. New York: Wiley.

Kane, T.R. 1968. *Dynamics*. New York: Holt, Rinehart, Winston.

Kreighbaum, E., and Barthels, K.M. 1995. *Biomechanics: a Qualitative Approach for Studying Human Motion*. New York: Macmillan.

Maquet, P.G.J. 1984. *Biomechanics of the Knee*. New York: Springer-Verlag.
Marone, P.J. 1992. *Shoulder Injuries in Sports*. Rockville: Aspen.
O'Malley, C.D., and Saunders, J.B. de C.M. 1952. *Leonardo da Vinci on the Human Body*. New York: Henry Schuman.
Marone, P.J. 1992. *Shoulder Injuries in Sports*. Rockville: Aspen.
McMahon, T.A. 1984. *Muscles, Reflexes, and Locomotion*. Princeton: Princeton University Press.
Mow, V.C., and Hayes, W.C. 1997. *Basic Orthopedic Biomechanics*. Philadelphia: Lippincott-Raven.
Nahum, A.M., and Melving, J.W. 1993. *Accidental Injury: Biomechanics and Prevention*. New York: Springer-Verlag.
Newton, I. 1999. *The Principia: Mathematical Principles of Natural Philosophy*. Berkeley: University of California Press.
Nigg, B.M. (ed.). 1986. *Biomechanics of Running Shoes*. Champaign: Human Kinetics.
Nordin, M., and Frankel, V.H. 1989. *Basic Biomechanics of the Musculoskeletal System*. Baltimore: Lea & Febiger.
Ozkaya, N., and Nordin, M. 1999. *Fundamentals of Biomechanics*. New York: Springer-Verlag.
Palastanga, N., Field, D., and Soames, R. 1989. *Anatomy and Human Movement*. Oxford: Heinemann.
Prescott, J. 1936. *Mechanics of Particles and Rigid Bodies*. London: Longmans, Green.
Spence, A.P. 1986. *Basic Human Anatomy*. Menlo Park: Cummings.
Taton, R. (ed.) 1964. *The Beginnings of Modern Science, From 1450 to 1800*. New York: Basic Books.
Winter, D.A. 1990. *Biomechanics and Motor Control of Human Movement*. New York: Wiley.
Zatsiorsky, V.M., Seluyanov, V., and Chugunova, L. 1990. Methods of determining mass-inertial characteristics of human body segments. *Contemp. Problems Biomech.* X:272–291.

B. Software

Articulated Total-Body Model (ATB), Armstrong Aerospace Medical Research Laboratory, 1994.
AUTOLEV3 by Online Dynamics, Inc., Sunnyvale, CA (Autolev@aol.com).
Complete Visible Human, National Library of Medicine's Visual Human Project.
MADYMO Model, TNO Road-Vehicle Institute, Department of Injury Prevention, *Software for Musculoskeletal Modelling (SIMM)*, Musculographics, Chicago, IL.
Working Model 3D, Knowledge Revolution, San Francisco, CA.

C. Journal Articles on Mechanics and Movements

Gymnastics and Jumping

Bobbert, M.F., and van Ingen Schenau, G.J. 1988. Coordination in vertical jumping. *J. Biomech.* 21:249–262.

Bobbert, M.F., Huijing, P.A., and van Ingen Schenau, G.J. 1986. A model of the human triceps surae muscle-tendon complex applied to jumping. *J. Biomech.* 19:887–898.

Cheetham, P.J., Sreden, H.I., and Mizoguchi, H. 1987. The gymnast on the rings—a study of forces. *SOMA* X:30–35.

Hay, J.G., Miller, J.A., and Canterna, R.W. 1986. The techniques of elite male long jumpers. *J. Biomech.*, 19:855–866.

McNitt-Gray, J.L., Yokoi, T., and Millward, C. 1994. Landing strategies used by gymnasts on different surfaces. *J. Appl. Biomech.* 10:237–252.

Pandy, M.G., and Zajac, F.E. 1991. Optimal muscle coordination strategies for jumping. *J. Biomech.* 24:1–19.

Pandy, M.G., Zajac, F.E., Sim, E., and Levine, W. 1990. An optimal control model for maximum-height human jumping. *J. Biomech.* 23:1185–1198.

van Ingen Schenau, G.J., and Cavanagh, P.R. 1990. Power equations in endurance sports, survey article. *J. Biomech.* 23:865–881.

Voigt, M., Simonsen, E.B., Dyhre-Poulsen, P., and Klausen, K. 1994. Mechanical and muscular factors influencing the performance in maximal vertical jumping after different prestretch loads. *J. Biomech.* 28:293–297.

Pole Vaulting and Aerial Movements

Dapena, J., and Braff, T. 1985. A two-dimensional simulation method for the prediction of movements in pole vaulting. *Biomechanics* 9B:458–463.

Hubbard, M. 1980. Dynamics of the pole-vault. *J. Biomech.* 13:965–976.

Ekevad, M., and Lundberg, B. 1995. Simulation of smart pole vaulting. *J. Biomech.* 28:1079–1090.

McGinnis, P.M. 1983. The inverse dynamics problem in pole vaulting. *Med. Sci. Sports Exercise* 15:112–117.

Morlier, J., and Cid, M. 1996. Three-dimensional analysis of the angular momentum of a pole-vaulter. *J. Biomech.* 29:1085–1090.

Yeadon, M.R. 1993. The biomechanics of twisting somersaults. Part I: Rigid body motions. *J. Sports Sci.* 11:187–198.

Yeadon, M.R. 1993. The biomechanics of twisting somersaults. Part II: Contact twist. *J. Sports Sci.* 11:199–208.

Yeadon, M.R. 1993. The biomechanics of twisting somersaults. Part III: Aerial twist. *J. Sports Sci.* 11:209–218.

Running

Bobbert, M.F., Yeadon, M.R., and Nigg, B.M. 1992. Mechanical analysis of the landing phase in heel-toe running. *J. Biomech.* 25:223–234.

Cavanagh, P.R., and LaFortune, M.A. 1980. Ground reaction forces in distance running. *J. Biomech.* 13:397–406.

de Clerqc, D., Aerts, P., and Kunnen, M. 1994. The mechanical characteristics of the human heel pad during foot strike in running: an in vivo cineradiographic study. *J. Biomech.* 27:1213–1222.

Gerritsen, K.G.M., van den Bogert, A.J., and Nigg, B.M. 1995. Direct dynamics simulation of the impact phase in heel-toe running. *J. Biomech.* 28:661–668.

Jacobs, R., Bobbert, M.F., and van Ingen Schenau, G.J. 1996. Mechanical output from individual muscles during explosive leg extensions: the role of biarticular muscles. *J. Biomech.* 29:513–523.

Jaeger, R.J., and Vanitchatchavan, P. 1992. Ground reaction forces during the termination of human gait. *J. Biomech.* 25:1223–1236.

MacMahon, T.A., Valiant, G., and Fredrick, E.C. 1987. Groucho running. *J. Appl. Physiol.* 219:709–727.

MacMahon, T.A., and Cheng, G.C. 1990. The mechanics of running: how does stiffness couple with speed? *J. Biomech.* 23:65–78.

Mann, R.A., and Hagy, J. 1980. Biomechanics of walking, running and sprinting. *Am. J. Sports Med.* 8:345–350.

Maronski, R. 1996. Minimum time running and swimming: an optimal control approach. *J. Biomech.* 29:245–249.

Novacheck, T.F. 1998. Running injuries: a biomechanical approach. *J. Bone Joint Surg.* 80:1220–1233.

Roberts, T.J., Marsh, R.L., Weyand, P.G., and Taylor, C.R. 1997. Muscular force in running turkeys: the economy of minimizing work. *Science* 75:1113–1117.

Salathe, E.P., Jr., Arangio, G.A., and Salathe, E.P. 1990. The foot as a shock absorber. *J. Biomech.* 23:655–659.

Skating

Arnold, A.S., King, D.L., and Smith, S.L. 1994. Figure skating and sports biomechanics: the basic physics of jumping and rotating. *Skating* 71:9.

de Boer, R.W., Ettema, G.J.C., van Gorkum, H., de Groot, G., and van Ingen Schenau, G.J. 1988. A geometrical model of speed skating the curves. *J. Biomech.* 21:445–450.

de Koning, J.J., de Groot, G., and van Ingen Schenau, G.J. 1992. Ice friction during speed skating. *J. Biomech.* 25:565–571.

van Ingen Schenau, G.J., de Groot, G., and de Boer, R.W. 1985. The control of speed in elite female speed skaters. *J. Biomech.* 18:91–96.

van Ingen, G.J. 1982. The influence of air friction in speed skating. *J. Biomech.* 15:449–458.

de Boer, R.W., Ettema, G.J.C., van Gorkum, H., de Groot, G., and van Ingen Schenau, G.J. 1987. Biomechanical aspects of push-off techniques in speed skating the curves. *Int. J. Sport Biomech.* 3:69–79.

White, J.D. 1992. The role of surface melting in ice skating. *Physics Teacher* 30:495–497.

Throwing and Hitting and Falling

Andrews, J.R., Dillman, C.J., and Fleisig, G.S. 1993. Biomechanics of pitching with emphasis upon shoulder kinematics. *J. Sports Phys. Ther.* 18:402–408.

Amis, A.A., Dowson, D., and Wright, V. 1980. Analysis of elbow forces due to high-speed forearm movements. *J. Biomech.* 13:825–831.

Jackson, K.M., Joseph, J., and Wyard, S.J. 1978. A mathematical model of arm swing during human locomotion. *J. Biomech.* 11:277–289.

Karlsson, D., and Peterson, B. 1992. Towards a model for force predictions in the human shoulder. *J. Biomech.* 25:189–199.
van der Kroonenberg, A.J., Hayes, W.C., and McMahon, T.A. 1996. Hip impact velocities and body configurations for voluntary falls from standing height. *J. Biomech.* 29:807–811.
Nightingale, R.W., McElhaney, J.H., Richardson, W.J., and Myers, B.S. 1996. Dynamic responses of the head and cervical spine to axial impact loading. *J. Biomech.* 29:307–318.
Putnam, C.A. 1993. Sequential motions of body segments in striking and throwing skills: descriptions and explanations. *J. Biomech.* 26:125–135.
Ryu, R.K.N., McCormick, J., Jobe, F.W., Moynes, D.R., and Antonelli, D. J. 1988. An electromyographic analysis of shoulder function in tennis players. *Am. J. Sports Med.* 16:481–485.
Welch, C.M., Banks, S.A., Cook, F.F., and Draovitch, P. 1995. Hitting a baseball: a biomechanical description. *J. Sports Phys. Ther.* 22:193–221.
Werner, S.L., Fleisig, G.S., Dillman, C.J., and Andrews, J.R. 1995. Biomechanics of the elbow during baseball pitching. *J. Sports Phys. Ther.* 17:274–278.
Yeadon, M.R. 1992. Comments on thoracic injury potential of basic competition Taekwondo kicks. *J. Biomech.* 25:1247–1248.

Weight Lifting

Donkers, M.J., An, K-N., Chao, E.Y.S., and Morrey, B.F. 1993. Hand position affects elbow joint load during push-up exercise. *J. Biomech.* 26:625–632.
Procter, P., and Paul, J.P. 1982. Ankle joint biomechanics. *J. Biomech.* 15:627–634.

Gravity—Weightlessness

Baldwin, K.M. 1996. Musculoskeletal adaptations to weightlessness and development of effective counter measures. *Med. Sci. Sports Exercise* 28:1247–1253.
Desplanches, D. 1997. Structural and functional adaptations of skeletal muscle to weightlessness. *Int. J. Sports Med. Suppl.* 4:S259–S264.

Physical and Geometric Properties of Human Body

Ackland, T.R., Blanksby, B.A., and Bloomfield, J. 1988. Inertial characteristics of adolescent male body segments. *J. Biomech.* 21:319–327.
Butler, D.L., Grood, E.S., Noyes, F.R., et al. 1978. Biomechanics of ligaments and tendons. *Exercise Sports Sci. Rev.* 6:125–181.
Clauser, C.E., McConville, J.T., and Young, J.W. 1969. Weight, volume, and center of mass of segments of the human body. AMRL-TR-69-70, pp. 59–60. Wright-Patterson Air Force Base, OH.
Dempster, W.T. 1955. Space requirements of the seated operator. WADC-55-159. Wright-Patterson Air Force Base, OH.
Hanavan, E.P. 1964. A mathematical model of the human body. AMRL Tech. Rep. 64-102. Wright Patterson Air Force Base, OH.
Hatze, H. 1980. A mathematical model for the computational determination of parameter values of anthropomorphic segments. *J. Biomech.* 13:833–843.

Jensen, R.K., and Fletcher, P. 1994. Distribution of mass to the segments of elderly males and females. *J. Biomech.* 27:89–96.

de Leva, P. 1996. Adjustments to Zatsiorsky-Selunayov's segment inertia parameters. *J. Biomech.* 29:1223–1230.

Matsuo, A., Ozawa, H.M., Goda, K., and Fukunaga, T. 1995. Moment of inertia of whole body using an oscillating table in adolescent boys. *J. Biomech.* 28:219–223.

Santschi, W.R., Dubois, J., and Omoto, C. 1963. Moments of inertia and centers of gravity of the living human body. AD-410-451. Wright-Patterson Air Force Base, OH.

Spoor, C.W., van Leuwen, J.L., Meskers, C.G.M., Titulaer, A.F., and Huson, A. 1990. Estimation of instantaneous moment arms of lower leg muscles. *J. Biomech.* 23:1247–1259.

Spoor, C.W., and van Leeuwen, J.L. 1992. Knee muscle moment arms from MRI and from tendon travel. *J. Biomech.* 25:201–206.

Wei, C., and Jensen, R.K. 1995. The application of segment axial density profiles to a human body inertia model. *J. Biomech.* 28:103–108.

Yeadon, M.R. 1990. The simulation of aerial movement. II. A mathematical inertia model of the human body. *J. Biomech.* 23:67–74.

Internal Forces

Alexander, R.M., and Vernon, A. 1975. The dimensions of knee and ankle muscles and the forces they exert. *J. Hum. Mov. Stud.* 1:115–123.

Collins, J.J. 1995. The redundant nature of locomotor optimization laws. *J. Biomech.* 28:251–267.

Collins, J.J. 1994. Antagonistic-synergestic muscle action at the knee during competitive weightlifting. *Med. Biol. Eng. Comput.* 32:168–174.

Crowninshield, R.D., and Brand, R.A. 1981. A physiologically based criterion of muscle force prediction in locomotion. *J. Biomech.* 14:793–801.

Dul, J., Johnson, G.E., Shiavi, R., and Towsend, M.A. 1984a. Muscle synergism. I: On criteria for load sharing between synergestic muscles. *J. Biomech.* 17:663–673.

Dul, J., Johnson, G.E., Shiavi, R., and Towsend, M.A. 1984b. Muscle synergism. II: A minimum fatigue criteria for load sharing between synergestic muscles. *J. Biomech.* 17:675–684.

Herzog, W., and Binding, P. 1992. Predictions of antagonistic muscular activity using nonlinear optimization. *Math. Biosci.* 111:217–229.

Herzog, W., and Leonard, T.R. 1991. Validation of optimization models that estimate the forces exerted by synergistic muscles. *J. Biomech.* 24:31–39.

Patriarco, A.G., Mann, R.W., Simon, S.R., and Mansour, J.M. 1981. An evaluation of the approaches of optimization models in the prediction of muscle forces during human gait. *J. Biomech.* 14:513–525.

Seireg, A., and Arkivar, R.J. 1973. A mathematical model for evaluation of forces in lower extremities of the musculoskeletal system. *J. Biomech.* 6:313–326.

Shultz, A.B., Alexanader, N.B., and Ashon-Miller, J.A. 1992. Biomechanical analysis of rising from a chair. *J. Biomech.* 25:1383–1391.

Yeo, B.P. 1976. Investigations concerning the principle of minimal total muscular force. *J. Biomech.* 9:413–416.

Index

f = figure

A

Abdominal crunch, 69, 69f
Abdominal wheel, 239
Abduction, 5, 5f, 68
Acceleration, 40
  angular, 94, 262
  of center of mass, 59
Actin, 18
Adduction, 5, 5f
Agonist, 20
Air resistance, 45, 49
Amphiarthrosis, 14
Anconeus, 25
Angular acceleration, 94
Angular impulse, 200
Angular momentum, 200
Angular velocity, 88
Annular ligament, 13
Antagonist, 21
Anterior, 4, 4f
Anterior cruciate ligament, 13
Appendicular musculature, 21
Appendicular skeleton, 7f, 8
Arabesque, 184
Arm swinging, 74
Articular capsule, 13f
Atlas, 16f
Auditory ossicles, 7f
Axial force, 136
Axial moment of inertia, 178
Axial musculature, 21
Axial skeleton, 6, 7f
Axis, 16f

B

Ball-and-socket joint, 16, 16f
Ballistic pendulum, 231
Bending, 140
Biarticular muscle, 21, 165
Biceps, 25
  curls, 50
  force, 154
Biceps femoris, 26
Bipennate muscle, 19, 20f
Body coordinates, 299
Body curls, 107
Bone
  cancellous, 10
  compact, 10
Bone fracture, 177
Bone matrix, 10
Brachialis, 25
Brachioradilis, 25
Bursa, 13

C

Cables, 127
Calcium phosphate, 10
Cantilever beam, 140
Car crash, 196
Carpals, 7f, 9
Cartilage, 12
Center of gravity, 59
Center of mass, 59

Chondrocyte, 12
Circular muscle, 20, 20f
Circumduction, 6
Clavicle, 7f, 8
Coccyx, 7f
Coefficient of friction, 209
Coefficient of restitution, 208
Collagen, 10
Collision, 198
Conservation
  of angular momentum, 71, 268
  of linear momentum, 57
  of mechanical energy, 226
Contact forces, 121
Contact inhibition, 180
Contraction, 174
Convergent muscle, 19, 20f
Conversion factors, 298
Cosine law, 161
Cranium, 7f

D
Dashpot, 250
Deltoids, 22, 22f, 24
Diarthrosis, 14
Distraction, 180
Diving, 104
Dorsiflexion, 5f
Drag coefficient, 55
Dumbell kickbacks, 183

E
Eccentric loading, 132f
Endomysium, 17
Epimysium, 17
Erector spinae, 23f, 25
Euler, 256
Euler's equations, 269
Extension, 4, 5f

F
Femur, 7f, 9
Fibers
  fast, 26
  intermediate, 26
  slow, 26
Fibroblasts, 14
Fibula, 7f, 9
Fibular collateral ligament, 13
Figure skating, 289
Filament
  thick, 18
  thin, 18
Flexion, 4, 5f
  of forearm, 162
Force, 43
  body, 43
  contact, 43
  couple, 67
  distributed, 144
  external, 46
  frictional, 45
  gravitational, 31, 44
  internal, 46
Free body diagram, 150
Free fall, 43–44
Friction, 46, 209
Frontal plane, 3, 4f
Fulcrum, 1, 2f

G
Gadd Severity Index, 214
Galileo, 30, 43
Gastrocnemius, 24f, 26
Geometric properties, 299
Glaneoid cavity, 17
Gluteus maximus, 25
Gluteus medius, 25
Gluteus minimus, 25
Gravity, 30, 43
Ground force, 205
Growth plate, 12, 180
Gyroscope, 290

H
Hamstrings, 26
Head injury, 218
Hinge, 15
Hip bone, 7f, 9
Hip fracture, 216
Hooke's law, 139
Hopping, 252
Human cable, 125
Hydroxyapatite, 176
Hyoid, 7f
Hyperextension, 4, 5f
Humerus, 7f, 9

I
Impact, 201
Impulse, 194
Impulsive force, 195f
Inferior section, 3, 4f
Initial motion, 211
Instantaneous center of rotation, 109
Internal forces, 135
Interroseous membrane, 13
Inverse dynamics, 156
Isometric contraction, 174
Isometric tension, 173f
Isotonic contraction, 175

J
Joint, 14
  angle, 161
  ball-and-socket, 16
  hinge, 14
  monoaxial, 15
  pivot, 15
  synovial, 15

K
Kinetic energy, 221

L
Latissimus dorsi, 22, 23f
Leg stiffness, 254
Lever, 1, 2f
Lever arm, 161
Ligaments, 13, 13f
Limb-lengthening, 178
Lumped mass, 72

M
Mass moment of inertia, 98, 265
Menisci, 15
Metabolic rate, 220
Method of reduction, 152
Moment arm, 161, 302
Moment of a force, 67
Moment of momentum, 70
Momentum
  angular, 97
  linear, 58
Monoaxial muscle, 21
Motor neurons, 18
Multibody systems, 232
Muscle, 1, 17
  activation, 174
  bipennate, 20f
  circular, 20f
  convergent, 20f
  externus, 21
  parallel, 20f
  skeletal, 17
  superficialis, 21
  unipennate, 20f
Myosin, 18

N
Newton, 30, 43
Newton's Laws of Motion, 30, 31, 43
Neutral equilibrium, 128

O
Obliques, 25
Optimization, 151
Oscillation, 47
  amplitude, 48
  of pendulum, 47
  period, 48
Osteoblasts, 11, 11f
Osteoclasts, 11, 11f
Oxygen consumption, 221
Oxygen uptake, 221

P
Parallel axis theorem, 155
Parallelogram law, 33, 34f
Patella, 7f, 9
Path coordinates, 42
Pectoralis major, 22
Pectoralis minor, 22
Pelvic girdle, 7f, 9
Pendulum, 47
Pennate muscle, 19, 20f
Perimysium, 17
Phalanges, 7f, 9
Pitching, 294
Pivot joint, 15
Planar motion, 85
Polar coordinates, 41
Pole vaulting, 235
Position vector, 40
Posterior, 4, 4f
Posterior cruciate ligament, 13

Potential energy, 129, 227
Power, 225
Precession of a football, 269
Primary planes, 3
Principia, 30, 31
Pronation, 5f
Pushups, 75, 115
Pythagoras' theorem, 36

Q

Quadriceps, 23f, 26
Quadriceps force, 152

R

Radius, 7f, 9
Radius of gyration, 155
Rate of shortening, 175
Reaction forces, 136
Rectus abdominis, 25
Rectus femoris, 24f, 26
Redundance, 127
Resistance force, 45
Resorption, 11
Rigid body, 84
Roller joint, 135
Rolling, 223
Rolling of abdominal wheel, 276
Running, 205, 233, 242

S

Sacrum, 7f
Sagittal plane, 4, 4f
Sarcomere, 18, 173
Scalar, 170
Scalar product, 65
Scapula, 7f, 8, 17
Seat belts, 196
Seated dumbell press, 261
Semimembranosus, 26
Semitendinosus, 24f, 26
Sesamoid bone, 9, 11
Shoulder joint, 16, 26f
Sideways fall, 217
Skeletal muscle, 18, 19f, 20f
Sliding filament theory, 18
Spinal cord, 6
Spinning top, 270
Spring constant, 190
Stable equilibrium, 128
Statically determinate, 133
Statically indeterminate, 133
Statics, 117
Sternum, 6, 7f
Stiff elbow, 178
Stiffness coefficient, 139, 176
Strain
  axial, 139
Stress
  axial, 139
  shear, 170
Structural stability, 127
Superior section, 3, 4f
Supination, 5f
Synarthrosis, 14
Synergist, 20

T

Tae kwon do kick, 204
Tendon, 13f, 17
Tension
  Achilles tendon, 156
  cables, 127
Tensor, 170
Terminal velocity, 45
Throwing, 283
Tibia, 7f, 9
Tibular collateral ligament, 13
Trajectory, 49, 49f
  golf ball, 49
Transverse plane, 3, 4f
Trapezoids, 23, 23f, 24
Tree of bodies, 267
Truss, 135
  method of joints, 136
  method of sections, 137
Turns of a dancer, 262
Turn table, 291
Twisting somersaults, 280

U

Ulna, 7f, 9
Unipennate muscle, 19, 20f
Units, 297
Unit vectors, 35
Unstable equilibrium, 128

V

Vastus intermedius, 26
Vastus lateralis, 26

Vastus medialis, 26
Vector
  absolute value, 36
  addition, 33
  direction, 36
  multiplication, 64–67
  product, 66
  projection, 34, 35f
  resultant, 37
  subtraction, 33
  time derivative, 39
Velocity, 40
  angular, 88, 259
  end point, 285
  path coordinates, 41f, 43
  polar coordinate, 41–42, 41f
  terminal, 45
Vertebrae, 7f
Vertebral column, 7
Vertical jumping, 61, 90, 253

W
Weight lifting, 157
  deltoids, 158
  erector spinae, 160
Work, 226
Work done
  by friction, 228
  by gravity, 227
  by springs, 228
Working model, 269

Y
Young's modulus, 139